T0245336

URBAN DC MICROGRID

URBAN DC MICROGRID

Intelligent Control and Power Flow Optimization

MANUELA SECHILARIU and FABRICE LOCMENT
Université de Technologie de Compiègne, France

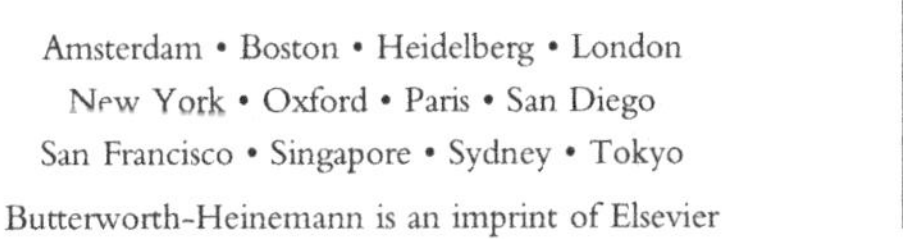
Amsterdam • Boston • Heidelberg • London
New York • Oxford • Paris • San Diego
San Francisco • Singapore • Sydney • Tokyo
Butterworth-Heinemann is an imprint of Elsevier

Butterworth-Heinemann is an imprint of Elsevier
The Boulevard, Langford Lane, Kidlington, Oxford OX5 1GB, UK
50 Hampshire Street, 5th Floor, Cambridge, MA 02139, USA

British Library Cataloguing-in-Publication Data
A catalogue record for this book is available from the British Library

Library of Congress Cataloging-in-Publication Data
A catalog record for this book is available from the Library of Congress

ISBN: 978-0-12-803736-2

For information on all Butterworth-Heinemann publications
visit our website at https://www.elsevier.com/

Publisher: Joe Hayton
Acquisition Editor: Lisa Reading
Editorial Project Manager: Peter Jardim
Production Project Manager: Sruthi Satheesh
Designer: Victoria Pearson

Typeset by TNQ Books and Journals

CONTENTS

Author Biographies *xi*
Foreword *xv*
Acknowledgments *xvii*
Abbreviations *xix*
General Introduction *xxi*

1. Connecting and Integrating Variable Renewable Electricity in Utility Grid **1**
1. Smart Grid—Solution for Traditional Utility Grid Issues 1
2. Microgrids 6
2.1 Alternating and Direct Current Microgrid 7
2.2 Research Issues in Microgrids 9
2.2.1 Control 9
2.2.2 Protection 11
2.2.3 Energy Management 12
3. Urban Direct Current Microgrid 14
3.1 Smart Grid, Smart City, and Smart Building 14
3.2 Smart Microgrids in Urban Areas 16
3.2.1 General Overview 16
3.2.2 Direct Current Microgrid for a Low-Voltage Direct Current Distribution Network 19
3.2.3 Dynamic Interactions Between the Microgrid and the Smart Grid 26
3.3 Urban Energy Management Strategies 26
3.4 Experimental Platform for Direct Current Microgrids 29
4. Conclusions 30
References 31

2. Photovoltaic Source Modeling and Control **35**
1. Photovoltaic Source Modeling 35
1.1 Photovoltaic Cell 35
1.1.1 Operating Principle of a Photovoltaic Cell 36
1.1.2 Electrical Characteristics of a Photovoltaic Cell 37
1.2 Photovoltaic Source Modeling 39
1.2.1 Photovoltaic Power Prediction 39
1.2.2 Equivalent Circuit Photovoltaic Model 40

5. Direct Current Microgrid Supervisory System Design **171**
1. Multilayer Supervisory Design Overview 171
2. Human-Machine Interface 174
3. Prediction Layer 175
3.1 Photovoltaic Power Prediction 176
3.2 Load Power Prediction 177
4. Energy Management Layer 178
4.1 Energy Cost Optimization Problem Formulation 179
4.1.1 Grid-Connected Mode 179
4.1.2 Off-Grid Mode 185
4.2 Solving the Problem 187
4.3 Interface for Operation Layer 189
5. Operation Layer 190
5.1 Control Algorithm for Grid-Connected Mode 192
5.2 Control Algorithm for Off-Grid Mode 193
6. Evaluation of the Supervisory System by Simulation 194
6.1 Simulation Results for Grid-Connected Mode 195
6.1.1 Power Flow Simulation Controlled by $K_D(t)$ 199
6.1.2 Power Flow Simulation Controlled by Constant K_D 201
6.1.3 Comparison and Discussion 201
6.2 Simulation Results for Off-Grid Mode 203
6.2.1 Power Flow Simulation Controlled by $K_D(t)$ 205
6.2.2 Comparison and Discussion 206
7. Conclusions 206
References 207

6. Experimental Evaluation of Urban Direct Current Microgrid **209**
1. Introduction 209
2. Considerations on Multilayer Supervisory Communication 211
3. Considerations on Power Control Algorithms Implementation 213
4. Direct Current Microgrid Operating in Grid-Connected Mode 214
4.1 Experimental Test Description for Grid-Connected Mode 215
4.1.1 Test 1 for Grid-Connected Mode 215
4.1.2 Test 2 for Grid-Connected Mode 221
4.1.3 Test 3 for Grid-Connected Mode 226
4.2 Results Analysis and Discussions for Grid-Connected Mode 229
5. Direct Current Microgrid Operating in Off-Grid Mode 232
5.1 Experimental Test Description for Off-Grid Mode 233
5.1.1 Test 1 for Off-Grid Mode 233
5.1.2 Test 2 for Off-Grid Mode 238
5.1.3 Test 3 for Off-Grid Mode 242

5.2 Results Analysis and Discussions for Off-Grid Mode 244
6. Conclusions 248
References 249

General Conclusions, Future Challenges, and Perspectives *251*
Index *269*

AUTHOR BIOGRAPHIES

BIOGRAPHY OF MANUELA SECHILARIU

Manuela Sechilariu received the Dipl.Ing. degree in electrical engineering in 1986 from the Institute Polytechnic Iasi, Romania, and the PhD degree in electrical engineering and automatic in 1993 from the Université d'Angers, France. In 2013 she obtained the HDR degree in electrical engineering from the Université de Technologie de Compiègne, France, the highest French academic title, and then the qualification required for full professor. The obtaining of HDR, accreditation to supervise research, confers official recognition of the high scientific level and capability to optimally manage a research strategy in a sufficiently wide scientific field (smart grid and microgrids). In 1989 she became an assistant professor with the Institute Polytechnic Iasi, Romania, and in 1994 she became an associate professor with the Université d'Angers, France. In 2002 she joined the Université de Technologie de Compiègne, France.

Manuela Sechilariu has more than 20 years of research experience. Her first research topic focused on the modeling and simulation of static converters by Petri Net, which quickly led to the study of hybrid dynamic systems. Contributions were made to the definition, classification, and optimal control of these systems. Since 2006 she has directed research in the study of decentralized renewable electricity production, urban microgrids, and energy management systems. She has delivered several invited lectures and has published more than 60 refereed scientific and technical papers in international journals and conferences, with more than 350 citations (SCOPUS), on topics such as renewable energy systems, including microgrids, photovoltaic-powered systems, economic dispatch optimization, supervisory control, and Petri Net and Stateflow modeling.

Her research has been funded by agencies and sponsors including the CNRS (National Center for Scientific Research), ADEME (The French Environment and Energy Management Agency), FEDER (European Fund for Regional Economic Development), and CRP (Picardie Regional Council). She has managed several national research projects and industrial research contracts.

She is a member of several professional bodies and academic boards, including the IEEE (Institute of Electrical and Electronics Engineers), the

French Research Group GDR SEEDS (Electric Power Systems in their Corporate Social Dimension), and the 63rd section of the French National Council of Universities. Manuela Sechilariu has reviewed projects of various scientific national research organizations (French and Czech) and articles for many international journals (active reviewer for several IEEE Transactions and Elsevier journals) and conferences. She has directed and co-supervised many dozens of MsEng. and PhD theses and dissertations. She has participated in many academic councils and committees either as a member or as a deputy member of the selection committee for candidates for associate professor position. For the last 10 academic years she has served as director of the Dipl.Ing. degree major "Systems and Networks for Built Environment" and then as a member of the PhD School Board.

Manuela Sechilariu's broad research interests focus on power and energy systems, the smart grid, microgrids, distributed generation, photovoltaic-powered systems, energy management, optimization, intelligent control, and Petri Net modeling.

Affiliations and Expertise

Professor and researcher on modeling, simulation, and power management applied to renewable energy in microgrids with AVENUES Laboratory, Université de Technologie de Compiègne, France.

BIOGRAPHY OF FABRICE LOCMENT

Fabrice Locment received the Dipl.Ing. degree in electrical engineering from Polytech Lille, Ecole Polytechnique Universitaire de Lille, France, in 2003, and MS and PhD degrees in electrical engineering from the Université des Sciences et Technologies de Lille, France, in 2003 and 2006, respectively. Since 2008 he has been an associate professor with the Université de Technologie de Compiègne, France. In December 2015 he obtained the HDR degree in electrical engineering from the Université de Technologie de Compiègne, France, the highest French academic title. The obtaining of HDR, accreditation to supervise research, confers official recognition of the high scientific level and capability to optimally manage a research strategy in a sufficiently wide scientific field.

His current research interests include designing, modeling, and control of electrical systems, particularly photovoltaic and wind turbine systems. He published more than 50 refereed scientific and technical papers in international journals and conferences, with over 450 citations (SCOPUS) on topics such as renewable energy systems, including microgrids, photovoltaic and wind powered systems, maximum power point tracking, and energetic macroscopic representation modeling.

Fabrice Locment was involved in several national research projects funded by agencies and sponsors including the CNRS (National Center for Scientific Research), ADEME (The French Environment and Energy Management Agency), FEDER (European Fund for Regional Economic Development), and CRP (Picardie Regional Council).

Fabrice Locment has reviewed projects of various scientific French national research organizations and articles for many international journals and conferences. He has directed and co-supervised many dozens of MsEng and PhD theses and dissertations. He has participated in many academic councils and department committees. During recent academic years he served as director of the Dipl.Ing. degree major "Integrated Technical Systems."

Affiliations and Expertise

Professor and researcher on designing, modeling, and control of electrical systems with AVENUES Laboratory, Université de Technologie de Compiègne, France.

FOREWORD

At a period when mankind is implementing an energy transition, Manuela Sechilariu and Fabrice Locment's book very aptly provides us with useful insights into electrical smart microgrids (for buildings, villages, districts, or cities) and into the exploitation of renewable resources on such a geographic scale.

Only renewable energy resources will be up to the task of reconciling the needs of a world population of 10 billion with the constraints of sustainable development. In such a context, electricity is to play a major role as is already demonstrated by its growing share in the global energy mix. Because it is now easily and economically converted from renewable resources, electric power is an undeniable vector of progress, but it is essential to continue improving the efficiency of its distribution and its uses. In this respect this book contributes to offering, with great scientific rigor, solutions to this wide-ranging issue.

In 2014 approximately 22% of global electricity was from renewable sources, and its share has been progressing at an average annual growth rate of almost 6% over the past decade. That same year the share of nonrenewable sources was in decline because it had dropped to a growth rate of 2.8% per year over the same period. Photovoltaic and wind sources have the greatest potential and play a major part in the growth of renewable electricity. To optimize performance these conversion chains now systematically use electronic power converters and, naturally, deliver direct current (DC). For the same reasons electricity storage systems are also well suited to DC. Likewise, all of the modern uses of electricity are much better suited to DC use. Under these conditions the use of alternating current (AC; 50 or 60 Hz), which is still widely dominant, contributes to the complexity of power architectures. AC also leads to an increase of losses in unnecessary conversion stages and to a waste of raw materials and embodied energy.

All of us have heard of the wars of the currents that happened in the late 19th century, particularly in Europe and America. Most famous among the advocates of DC were Marcel Deprez in Europe and Thomas Edison in America, whereas among the defenders of AC there were engineers from Siemens and Nikola Tesla (Westinghouse). AC finally took over because there were, at the time, very good technological reasons to justify its supremacy. However, since the late 20th century a revolutionary technology has gradually come to the forefront—power electronics (solid-state conversion with power semiconductors). This technology is now almost everywhere and will now allow DC to regain ground over AC power. Of course the inertia of standards is a major obstacle, and it may be long before DC surpasses AC, but I am sure this will eventually happen!

DC power distribution, especially in buildings and urban areas, is to play a key role in an efficient use of renewable resources as well as in the securing of greater resilience from electrical systems. DC will produce better performing smart grids, which will be more reliable and more efficient all along their life cycle while saving energy resources and raw materials.

This book, which is based on scientific and technological research performed by the team of Manuela Sechilariu and Fabrice Locment, presents a very relevant synthesis of DC electrical architectures and power management methods. Technological aspects are thoroughly examined and give greater credibility to the book. The authors provide numerous energy models as well as management strategies and control laws at the different stages of power conversion. They focus on the conversion of solar renewable resources (photovoltaic conversion); energy storage systems; backup generators; and, of course, smart microgrids, which combine all of these aspects. Moreover, the numerous experimental results and associated simulations strongly contribute to the high quality of this book.

I hope this book will have many readers who, whether they are scientists or students, will no doubt appreciate the excellent quality of the work performed by the authors. Finally, I hope that this book will contribute to accelerating the sustainable energy transition that mankind so urgently needs.

Rennes, December 7, 2015

Bernard Multon
Professor at Ecole Normale Supérieure de Rennes
SATIE CNRS Laboratory

ACKNOWLEDGMENTS

We are heartily thankful to Professor Bernard Multon, French forward-thinking leader in renewable energy, whose leading-edge research in electrical engineering is making huge contributions to renewable energies field development. We are very grateful for his permanent encouragement and especially for his eloquent foreword, which introduces this book and highlights our expertise field, despite his very busy schedule. Thank you, Professor Multon, for inspiring all of us.

We would like to express our gratitude to several PhD students at Avenues Laboratory, Microgrid Research Team, for their scientific and technical contributions as well as some experimental results. Many of our scientific and technical papers in international journals and at conferences on the field of microgrids were coauthored with these PhD students whose theses were supervised by us.

We would like to thank and acknowledge the valuable support of CNRS (National Center for Scientific Research), ADEME (The French Environment and Energy Management Agency), FEDER (European Fund for Regional Economic Development), and CRP (Picardie Regional Council) that funded some of the research included in this book.

Our thanks are extended to the Université de Technologie de Compiègne for creating and maintaining an excellent academic environment that promotes innovation and technology; this had a positive impact on this research. Thanks also to our colleagues of the Urban Systems Engineering Department, Avenues Laboratory, and the LEC laboratory for a friendly and interesting working environment. We would like to extend special acknowledgement to our academic staff.

Special thanks to the team at Elsevier—in particular Raquel Zanol, Lisa Reading, Peter Jardim, and Natasha Welford—for their persistent proposition, careful consideration, and dedication to bring the idea of this book to its publication.

Lastly, we are thankful to all of those who provided support, read, wrote, and offered comments and review on this research work; we offer our regards to all of those who played a part and who supported them in any respect during the completion of this book project.

ABBREVIATIONS

AC	Alternating current
ACR	Automatic current regulator
AGM	Absorbent glass mat
AVR	Automatic voltage regulator
CEC	Californian Energy Commission
DC	Direct current
EDF	Electricité de France
FL	Fuzzy logic
HMI	Human-machine interface
HVDC	High-voltage direct current
IEEE	Institute of Electrical and Electronics Engineers
IGBT	Insulated gate bipolar transistor
ImP&O	Improved perturb and observe
InC	Incremental conductance
IoT	Internet of Things
IP	Integral proportional
LED	Light-emitting diodes
Li-ion	Lithium ion
LUT	Look-up table
M2M	Machine to Machine
MPP	Maximum power point
MPPT	Maximum power point tracking
NiCd	Nickel cadmium
NiMH	Nickel metal hydride
NOCT	Nominal operating cell temperature
P&O	Perturb and observe
PCC	Point of common coupling
PEL	Programmable electronic load
PI	Proportional integral
PLL	Phase-locked loop
PN	Petri Net
PV	Photovoltaic
PVA	Photovoltaic array
PWM	Pulse width modulation
SD	Science Direct
STC	Standard test condition
V2G	Vehicle-to-Grid
V2L	Vehicle-to-Load

GENERAL INTRODUCTION

1. CONTEXT AND MOTIVATION

Currently, the global environmental issue, in part because of the use of fossil/fissile fuels for electricity production, is a key concern in the various strata of society in many countries. To avoid an ecological crisis that will no doubt be more severe than the economic one, reduction of the environmental footprint, greenhouse gas emissions, and consumption of fossil/fissile fuels in favor of alternative energy is a mandatory crossing point. This is the global energy transition that means the passage of the current energy system using nonrenewable resources to an energy mix based mainly on renewable resources. This means developing alternatives to fossil and fissile fuels, which are finite and nonrenewable resources at the human scale. The energy transition provides for their gradual replacement by renewable energy sources for almost all human activities (transport, industry, lighting, heating, etc.).

The international community is becoming aware of the major environmental problems caused by human activity. The World Energy Council is an international organization supporting accessible and sustainable energy development across the planet. It highlights that to provide sustainable energy policies it is important to take into account the three following dimensions:

- *Energy security*: The effective management of the primary energy supply from domestic and external sources, the reliability of the energy infrastructure, and the ability of energy providers to meet current and future demand.
- *Energy equity*: Accessibility and affordability of the energy supply across the population.
- *Environmental sustainability*: The achievement of supply- and demand-side energy efficiencies and the development of energy supply from renewable and other low-carbon sources.

Thus the energy transition also induces a behavioral and sociotechnical transition, involving a radical change in energy policy as moving from demand-oriented policy to a policy determined by supply along the possibilities of distributed production. This is also to avoid overproduction and unnecessary consumption to save more energy and benefit from better energy efficiency.

The public power grid that operates today is confronting the demands to improve reliability, reduce costs, increase efficiency, comply with policies and regulations concerning the environment, integrate renewable energy sources and electric vehicles to the power grid, etc. The promising smart grid can meet these priorities. This network is designed primarily for information exchange concerning the requirements and availability of the power grid and for help balancing power by avoiding an undesirable injection and performing smoothing of loads during peak hours. The smart grid is defined as the power grid that uses innovative monitoring, controls the transmission of information, and uses self-healing technologies to provide better services to electricity producers and distributors, flexible choice for end users, good reliability, and security of supply. This very complex smart grid, with bidirectional power flow and communication, requires much work to implement it in reality.

On the other hand, the electricity production seeks to produce more and more energy from renewable sources (wind, solar, biomass, and geothermal sources), but integrating power from renewable resources into the utility power grid (ie, public grid) can be a huge challenge. The intermittent and random production of renewable sources is always a problem for their large-scale integration into the power grid. There is not yet a worldwide standard for smart grid topology, but regarding better integration of renewable sources of low and middle power, microgrids seem to have an important place. A microgrid consists of renewable and traditional sources, energy storage systems, and controllable loads that can be adjusted. A microgrid allows the connection with the public grid and ensures ancillary services (control of the voltage and frequency fluctuations), energy flow, load sharing, and load shedding during islanding, and it takes into account the constraints of the public grid transmitted by the smart grid through the smart grid communication bus. Thus around the world researchers and engineers are deploying increasing efforts to design and implement intelligent microgrids to achieve the energy goals of the 21st century, such as improved reliability based on diversification of sources of electricity production. Nevertheless, ensuring reliable distribution of electricity based on a microgrid and realizing its integration into the centralized larger production of the power grid are not easy to achieve.

Regarding environmental sustainability, one of the most energy-intensive sectors is the construction sector, representing in the near future almost a half of total energy consumption and a quarter of greenhouse gas emissions released into the atmosphere. Regarding environmental challenges, the building sector is now positioned as a key player to achieve the energy transition. It could be the only one that provides opportunities for considerable progress to meet the international community commitments to reduce greenhouse gas emissions. In fact, it is found that progress pathways in the building sector can be identified much better now than in previous years, especially because of the aspect that buildings can use several energy sources, including renewable energy. In addition, the buildings' occupants have energy use behaviors relatively constant over time; their needs change over long cycles, with no abrupt break, and can be reasonably anticipated.

Therefore it is essential today to reduce the environmental impact of existing and future buildings and to find solutions to reduce energy consumption and increase the share of renewable energies. The trend is actually to give more and more "local power" to urban areas to control the energy distribution and production. Everything should be set up, throughout the territory of the city, to provide the opportunity and the desire to produce its own energy. Thus many calls for project proposals are launched for the creation of positive energy buildings and territories, which undertake a path to achieve the balance between consumption and production of energy at the local level aiming for renewable energy source deployment. To mitigate the intermittency and randomness of different renewable energies, the engineering and technology of the smart grid are being developed at full speed, representing a new industry. The occurrence of the smart grid is launched at all scales: building, block, neighborhood, city, and between territories. Many countries are currently dealing with the formidable problem of financing of the smart grid and microgrids—from research project to implementation of experimental facilities that can become a demonstration and pilot site.

In this context, in urban areas and for buildings equipped with renewable energy resources, the building-integrated microgrid, with off-grid/grid-connected operating modes, can become an answer to these technical difficulties. This microgrid represents a form of local power generation, often multisource, and it can operate in grid-connected and in off-grid operating mode. The off-grid aspect is given by the fact that the energy production is intended mainly for self-feeding. Because of the grid connection, the microgrid can receive power from the public grid. Moreover, excess power can be traded back to the utility grid or directed toward other urban microgrids.

2. BOOK OVERVIEW

On the basis of a representative microgrid in an urban area and integrated in a building, this book focuses on increasing integration of renewable electricity sources to obtain a robust electricity grid, to solve consumption peak problems, and to realize optimal energy and demand-side management. Assuming that the locally generated renewable electricity is consumed where, when, and in the form in which it is produced, with a public grid seen as a backup source, the building-integrated microgrid is a solution for self-feeding and injection-controlled electricity.

Research works conducted on microgrids have greatly increased in the last years. However, systemic study of microgrids integrated in urban areas is still rare. Two important aspects must be particularly noted:

- the control aiming the power balancing and power flow optimization are often studied separately; and
- regarding the power flow optimization based on a predictive model, its validation is often only demonstrated by simulation.

Thus, knowing that optimization is based on forecasting data, the main scientific lock is implementing the optimization method in real-time operation. However, the real operating conditions are usually different from those of the prediction. The uncertainties can degrade or cause failure of the operating system, and it is believed that innovation is still needed to propose an implementation of robust optimization to deal with the prediction uncertainties.

Thus the objective is to study, design, analyze, and develop an urban direct current (DC) microgrid that integrates photovoltaic (PV) sources, which represent the most common renewable energy sources for urban areas. This system should be able to extract maximum power from the PV plants and manage the transfer of power to the load (ie, building distribution network) while taking into account the connection with the public grid, the available storage elements, and other backup sources. Highlighting the scientific issue on the implementation of an optimization in real-time operation, the ultimate goal is to achieve, through an intelligent control, an optimized local electricity production-consumption, in secure mode, with controlled power injection while taking into account the public grid vulnerabilities. The main application is represented by passive and positive energy buildings as well as positive energy territories.

Moreover, this book intends to make a further contribution toward the conceptualization of microgrid control in which smart grid communication is integrated. The goal is to design an intelligent control of local energy by a hierarchical control that allows a decentralized and cooperative architecture for power balance as well as for power flow optimization. Specifically, this advanced local energy management and control takes into account forecasting data related to PV power production and load power demand while satisfying constraints such as storage capability, grid power limitations, grid time-of-use pricing, and grid peak hours. Optimization, the efficiency of which is related to the prediction accuracy, is performed by mixed integer linear programming. Experimental results show that the proposed urban DC microgrid is able to implement optimization in real-time power control and ensures self-correcting capability. The power flow can be controlled near optimum cost when the prediction error is within certain limits. Even if the prediction is imprecise, power balancing can be maintained with respect to rigid constraints. The proposed supervisory control can respond to issues of performing peak shaving, avoiding undesired injection, and making full use of locally produced energy with respect to rigid element constraints.

The purpose of this book is to provide a clear and concise overview of the DC microgrid field and covers in particular a discussion of building-integrated microgrids able to provide smart grid communication in urban areas. This book is based primarily on research works conducted by the authors in the last 8 years at the Université de Technologie de Compiègne, France, aiming at a contribution to advanced energy management and optimization. To validate the proposed models and associated controls, the development of these research works led to the realization of an experimental platform installed in the Center Pierre Guillaumat 2 of the Université de Technologie de Compiègne. Using this experimental platform several tests were conducted and the results validate the design and choice of models and the technical feasibility of the proposed microgrid system.

3. BOOK CHAPTER ORGANIZATION

The content of this book is organized into six chapters, which explore, by adopting a systemic approach, the DC microgrid subject from the modeling and command of each DC microgrid component to the needed DC microgrid interface control. The systemic approach allows for generating evaluation criteria of the proposed models and controls that respond to specific needs such as an urban DC microgrid requires. Another important aspect concerns some considerations about the quality of the energy produced by the energy sources: to meet signal quality requirements of the public grid with any power injection or power absorption and to satisfy the DC load while ensuring the stability of the DC bus voltage. Moreover, the importance of component modeling is highlighted in the design of the interface needed to connect the microgrid with the smart grid. This interface, also called supervisory control, aims to manage the balance of the instantaneous power based on an optimization of total energy costs for a determined time period with respect to several constraints. The main issue is the difficulty of global optimization related to offset risks between forecasting, planning, and operational reality on the one hand and the need to take into account the criteria, constraints, and requirements of the public grid on the other hand.

Chapter "Connecting and Integrating Variable Renewable Electricity in Utility Grid," first provides a review of current problems in the utility power grid and defines the smart grid concept as a solution for traditional utility grid issues. Then, in the context of variable renewable energy integration, the alternating current (AC) and DC microgrids are presented and described, and their general issues are analyzed. The urban context places the microgrid concept in association with the smart building and smart grid, resulting in forming a smart city. Regarding improved energy efficiency at the local level, the DC microgrid is presented versus the AC version. The overview concerning the dynamic interactions between the microgrid and the smart grid, on the one hand, and the context of major preoccupation in urban areas, on the other hand, enable charting future urban energy strategies. Finally a brief description of the experimental platform is provided.

The variable renewable source integrated in the urban DC microgrid, which is the PV generator, is completely explored in chapter "Photovoltaic Source Modeling and Control." This chapter is devoted to the modeling and control of the PV generator. PV modeling aims at supporting the achievement of numerical simulations of the PV source system, which must be valid for all weather conditions associated with the operation. To meet the needs of developing an energy microgrid model, the reliability of the PV model must be high enough to provide credible PV production forecasts. Two classic mathematical models presented in the literature are studied and analyzed. Then, to overcome the weaknesses of previous models, a third model, the purely experimental model, is proposed. The experimental comparison of three models allows for highlighting the application limits of each model against the criteria of choice. Regarding the control of the PV source, it must meet the maximum energy performance requirements at a reasonable cost, focusing not only on maximum power point tracking (MPPT) methods but also on constrained power control algorithms. Four MPPT algorithms and their energy performance are explored and proposed for experimental validation under different weather conditions. The analysis resulting from this experimental comparison validates the choice of the control method. In addition, an algorithm for PV constrained power limiting control is proposed and experimentally validated.

In chapter "Backup Power Resources for Microgrid," the possible backup power elements to be integrated in the microgrid system are introduced. Depending on the microgrid operating mode, grid-connected or off-grid, these backup elements must be able to provide continuous energy for the microgrid internal load. After a brief review of different backup resources, the electrochemical storage based on lead-acid batteries is presented as the most common storage for stationary application including microgrids. Taking into account the main characteristics of electrochemical batteries, this section presents the modeling of a lead-acid accumulator. Several electric models are studied, and then an experimental lead-acid battery model is provided. These models allow for increasing knowledge on batteries and to analyze and study their relevance to microgrid modeling needs, especially for state of charge control. For the off-grid operating mode, the diesel generator is proposed as a traditional source to secure microgrid operation. To better understand the complex problem of the sluggish dynamic of a prime mover and to decide the realistic working hypothesis, the characteristics, operating principle, and operating cost analysis for a diesel generator are presented. Regarding the public grid connection, a linear control based on pulse width modulation and a conventional structure for the voltage and current synchronization are provided. The experimental validation of this grid connection allows for emphasizing the problems of absorption or injection at low power and an improvement is proposed and detailed.

Chapter "Direct Current Microgrid Power Modeling and Control," presents and provides details on DC microgrid power system modeling and power balancing control. The interaction between the smart grid messages and the DC microgrid is taken into account. The overall power balancing control strategy requires identifying system constraints and operating mode coordination between different sources. Thus the DC microgrid power system model is designed by interpreted Petri Net formalism and the Stateflow tool provided by MATLAB-Simulink software. On the basis of power system behavior modeling, the power system control strategy is designed with consideration of each element's constraints and their behavior. Simulation and experimental results of several day tests validate the basic power balancing control for grid-connected and off-grid operating mode. The proposed control algorithm ensures power balancing while respecting all element constraints.

The DC microgrid supervisory system design is presented in chapter "Direct Current Microgrid Supervisory System Design." On the basis of the hybrid dynamic systems and designed as a multilayer structure, the proposed supervisory control combines power balancing, energy management, and smart grid interaction. The aim of supervisory control is to provide a continuous supply to the load with an optimized energy cost under multiple constraints. The entire supervisory control is developed for grid-connected mode and off-grid mode, which includes power balancing, optimization, prediction, and interactions with the smart grid and end user. This control system is presented layer by layer: the human-machine interface allowing for end-user interaction; the prediction layer regarding the PV and load power prediction calculations based on metadata communication; the energy management layer that proceeds to the energy cost optimization and proposes a day-ahead optimized source scheduling; and the operational layer, which maintains power balancing, taking into account multiple constraints so that they together handle the power balancing control, PV MPPT, and constrained power commands as well as load shedding control. The supervisory system is evaluated by simulation tests for different cases. The obtained results show the control system ability to efficiently manage an optimized power flow while maintaining power balancing in any case. The effect of optimization on total energy cost is proved by comparison with a non-optimized power flow control. Nevertheless, the optimization effect relies on prediction precision.

Experimental tests for different cases and for grid-connected mode and off-grid mode are performed and described in chapter "Experimental Evaluation of Urban Direct Current Microgrid." This chapter presents the assumptions under which the tests were performed and provides details on the experimental platform. To analyze DC microgrid validation and its technical feasibility, three case studies, corresponding to three different types of solar irradiance evolution, are proposed and discussed following the two operating modes (ie, grid-connected and off-grid modes). The obtained results show that supervisory control can maintain power balancing while performing optimized control, even with the uncertainties of prediction and arbitrary energy tariffs. In addition, these experimental results highlight the strong influence of the solar irradiance evolution and induced prediction errors. To summarize, the obtained experimental results validate the feasibility of DC microgrid control, which provides an interface for energy management and smart grid interaction, in real-time operation with respect to rigid constraints. The feasibility of implementing optimization in real-time operation is also validated even with uncertainties.

The last part of this book is represented by "General Conclusions, Future Challenges, and Perspectives." The first challenge was to develop an algorithm for the balance of powers in accordance with all requirements. A bottom-up approach is proposed using modeling by Petri Net to analyze the behavior of each element and constructing an algorithm for the power balance. This approach facilitated the identification of an interfacing variable for coupling the operating layer and the energy management layer. Regarding optimization, the scientific challenge has been to conduct optimization under multiple constraints. For the public grid it is necessary to reduce consumption during peak hours and avoid undesired energy injections. In addition, limitation of the PV output power and partial load shedding should be minimized by optimization. The optimization is formulated to minimize the energy cost while respecting the imposed constraints. For this the linear mixed integer programming is applied. Grid-connected and off-grid procedures are explored and experimentally validated. In on-grid mode the microgrid answers the questions related to peak shaving and reduces undesired energy injections. In off-grid mode the microgrid is able to minimize the fuel consumption of a diesel generator. Second, the technical challenge was to process data, constraints, and measures 1000 times during a test day. At this scale the optimization problem is difficult to solve by an already programmed function in MATLAB. The solution is to formulate the problem with the C++ programming language and using the CPLEX solver, which is very effective in solving this type of problem. Despite the uncertainties on the calculated predictions, the experimental results show that the proposed microgrid structure is able to effectively manage the power flow, ensuring the balance in all cases through the proposed operational control algorithm, which provides autocorrective actions. It can be concluded that the proposed optimization leads to good predictive energy management while minimizing load shedding and/or limiting PV production. Despite the uncertainties the optimization implementation feasibility in real operation, through a simple interface while respecting the constraints, is validated. The uncertainties do not disturb the balance of powers, but the optimization of performance is related to the prediction information. This can be improved by updating the real-time optimization model.

This research study is ongoing, and many future challenges are presented as works in progress or mid- and long-term perspectives on the potential of microgrids to better meet the needs of the end user and the public grid, which must facilitate the implementation of the future smart grid.

CHAPTER 1

Connecting and Integrating Variable Renewable Electricity in Utility Grid

1. SMART GRID—SOLUTION FOR TRADITIONAL UTILITY GRID ISSUES

In the new energy landscape, the increasing power consumption requires maintaining power grid safety and reliability with permanent innovations in electricity flow regulation, with less mismatching between electricity generation and demand and integration of renewable energies. In addition to the performance of load demand management, optimizing scheduling, improving energy quality, improving assets efficiency, integrating dynamic pricing, and incorporating more renewable electricity sources, the continuous challenge of the traditional utility grid is power balancing. Even if the supply interruption rate and accumulated duration is very weak today, the power generation, transmission, and distribution remain vulnerable because of major changes undergone by this system in the context of current environmental, technical, and economic constraints. Power grid fluctuations in power demand and power generation, even for few seconds, induce an effect causing the commissioning of additional conventional production units. These conventional production units are based on fossil primary energy (gas, oil, coal) and form the spinning reserve of the utility grid. Thus, to ensure the balance between power generation and increasing power demand, the number of conventional production units in operation must grow. To reduce the spinning reserve, power fluctuations could be minimized by better integration of renewable energy generation and increasing the power demand response (temporary changes to electric loads in response to supply conditions).

Facing the increase of energy demand, environmental problems, and decreasing fossil energies, the renewable energies have to be integrated in the utility grid. Indeed, to reduce the greenhouse gases of power generation, the existing utility grid has already incorporated renewable energy

Urban DC Microgrid
ISBN 978-0-12-803736-2
http://dx.doi.org/10.1016/B978-0-12-803736-2.00001-3

resources as the necessary complement to traditional electricity generation. Nowadays, the distributed power generation is based on systems that may be classified as:

- a grid-connected system, with a total and permanent power injection;
- a stand-alone system, seen as a substitute of utility grid connection, usually for remote sites; or
- an off-grid/grid-connected and safety network system.

Because of the renewable energy purchase conditions, the grid-connected system for permanent energy injection is proposed in most applications, especially for variable renewable electricity generation such as wind turbine generators and photovoltaic (PV) sources. However, knowing that this kind of renewable power generation is very intermittent and random, this increased permanent injection of energy tends to cause grid-connection incidents, which have become true technical constraints. If such continuously growing production is injected into the grid without control, regardless the spinning reserve expanding, then it will increase the power mismatching in the utility grid and cause fluctuations in voltage and frequency [1]. Therefore, the vulnerability of the utility grid could drastically increase. This is because the variable renewable electricity generation, which is hardly predictable and very unsettled, is not participating in technical regulations for grid connection (setting voltage and frequency, islanding detection, etc.) and behaves as passive electric generators [1]. In response to these technical constraints, research is being performed on grid integration of decentralized renewable energy generation [2] or developing new supervision strategies as high-level energy management control [3].

Concerning grid-connected systems, many studies have been performed and solutions have been proposed on power electronic converters [4], a complex systems approach [5], and grid system connection [6]. However, because of the absence of the grid-integrated energy management, the development of renewable energy grid-connected systems could be restrained, especially by the power back grid capacity in real time [7].

Energy storage seems to be a perfect solution to handle the intermittent nature of renewable energy, but it has limitations based on available technologies, capacity, response time, life cycle cost, specified land form, and environmental impact [8—10]. For a large-scale renewable energy plant, such as a wind farm, the pumped-storage hydroelectricity station is a promising technology to deal with the random production of renewable sources [9]. This technique is the most cost-effective form of current available grid energy storage. However, capital costs and the requirement of appropriate terrain cannot generalize this solution. Recent progress in grid energy storage makes hydrogen technologies (combined fuel cells and electrolyzers with hydrogen tanks) an alternative to pumped storage [11]. In contrast, for a small-scale plant such as building-integrated renewable generators, there is little innovation to overcome the lack of grid-interactive control for grid-connected systems. For PV systems, lead-acid batteries are commonly used as storage because of the low cost with regards to their performance. However, considering a limited storage capacity, an energy management strategy needs to be developed to optimize the use of variable renewable energy for high penetration level.

Given the intermittent nature of renewable sources, the major problem associated with the stand-alone systems is the service continuity, from whence the energy storage and the number of conventional sources are required. The studies in this axis concentrate more on the techno-economic feasibility conditions, optimized storage sizing, and load management, as in [12—14].

Therefore the distributed energy generation shows a very rapid growth and reveals an increasing complexity for grid managers due mainly to prosumer sites (ie, producer and consumer sites). The intermittent nature of renewable energy sources (eg, PV and wind turbine generators) remains an issue for their integration into the public grid, resulting in fluctuations of voltage and/or frequency, harmonic pollution, difficulty for load management, etc. This leads to new methods for power balancing between production and consumption [5].

Fig. 1.1 shows that the electricity landscape includes electricity production sites on the one hand and electricity production/consumption sites on the other hand.

As mentioned previously, the intermittent nature of renewable sources leads to new methods for balancing of production and consumption. Indeed, the production/consumption sites, also called energy prosumer sites, involve a bidirectional power flow that was nonexistent in the last decades and for which the traditional utility grid was not designed.

Therefore in this context and in terms of energy and territory scale, the power balancing, such as main regulation, does it have to remain centralized or could it also become a local regulation? Which is the best way to integrate variable renewable energy sources to accommodate the needs of

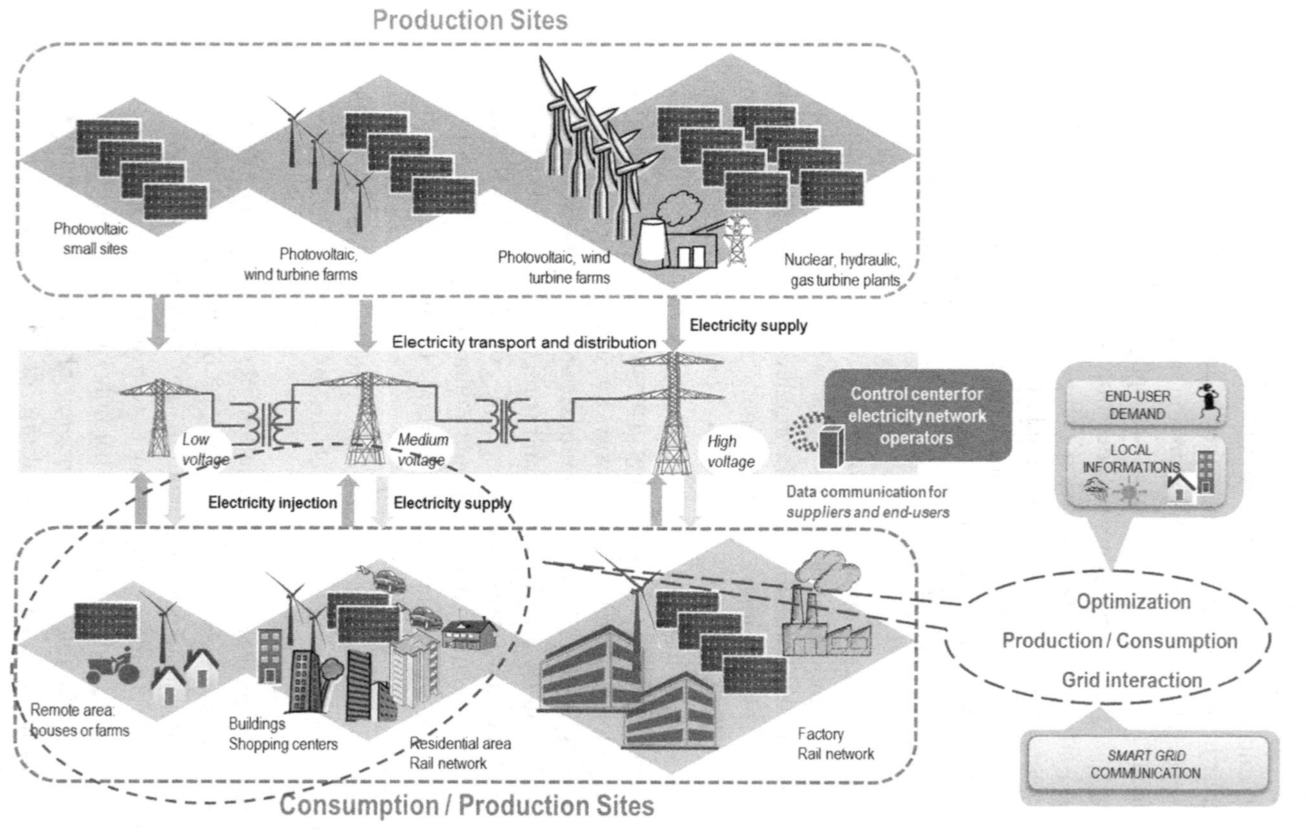

Figure 1.1 General overview of the electricity landscape.

utility grid in real time? On the other hand, information of grid needs and availability could assist in power balancing by avoiding undesired injection and performing load shaving during peak hours. For this, the smart grid is being created to facilitate information exchange [15]. The concept of the smart grid, born in recent years, seems to be the solution for energy prosumer sites to reduce losses and mitigate energy demand peaks in the territory scale and to operate a local grid regulation through data communication, energy management optimization, and interaction with the whole utility grid.

So, what is the smart grid?

- "The smart grid is ultimately about using megabytes of data to move megawatts of electricity more efficiently and affordably." (Definition given by Ontario Smart Grid Forum report, May 2011.)
- "Smart grid" generally refers to a class of technologies that people are using to bring utility electricity delivery systems into the 21st century, using computer-based remote control and automation. These systems are made possible by two-way digital communications technologies and computer processing that has been used for decades in other industries. They are beginning to be used on electricity networks, from the power plants and wind farms all the way to the consumers of electricity in homes and businesses. They offer many benefits to utilities and consumers—mostly seen in big improvements in energy efficiency and reliability on the electricity grid and in energy users' homes and offices." (Definition given by the US. Department of Energy.)

To summarize, smart grid could be defined as the electricity delivery system, which transports, converts, and distributes the power efficiently (from producers to consumers), integrated with communication and information technology.

The main goal of smart grid communication is to assist in balancing the power generation and the power consumption. The smart grid is a very complex network with nonlinearity, randomness, bidirectional power flow, and bidirectional communication. Consequently, supervising the status of the whole system and dealing with the large-scale real-time data remain an open problem despite the technologies of smart devices and communication protocol.

On the other hand, to achieve a high level of renewable energy penetration into a grid, strategies and means of power management should be developed to build a more robust utility grid. Moreover, to avoid undesired injection and performing load shaving during peak hours, information on grid needs and availability are very important. For this the smart grid is supposed to facilitate information exchange.

2. MICROGRIDS

Using sensors, communications, and monitoring technologies, the smart grid will assist in power balancing and the power consumption scheduling. In this context and turning to the necessary integration of variable renewable energies, microgrids are proposed as a key component of the grid power energy mix [5]. A microgrid includes a multisource system consisting of renewable and conventional energy sources, storage systems, and controllable loads. A full interface controller between the utility grid and the microgrid is used to interact with the smart grid; it provides voltage control, power balancing, load sharing, or load shedding, and it takes into account the constraints of the utility grid provided by smart grid communication. Therefore concerning ancillary services (power grid technical regulations), for better decentralization of production, microgrids play an important role.

According to recent studies, a microgrid is a form of distributed energy generation, able to operate grid connected and off-grid. In grid-connected operating mode, a microgrid can exchange power with the utility grid (to absorb or to inject power); when the power generation and demand are equal, the power transferred between the microgrid and the utility grid is zero. If the microgrid can be connected with the utility grid but the system is working independently, then it is named islanded operating mode. However, the islanding operation is highly sensitive and requires an important control because of the lack of microgrid inertia. In such case, even the slightest perturbation coming from sources or loads can produce large, transient deviations in voltage and frequency.

Otherwise, if the physical connection is absent, it is called isolated operating mode. During off-grid operating mode, a microgrid should be able to continuously provide enough energy to feed the majority part of its internal load as self-supply or energy sharing with other microgrids. Therefore storage participation and demand-side management are used to increase the security of loads supply. However, the microgrid systems are usually prepared to work in all of them, grid-connected, islanded, and isolated operating mode, the choice of the operating mode will depend on the state of the utility power grid and the microgrid system. Each of these operating modes has its own control scheme that follows the requirements of the utility grid and microgrid.

2.1 Alternating and Direct Current Microgrid

It is generally considered that a microgrid can operate as a single controllable system; it controls on-site generation and power demand to meet the objectives of providing local power, ancillary services, and injecting power into the grid if required. Thus microgrids combine power balancing control and energy management to provide the ability to adjust the grid power level at the point of common coupling (PCC). Considering the common bus in which sources and loads are connected, microgrid systems can be distinguished between alternating current (AC) bus or direct current (DC) bus.

Typical microgrids interconnected with the main grid through PCC are shown in Fig. 1.2.

Renewable energy sources and loads are connected to the common bus through converter interfaces. Depending on bus voltage value, AC loads for AC bus and DC loads for DC bus may eliminate the converter interface.

During normal operation, an AC microgrid is connected to the utility grid and the local loads are supplied mainly by the power produced by the microgrid and, depending on grid availability and energy tariffs, by the utility grid. However, the excess power produced by renewable energy sources may be injected in the utility grid. In most cases, the AC microgrids adopt the voltage and frequency standards applied in most conventional AC distribution systems.

A DC microgrid, as a typical system, is presented in Fig. 1.2b. The DC microgrid operation is similar to the AC microgrid. The common DC bus approach is applied to avoid some necessary energy conversions in the AC architecture. However, the DC bus architecture requires an energy

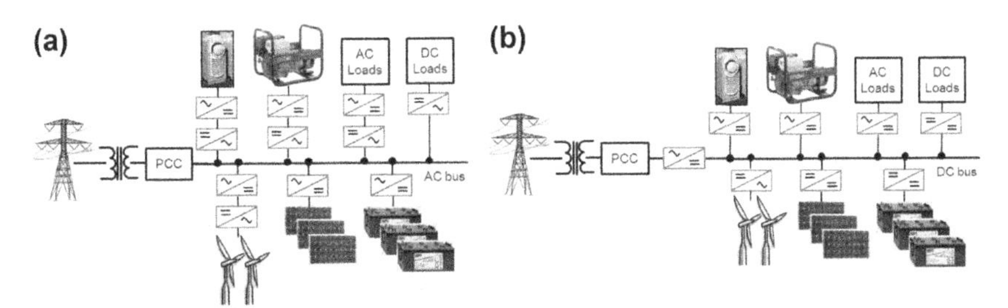

Figure 1.2 (a) AC microgrid architecture and (b) DC microgrid architecture.

conversion in the PCC to be able to exchange power with the utility grid. The DC microgrid approach is more interesting from the efficiency point of view; it is presented in detail in Section 3.2.2.

By combining AC microgrid with DC microgrid, a hybrid AC/DC microgrid can be also defined. It consists of an AC sub-microgrid directly connected to the utility grid, a DC sub-microgrid connected to the utility grid through an interlinking converter, and power electronics interfaces between AC and DC buses [16]. In hybrid AC/DC microgrids the storage systems can be theoretically installed in either AC sub-microgrid or DC sub-microgrid, but its implementation is not trivial and should be highly studied and designed by considering load types, power flow, operational reliability, and cost. In addition, the design of a hybrid microgrid control able to handle all operating modes is still a challenge.

Turning to DC microgrid architecture, the AC microgrid presents some advantages. For AC systems, the AC voltage can be easily and economically transformed either increasing or decreasing by electromagnetic transformers without control strategy. The DC voltage magnitude conversion requires power electronic devices associated with measurements and complex control. Regarding the electrical circuit protection for DC microgrids, the existing technologies are not yet fully developed and still remain immature.

In contrast, because of the nature of the incorporated renewable energy sources, the DC microgrid can offer several main interests compared with AC microgrids. The DC native renewable energy generators (ie, fuel-cells and PV sources) and the electrochemical storage (ie, batteries and super-capacitors) can be more easily and efficiently integrated into the DC microgrid, and the DC/AC energy conversion stage is not involved. Knowing that the AC microgrids use AC/AC converters, which are generally based on AC to DC and then DC to AC conversions, DC microgrids may save up to 20% of energy losses by using mainly DC/DC conversions, depending on various devices and/or their rated power [17]. In addition, DC systems do not suffer from skin effect, therefore thinner cable can be used with improved material efficiency.

Coupling AC sources on a common AC bus, which is the case for AC microgrids, requires synchronization; that is, all conversion stages have to operate at the same voltage, frequency, phase sequence, and phase angle. Concerning DC microgrid, which is a zero-frequency system, only the voltage amplitude should be regulated and the synchronization is not required when connecting DC sources on the common DC bus. In addition, because no reactive power is present in the DC bus, connecting

AC sources on the common DC bus that can run only with active power increases the power efficiency and power transfer ability.

DC microgrid involves better current control because there is no negative and zero sequence currents, which cause problems in an AC microgrid.

AC microgrids currently have more advantages in certain high-voltage and high-power applications whereas DC microgrids may provide advantage in low-voltage levels of the power delivery network.

2.2 Research Issues in Microgrids

Nowadays, microgrids are very frequently proposed in distribution power networks, and this increasing use may certainly change the traditional topology of the network. Much more recently, microgrids are also studied and conceived to be connected to the transportation power network. DC microgrids are expected to bring energy efficiency, especially for the high-voltage DC (HVDC) network section. Nowadays, there are many research works on different aspects of microgrids and microgrid studies vary from one application to another. Nonetheless, it could be highlighted that the main research issues in microgrids are linked with power quality, protection, energy management, communication, dynamics, control, economy, secure operation, and so forth. However, control, protection system and devices, and energy management seem to be the most important.

2.2.1 Control

The utility grid is traditionally able to absorb large, transient, active and reactive power imbalances to maintain the voltage and frequency by means of a physical prime mover of AC generators; in general, a synchronous generator reduces the frequency when the real power increases. This method, called droop control, allows operation, when several parallel connected generators supply an electrical variable load, so that loads are shared among generators in proportion to their power rating.

Some microgrids, especially AC microgrids that include a physical prime mover, operate following the same principle. Because of the increased penetration of distributed generation, other microgrids operate based on a virtual prime mover; voltage and frequency should be controlled by other types of appliances such as a converter interface. In this case, the converter imitates the behavior of a synchronous generator and regulates the frequency and the magnitude of voltage reference to achieve proper power sharing.

Thus, to achieve the microgrids' control goal, many power sharing control schemes have been proposed according to the following practical methods: central control, decentralized control, and hierarchical control.

By using a powerful and high-speed communication system, the central control basically measures microgrid state variables, calculates, and dispatches power references to each source so that all sources can simultaneously generate proper real and reactive powers to maintain the microgrid as stable. To go further, to enable functions as economic dispatching, active energy management, and other optimization calculations, a microgrid central decision-making controller is required [18]. Therefore the control functions become similar to hierarchical control systems. However, the central character of this control lies in the fact that it has to gather all data available in the microgrid, to make decisions based on these data, and to transmit power references to sources and sometime even to loads. Depending on the power scale and distances between sources, this control is very complex, sensitive, and difficult to implement; it is also expensive because of the powerful communication system.

The decentralized control, supposedly without communication, is based on individual converter interfaces able to respond automatically to variations in local state variables and to guarantee stability on a global scale. The converters are supposed to actively damp oscillations between the output filters and to prevent any voltage offsets on the microgrid. The microgrid decentralized control may be applied to radial microgrid topologies rather than meshed topologies for which the control concept is still immature.

The aforementioned controls, considered mostly for islanding and grid-connected microgrid applications, result in problems associated with the high-speed communication and local measurements, respectively.

The hierarchical control, which is often applied for microgrids to operate in grid-connected and islanded modes, is based on the above methods and three control levels that are defined as primary control (ie, low-level or bottom-level controller), secondary control, and tertiary control (management level). For AC and DC microgrids, the hierarchical control requires communication between the sources and the central controller and it can be implemented in grid-connected and islanding operational conditions.

For AC microgrid hierarchical control, the primary control is indeed a droop control method, including an output virtual impedance loop. The secondary control allows for the restoration of deviations produced by the primary control whereas the tertiary control manages the power flow between the microgrid and the utility grid at the PCC.

Concerning the DC microgrids, the primary control handles only the DC bus voltage stability when variable local loads are connected to the DC bus so that the controller obtains an equal or proportional DC load current sharing. Because the secondary control aims to restore the deviation induced by the primary control, in this case it has to eliminate the DC bus voltage deviation. Being similar to the AC microgrid, the tertiary control regulates the power flow at the connection point to external DC microgrids.

The tertiary control is used mostly for the grid-connected operating mode to adjust the power flow of AC or DC microgrid following some optimization objectives: minimizing total energy cost for the end user and the microgrid owner, improving demand-side management, improving demand response, reducing pollutant emissions, and improving microgrid security.

2.2.2 Protection

The protection system for microgrids consists of protection devices, protective relays, breakers, measurement equipment, and grounding methods. Because microgrids introduce bidirectional current, new fault detection and protection control schemes and algorithms have to be conceived and developed according to the microgrid architecture and operating mode (ie, grid-connected, islanded, or isolated modes).

Conventional relays, breakers, and other protection devices may still work in AC microgrids for a traditional operating mode. It is obvious that the protection system for an AC circuit is more mature than for a DC circuit. This is mainly because for AC systems the line impedance is much higher than for DC systems. Furthermore, for AC circuits the use of an electromagnetic transformer allows for withstanding more overloading than for DC circuits, for which a slight overloading for milliseconds over the power electronic device rating could permanently damage the power converter. Concerning the fault current control, according to its high impedance, the AC current can be better limited in proper time through the protection devices. In DC systems the low DC "impedance" induces high rising rate in the fault current, resulting in more challenges for the design of a fast response DC circuit breaker.

An AC circuit breaker is much more mature than a DC one. Forced by AC voltage, AC current has natural zero-crossing features whereas DC current is always persisting. Breaking the DC circuit is much more difficult, especially in a high current rating, than breaking AC current.

Regarding the DC circuit breaker, the main issue is related to the lack of a natural zero crossing of the current; hence, the formation of a significant electric arc could occur during the breaking procedure. Several solutions based on the static switches (especially insulated gate bipolar transistor [IGBTs]) have the advantage of speed and reliability; however, the losses are large. Other studies show that a hybrid structure, static and electromechanical, would be the best technical compromise, but this remains an expensive solution. Thus DC microgrid protection faces the challenges posed by the lack of standards, guidelines, and practical experience.

In the coming years, new protection solutions are expected to be provided to ensure safe and secure operation of microgrids.

2.2.3 Energy Management

Microgrids are able to drive smart energy management. The objective is to fully use each renewable energy source while respecting their constraints on capacity and power by managing power flow in the microgrid and power flow exchanged with the utility grid during the grid-connected mode. Using data communication by a smart grid, this operation can be optimized and the energy global cost can be reduced; thus the overall utility grid performance is enhanced.

To build strategies for better energy management, information on power and energy requirements and the availability of the utility grid are very important. This could help balancing power by avoiding an undesirable injection and/or by performing a partial load shedding during critical hours. Thus the concept of a smart grid that involves innovative and intelligent monitoring, transmission of information, and self-healing technologies has to provide better services to producers and electricity distributors, flexible choice for end users, reliability, and security of load supply. Furthermore, the electricity market will become more responsive and treatment of dynamic pricing over time will become possible. Because the smart grid is a very complex network with nonlinearity, random flow of bidirectional power flow, and bidirectional communication, significant research work needs to be done. Despite the development of intelligent systems technologies and various communication protocols, managing the status of the entire utility grid, from the producer to the end user, and processing Big Data on a large scale in real time remains an open problem. In response to this global issue, recent research involves knowledge of the messages transmitted by the communication network of a smart grid. In Ref. [19], the authors propose a power management, situated at a high level in the hierarchical architecture of the public grid, using an approach based on multiagent systems. They validate this supervision for an application of a hybrid distributed generation combining renewable and conventional sources. Other studies, such as Ref. [20], propose local energy management based on real-time tracking of distributed generation sources depending on operating modes.

In general, microgrid energy management aims at economic power dispatching, often operated a day-ahead, and online optimization while performing power balancing at the local level and ensuring load supplies, and it improves reliability and power quality. The energy management can be classified into rule-based and optimization-based approaches. A rule-based approach manages the system according to prefixed rules, such as simple rules, defined by a multiagent system and based on fuzzy logic approaches. An optimization-based approach manages the system by mathematical optimization, performed with objective function and constraints. The optimization methods include the artificial intelligence joint with linear programming, linear programming or dynamic programming, and genetic algorithms.

The rule-based method is simple and robust, but it does not guarantee the optimal performance with given operating conditions. Moreover, rules become complex when facing different scenarios. The optimization method gives an optimal solution within given constraints and operation conditions. Nevertheless, optimization is usually treated as a separated problem from the power balancing strategy on the one hand, and it requires a priori information on energy production and energy consumption on the other hand. Error between the forecasted and real condition could result in degradation of optimization performance or even undesired operation that may violate certain constraints, and then the system would no longer be able to operate. Hence, a requirement is to develop the power balancing and optimization together and taking into account that when errors of a priori information occur, the power balancing strategy should be affected as little as possible. For this, a multiobjective cost function may be formulated, avoiding load shedding and respecting the energy sources limits, utility grid availability, and dynamic pricing constraints. Moreover, the optimization takes into account the forecast of variable renewable power generation and load power demand while satisfying many constraints. This local control handles instantaneous power balancing following the estimated power flow day-ahead optimization and provides interface for metadata communication (forecasting data, smart grid data, etc.). This control is developed and implemented for the grid-connected mode in [21] and for the off-grid mode in [22].

3. URBAN DIRECT CURRENT MICROGRID

Urban areas have great potential for intensive development of PV sources; however, the intermittent and unpredictable nature of PV energy remains an issue for its utility grid integration resulting in fluctuations of voltage and/or frequency, harmonic pollution, difficult load management, etc. To increase their integration level and obtain a robust power grid, the smart grid could solve problems of peak consumption, optimal energy management, and demand response. The smart grid is being designed primarily to exchange information on grid needs and availability, help in balancing power, avoid undesirable injection, and perform peak shaving [7]. Concerning ancillary services (power grid technical regulations), for a better decentralization of production, microgrids play an important role. Thus, at the urban scale, a microgrid integrated into the building can become an answer to these technical difficulties. Assuming that renewable locally generated electricity is consumed where, when, and in the form in which it is produced, with a utility grid connection seen as backup, this microgrid is a solution for building self-feeding and the controlled injection of surplus electricity.

3.1 Smart Grid, Smart City, and Smart Building

Following the example of the smart grid concept, communication broadband networks and smart systems serving cities and territories are nowadays developed more and more. Smart grid; smart city; smart building; smart transportation; smart metering; smart water; and smart services combined with concepts such as machine to machine (M2M), Internet of Things (IoT), Big Data, etc., are smart technologies to make cities more attractive, intelligent, sustainable, connected, and based on sustainable territories at the service of citizens. Indeed, many applications; solutions; and innovations in broadband and ultra-broadband, intelligent networks for development of cities and territories, energy efficiency, and smart buildings are currently developed and proposed.

A smart or intelligent system is an automatic system based on the technology of acquiring and processing data. Artificial intelligence is implemented in the devices aiming at calculation and formatting of input and output signals. A smart system covers logical machines and systems developed to help and/or assist humans in several tasks. Some smart systems help to strengthen the technical aspects by the process control, and others help to strengthen the systems and the human-machine interaction aspects. The general problem is to design and implement a reliable control system capable of reacting quickly enough to external stimuli to ensure the asked operation while respecting the safety of physical assets and the people using these facilities.

Smart and connected cities will be built around many sectors such as health care, public services, smart commercial buildings, smart homes, transportation, utilities, etc., which implies a safety at the highest level of end users, suppliers, and collective data. A smart city network, which is often related to urban system engineering, is a set of solutions and systems allowing city network operators to monitor and diagnose problems, continuously and remotely prioritize and manage maintenance operations, and use the data provided to optimize all aspects of smart network performance.

To transform traditional urban system engineering into smart networks, current urban systems should be equipped with four main abilities: (1) to accept unconventional operations; (2) to improve operation, safety, and continuity of service (remote reading and remote management of infrastructure); (3) to generalize the communicating count that will allow a better knowledge and management of network operation and energy consumption; and (4) to increase the flexibility of the needed energy system by becoming a place of smart consumption, production, and storage.

Today there are mainly three technologies (ie, IoT, M2M, and Big Data) closely linked so that cities consume less or better and making the citizen an actor of its own city life. The smart grid interacts with the smart city because Big Data allows a detailed analysis of the information reported by the communicating objects via the Internet from users (IoT) and machines (M2M). With the maturity of technologies combined with cloud computing and sensors, cities began their revolution to greater intelligence, capable of generating financial and time savings as well as better quality of life.

Through many market reports, technological innovations, and scientific studies, it is observed that individual houses, residential buildings, and business buildings constitute the bulk of the smart city. In fact, the smart grid would thus intend to serve as a backbone for smart cities of tomorrow composed of smart buildings. This paradigm is possible thanks to the massive arrival of digital technologies in the buildings.

On the other hand, a smart building is a building for which the decision, command, and control system detects and predicts the need to change an aspect of building operation and drives the technical equipment. Among the technology that helps in the realization of a smart building, there are the active materials, controllers, automata, specialized calculators, local process networks, and communication networks. Nevertheless, to reduce pollutant emissions, renewable energy and passive as well as active energy efficiency are significant efforts that will have positive effects, but they are limited if they are not integrated into an overall system. At present, each building, especially smart building, has its own system and is still too isolated from others. To make that energy efficiency optimal, it must operate in an environment where smart grid systems are interoperable and interconnected buildings. Furthermore, to facilitate the analysis of needs and ensure consistency and interoperability of deployed systems today is not so much to use new technologies as it is knowing how to "speak" to the buildings, to define a common language, to make them "ready to grid" and therefore "ready for services" that are supposed to guarantee interoperability and the optimal operating mode in connection with the other intelligent systems.

With the development of current and future energy needs, in particular the energy transition, cities must gradually integrate intelligence into their networks to finely manage the generation, transmission, and use of electricity. Presently, with the smart grid concept, around the world there is an increasing dynamic toward the smart city so that the urban building-integrated smart microgrids become the key system.

3.2 Smart Microgrids in Urban Areas

As renewable and distributed electricity sources increase in urban areas, their grid integration associated with an energy management system is more necessary than ever. The association and interconnection of the smart grid and the smart building are supposed to bring innovative solutions.

3.2.1 General Overview

The microgrid system is considered as one promising approach to facilitate the smart grid. By organizing a set of microgrids with several grid connections, through an adequate interface controller, power system balance becomes rather a local issue than a region-wide issue [7]. In urban areas, the

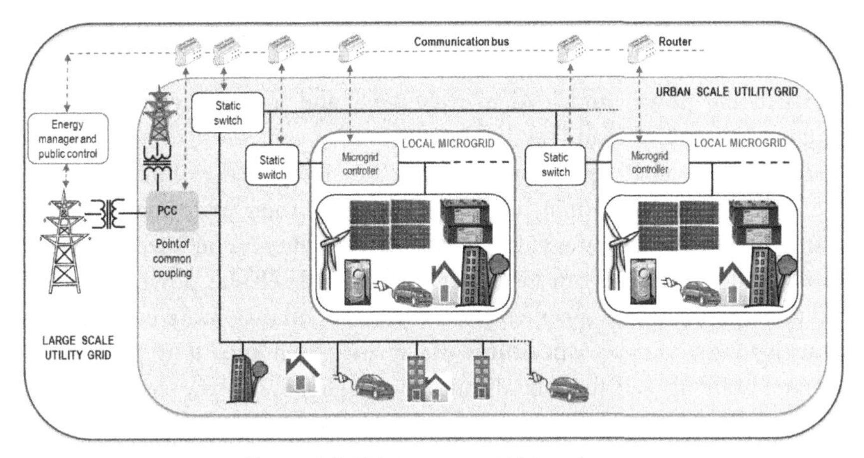

Figure 1.3 Urban smart grid topology.

possible smart grid topology evolution, based on building-integrated microgrids as shown in Fig. 1.3, could be presented following three levels.

In urban areas, at the local level, the microgrid is integrated to the building producer-consumer and connected to the utility grid by an adapted controller. The microgrid could refer to different power scales from a few kilowatts to megawatts and is able to maintain a basic power balancing and to exchange power with another microgrid as well as with the traditional utility grid through a specific interface controller capable of exchanging data.

At the urban scale there are several building-integrated microgrids and parts of a traditional utility grid as a single system capable of exchanging power and data through a specific interface at the PCC.

The large scale consists of numerous microgrids implemented in the power distribution network as well in the power transport network, combined with the traditional utility grid and a communication network to transform the traditional power grid into a smart grid.

Intelligent static switches allow grid connection and islanding of microgrids.

The communication network is mainly composed of communication bus and routers, which are dedicated to direct messages following energy management priorities or special management areas. All control interfaces and static switches are supposed to be able to generate and receive messages.

The urban DC microgrid described in this book is building integrated and connected to the smart grid as previously described. It is considered that microgrid controls on-site generation and power demand to meet the objectives of providing local power, ancillary services, and injecting power into the utility grid if required. The microgrid controller becomes essential for balancing power and load management, and it facilitates the sources pooling during grid islanding.

As many small PV plants are associated or integrated to buildings, it is essential to restructure their use and to improve their performance by an energy management strategy. For distributed PV energy, on-site generation through the microgrid can be better scaled to match the power needs of end users, who require specified power services and may more easily accept some load shedding. Concerning the tertiary building, it is possible to operate building self-feeding during the daylight and the building can become self-sufficient. In case of any energy excess, it can even inject part of its production into the utility grid or in another microgrid in which demand would exceed production (pooling of resources operating mode).

In this context, at the urban scale, the proposed system is a building-integrated DC microgrid that provides a solution for the self-supply of buildings and grid-interaction control. It consists of a physical power system and a supervisory control system. The power system includes a DC load, which is the building as producer-consumer, and sources. The considered sources (ie, PV generator, electrochemical storage, diesel generator, and grid connection) are connected on a common DC bus through their dedicated converters, whereas the DC load demands power directly from the DC bus. The microgrid supervisory control is designed and developed similar to an intelligent energy management system that optimizes power transfer, adapts to conditions imposed by the public grid through the smart grid bus communication, and takes into account the various constraints to minimize the energy consumption from the public grid and to make full use of local production. The interface between the smart grid and the proposed microgrid offers strategies that ensure, at the same time, local power balancing, local power flow optimization, and response to grid issues such as peak shaving and avoiding undesired injections.

The main scientific issue is the difficulty of global optimization due to the risk of mismatch between production/consumption predictions and real-time operating conditions on the one hand and the need to take into account the constraints imposed by the public grid on the other hand.

Therefore, the urban DC microgrid integrated into the building can become an answer to some current technical difficulties. Assuming that renewable locally generated electricity is consumed where, when, and in the form in which it is produced, with a utility grid connection seen as a backup, this microgrid is a solution for building self-feeding and the controlled injection of the excess electricity production.

3.2.2 Direct Current Microgrid for a Low-Voltage Direct Current Distribution Network

One of the original aspects of this study is the DC current aspect. It is well known that for more than a century the AC current has established itself as the worldwide standard in electrical power distribution. During the last 10 years, several research works propose the study of DC current applications, especially for buildings.

Regarding the urban DC microgrid, the common DC bus architecture is chosen for an efficient integration of other renewable sources and storage that are technologically in DC current. This is for the absence of phase synchronization involved in the case of the three-phase AC current because only the voltage must be stabilized, and only a single inverter is required to connect an AC load, if any.

In addition, considering a common DC bus and a DC load directly connected, the overall performance is improved by removing multiple energy conversions. Indeed, a DC network building distribution may use the existing cables with the same power transfer as in AC distribution. The DC bus can directly supply many building appliances (lighting, ventilation, electronic office equipment, etc.) as well as an electric vehicle.

Thus a new debate on AC versus DC emerges. In a tertiary building, should a DC electrical system replace or complete the AC electrical system? In other words, is it possible to directly feed electric loads with DC power?

An example of a building-integrated microgrid system, in grid-connection operating mode, is presented in Fig. 1.4.

PV generators, electricity storage, public grid connection, and electric building loads are coupled, through their dedicated converters, on a common DC bus. The PV generators are considered as a source controlled to extract a maximum power but also able to output a limited power if necessary. The storage system is an electrochemical system that is technically and economically well adapted for a building-integrated microgrid system. The storage is required to smooth the power output from renewable sources. The utility grid connection and the building distribution bus connections are made by static-state or hybrid switches. The microgrid should be able to optimize the power flows on the bus to obtain a minimized daily cost for end users [21]. In comparison with an AC bus, the presented microgrid is based on a common DC bus for efficient integration of renewable sources and storage and for the

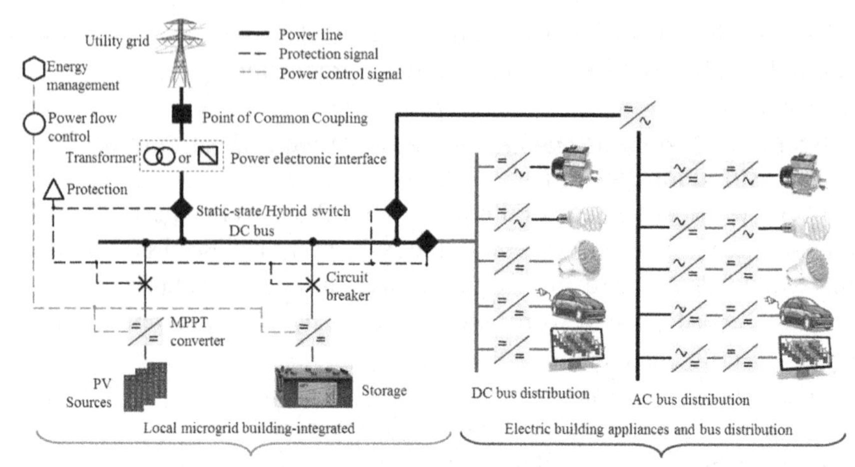

Figure 1.4 Example of building-integrated microgrid system.

absence of the frequency and phase synchronization; therefore only the DC bus voltage needs to be stabilized [21].

Regarding the electric building loads, there are three possible connections: (1) using an inverter at the output of the microgrid and an AC bus distribution, (2) considering a DC bus distribution directly connected to the DC bus of the microgrid, and (3) an AC/DC separated distribution. Concerning the building electrical loads, an increase of the number of appliances that work internally with DC power is observed, but this DC power is converted inside of the device from a standard AC supply. On the other hand, building energy performances are improved by using frequency converters that, in their last level, transform a DC signal into an AC signal at different values of voltage and frequencies. Thus many buildings' electrical appliances could be fed directly with DC power, such as devices based on microprocessors; computer system power supply; switched-mode power supply; variable-frequency drives for the speed variation of the motors that equip the systems of heating, ventilation, and air conditioning; and lighting based on light-emitting diodes. These examples represent a very important percentage of tertiary buildings' electric appliances, and several projects already show the feasibility of its supply directly in DC and the trend of certain manufacturers to make this change. This is the context that justified the choice to study an urban DC microgrid and a DC load. Furthermore, the energy efficiency improvement could increase the advantage of the positive-energy building and make better electric vehicle utilization.

DC bus distribution maximizes the efficiency of PV generation by avoiding some energy conversion losses (elimination of one or two energy conversion stages, absence of reactive power and harmonics) [23]. In addition, DC bus distribution is highly compatible with electricity storage, which may increase the efficiency of plug-in hybrid electric vehicles and electric vehicles, for which the number is supposed to increase in a few years. In traditional AC bus distribution, the local microgrid produces DC power that is converted to AC power to supply a building's electric system; this power then has to be reconverted to DC for many end uses as previously cited. The AC/DC separated bus combined the advantage of the two kinds of distribution by consisting of a DC bus distribution and an AC bus distribution in parallel. It may save the conversion step and be compatible with all kinds of electrical devices. However, the DC bus distribution is currently not compatible with the existing plug-and-play appliances and its use is more complicated.

Microgrid Global Power Transmission Efficiency: Alternating Current Bus versus Direct Current Bus

On the basis of the presented local microgrid, the comparison of power transmission efficiency between the AC bus distribution and the DC bus distribution is presented for steady-state analysis [23].

The behavior of a low-voltage distribution line is often modeled by a resistor, an inductor that is the ratio between the total magnetic field of the leakage flux and the current flowing through the line, and a capacity translating the impact of the created electrical field. In the case of a short line (less than a dozens of kilometers), the capacity can be ignored. For a low-voltage electrical installation, the reactance of the line is negligible [24], and at 50 Hz, the skin effect is not significant.

Assuming that the power transmission is implemented on the same cables (ie, same cable length l and same resistance r per unit length), the line losses for the same load power P_{load} are estimated, in AC and DC cases, by Eqs. [1.1] and [1.2], respectively.

$$p_{\text{AC}} = 2\cdot r\cdot l\cdot I_{\text{AC}}^2 = 2\cdot r\cdot l\cdot P_{\text{load}}^2/\left(V_{\text{AC}}^2\cdot\cos^2\varphi\right) \qquad [1.1]$$

with p_{AC} the line losses, I_{AC} the RMS value of phase current, V_{AC} the RMS value of the voltage between phase and neutral, and φ the phase difference between the current and the voltage for the AC bus distribution case.

$$p_{\text{DC}} = 2\cdot r\cdot l\cdot I_{\text{DC}}^2 = 2\cdot r\cdot l\cdot P_{\text{load}}^2/U_{\text{DC}}^2 \qquad [1.2]$$

with p_{DC} the line losses, I_{DC} the current, and U_{DC} the voltage for the DC bus distribution case.

In general, for a building with a low-voltage distribution of 400/230 V AC, the network is made by cables for which the operating voltage value is 600 V and even up to 1000 V. Thus the DC bus voltage U_{DC} can be equal to the value $V_{AC} \cdot \sqrt{2}$ without taking risks for the cables, which are assumed to be included in an existing installation. Therefore for the same load power P_{load}, the line losses in DC bus distribution are less than the losses in the AC bus distribution, as expressed by Eq. [1.3].

$$p_{DC}/p_{AC} = \cos^2 \varphi/2 \qquad [1.3]$$

To be economically profitable, the low-voltage electrical distribution within buildings requires an optimal value of DC bus voltage that may be between 325 V DC and 400 V DC [25].

Aiming to use existing building cable infrastructure, Fig. 1.5 shows the transition possibilities from AC to DC for the three-phase case and single-phase case, respectively, in which PE is the underground electric cable protection.

The power transmission rates, which are compared in three-phase and single-phase cases, are expressed by Eqs. [1.4] and [1.5], respectively.

$$\frac{P_{DC\_5\ wires}}{P_{AC\_5\ wires}} = \frac{2 \cdot U_{DC} \cdot I}{3 \cdot V_{AC} \cdot I \cdot \cos \varphi} = \frac{2 \cdot \sqrt{2} \cdot V_{AC} \cdot I}{3 \cdot V_{AC} \cdot I \cdot \cos \varphi} = \frac{2 \cdot \sqrt{2}}{3 \cdot \cos \varphi} \qquad [1.4]$$

$$\frac{P_{DC\_3\ wires}}{P_{AC\_3\ wires}} = \frac{U_{DC} \cdot I}{V_{AC} \cdot I \cdot \cos \varphi} = \frac{\sqrt{2} \cdot V_{AC} \cdot I}{V_{AC} \cdot I \cdot \cos \varphi} = \frac{\sqrt{2}}{\cos \varphi} \qquad [1.5]$$

For the three-phase cable and for a power factor less than 0.942, the transmitted DC power $P_{DC\_5\ wires}$ is greater than the transmitted AC power $P_{AC\_5\ wires}$. For the single-phase cable and for any power factor, the

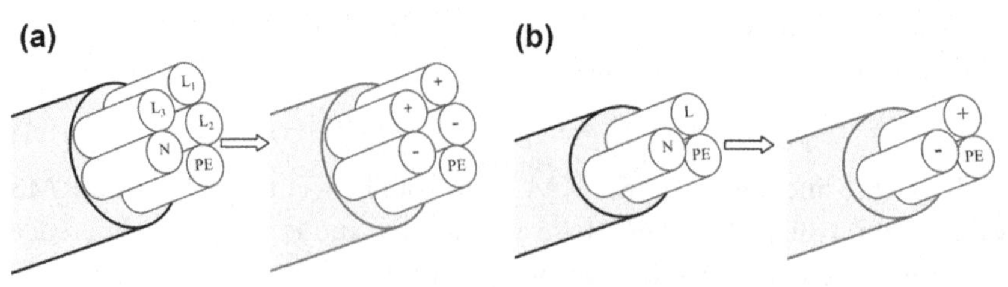

Figure 1.5 Cable transition from AC to DC for (a) three-phase case and (b) single-phase case.

transmitted DC power $P_{DC\_3\ wires}$ is always greater than the transmitted AC power $P_{AC\_3\ wires}$.

Microgrid Global Energy Efficiency: Alternating Current Bus versus Direct Current Bus

Regarding the global energy efficiency comparison between the AC bus distribution and the DC bus distribution, the energy conversions stages implemented for the AC and DC bus distribution cases are highlighted in Fig. 1.4 [24]. Considering that the produced PV power at its maximum power point is P_{PV_MPP}, the average of global energy efficiency for the AC bus distribution can be expressed as in Eq. [1.6] whereas the average of global energy efficiency for the DC bus distribution can be expressed as in Eq. [1.7]:

$$\eta_{AC} = \eta_{chopper\_MPPT} \cdot \eta_{inv\_microgrid} \cdot \left(\eta_{rect} \cdot \eta_{inv/chopper}\right)_{load} = P_{load}/P_{PV\_MPP} \quad [1.6]$$

$$\eta_{DC} = \eta_{chopper\_MPPT} \cdot \left(\eta_{chopper/inv}\right)_{load} = P_{load}/P_{PV\_MPP} \quad [1.7]$$

with η_{AC} the power efficiency for AC bus distribution case, $\eta_{chopper\_MPPT}$ the maximum power point tracking (MPPT) converter efficiency, $\eta_{inv\_microgrid}$ the microgrid inverter efficiency, $(\eta_{rect} \cdot \eta_{inv/chopper})_{load}$ the product of the power converter efficiencies at the load level, and η_{DC} the power efficiency for the DC bus distribution case with $(\eta_{chopper/inv})_{load}$ the product of the power converter efficiencies at the load level. Following Fig. 1.4, because of the conversion stages elimination and taking into account any arbitrary average efficiency on duty cycle, the global energy efficiency of the DC bus distribution case is higher than the AC bus distribution case efficiency.

The impact of power converters' efficiency on a building-integrated microgrid is obvious. Hereafter, some theoretical aspects of the efficiency of power converters are discussed, and the case of the microgrid inverter behavior is highlighted [24]. Fig. 1.6 represents average power converter efficiency and the optimal operational window within which it performs at its most efficient operating points.

This optimal operational window (between P_1 and P_2) depends on the average point of operation. The power converter's efficiency is defined as a power transfer function during normal operation, depending on the instantaneous power, either of the input or of the output. This power transfer function is nonlinear with a threshold input power as the starting operating point, often understood as the converter's own consumption.

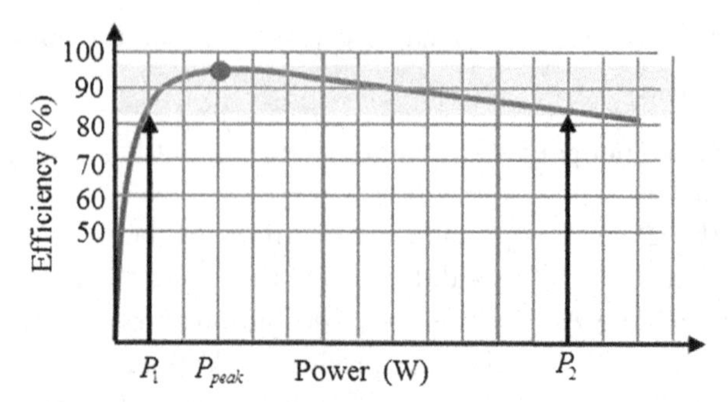

Figure 1.6 Typical power converter efficiency profile.

The ohmic effect of the output circuit induces a loss that increases quadratically with the current (or power), leading to a decrease of efficiency after a maximum. The peak efficiency of a power converter indicates the efficiency of the device at one operating point, considered the optimal performing point (corresponding to the peak power P_{peak} in Fig. 1.6; that is, when the power converter is operating (usually) at its rated capacity). Although this peak efficiency is almost always higher than 95% for the newest devices, it is important to note that the converter may only operate at its peak efficiency range for a very small proportion of the duty cycle or not at all. The power converter efficiency curve is rarely explicitly given by the manufacturers or measured by independent institutes. Fig. 1.7 gives the efficiency curves of an industrial PV inverter and the PV inverter main data sheet.

The inverter efficiency depends on the output power as well as the applied DC voltage of the PV plant. On the one hand the DC voltage varies to impose the maximum power point at any time at the PV plant; on the other hand the power demanded by the load also varies, especially for electric building appliances. Thus the optimal sizing of the inverter may become very complex.

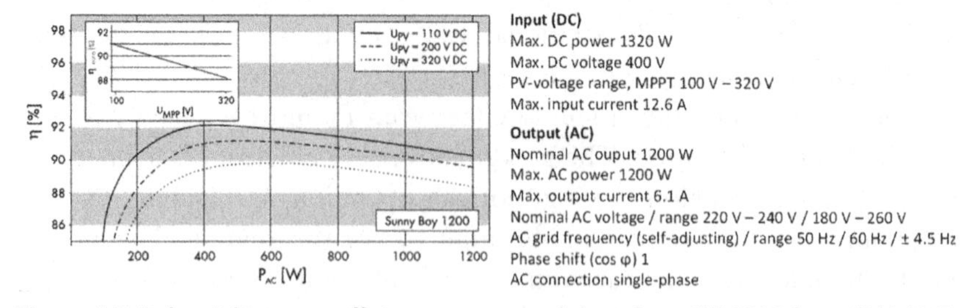

Figure 1.7 Industrial inverter efficiency curves and data sheet (SB 1200 from SMA [26]).

For the PV inverter, its efficiency relies on the average PV power received on-site, which could be calculated by the meteorological data and estimated PV panel efficiency at such conditions [27]. The European weighted efficiency benchmark is expressed by Eq. [1.8], where η_i is the conversion efficiency at *i*% of the rated output power of the inverter. It is an averaged operating efficiency over a yearly power distribution corresponding to the middle-European climate and is referenced on nearly every inverter data sheet.

$$\eta_{\mathrm{EU}} = 0.03 \cdot \eta_{5\%} + 0.06 \cdot \eta_{10\%} + 0.13 \cdot \eta_{20\%} + 0.10 \cdot \eta_{30\%} + 0.48 \cdot \eta_{50\%} + 0.20 \cdot \eta_{100\%} \quad [1.8]$$

One can note that the SB 1200 efficiency presented in Fig. 1.7 is quoted by the manufacturer at 92.1% whereas its European weighted efficiency is given as 90.7%. According to the climate of higher solar irradiance, the Californian Energy Commission weighted efficiency is given by [1.9].

$$\eta_{\mathrm{CEC}} = 0.04 \cdot \eta_{10\%} + 0.05 \cdot \eta_{20\%} + 0.12 \cdot \eta_{30\%} + 0.21 \cdot \eta_{50\%} + 0.53 \cdot \eta_{75\%} + 0.05 \cdot \eta_{100\%} \quad [1.9]$$

It is obvious that efficiency depends on many parameters, including the applied voltage, the current, the switching frequency, temperature, output circuit, etc. Because of the nature of renewable energy, it is almost impossible to set an operating point at some point. Therefore, it is not enough to just know the optimal operational window around its peak efficiency; it is also necessary to study the changes of the converter efficiency over the entire operating range. Therefore further converter efficiency characterization and analysis must be performed to propose a global energy efficiency optimizing method well adapted to local DC microgrids based on PV sources.

From an energetic point of view, the DC distribution bus presents advantages because it saves power losses by removing the inverter, by minimization of power line losses, and by removing the power conversion stage in the electric loads. In conclusion, the DC microgrid with DC bus distribution is superior in efficiency, cost, and system size. It can supply the same power as in AC distribution, and it is more environmentally friendly than AC distribution. To implement a DC bus distribution network, DC/DC converters are required; therefore advancements in DC/DC converter technology and efficiency are significantly necessary.

3.2.3 Dynamic Interactions Between the Microgrid and the Smart Grid

The concept of the smart grid in urban areas leads to microgrids for producer-consumer sites to reduce losses and peak energy demand, and to play a role in local regulation, through the data communication. The microgrid controller must provide the interface between the utility grid and the loads (eg, buildings' electrical appliances and electric vehicles), aiming at an optimal power management.

Fig. 1.8 illustrates the power management interface principle based on the main data, which have to be exchanged between the microgrid and utility public grid.

Thus the microgrid controller must take into account information about the public grid availability and dynamic pricing, inform the smart grid on injection intentions and power demand, meet the demand of the end user with respect to all physical and technical constraints, and operate with the best energy cost for the public grid and for the end user. To meet these objectives as well as forecasting, smart metering, monitoring, and other actions described in Fig. 1.8, a specific interface associated with the urban microgrid has to be designed; one of the solutions of this kind of interface is proposed later in this book.

3.3 Urban Energy Management Strategies

Currently, some major preoccupations in urban areas are the buildings' energy performances and the construction of charging stations for electric vehicles, the number of which is predicted in future years. Turning to the emergence of the smart grid combined with microgrids on the one hand and the increase of the positive-energy buildings on the other hand, one of the solutions is the urban building-integrated DC microgrid.

For urban DC microgrids, several operating strategies may be developed based on sources that make up the microgrid: renewable sources such as PV generators and wind turbines; electrochemical storage; utility grid connection; traditional sources such as microturbines or (bio)diesel generators; and mainly two kinds of load, buildings loads (electric appliances) and electric vehicles charging stations. Fig. 1.9 presents the main possible strategies.

Renewable energies and the storage supply the building and charge the electric vehicles. The renewable excess energy could be stored and/or

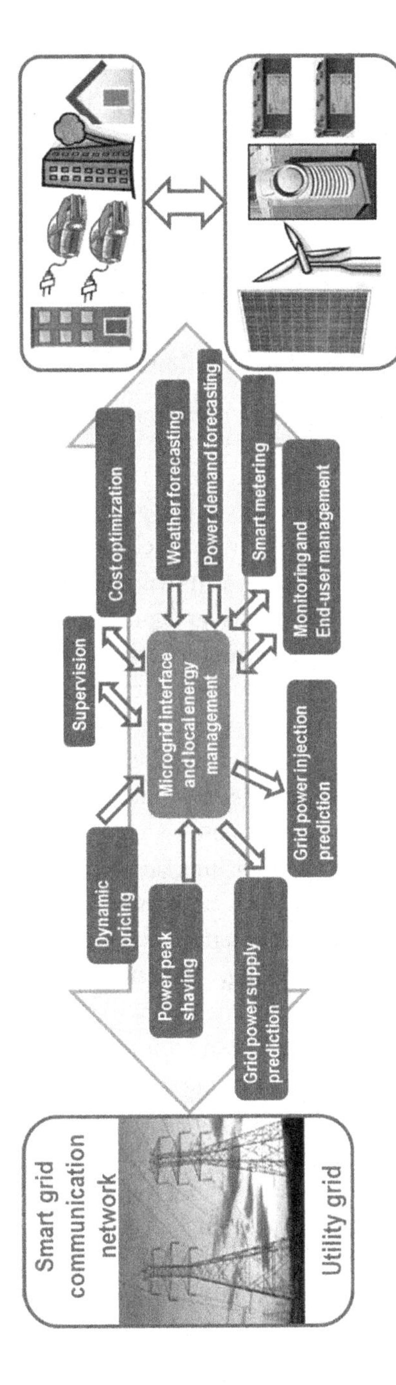

Figure 1.8 Interactions between the microgrid and the smart grid.

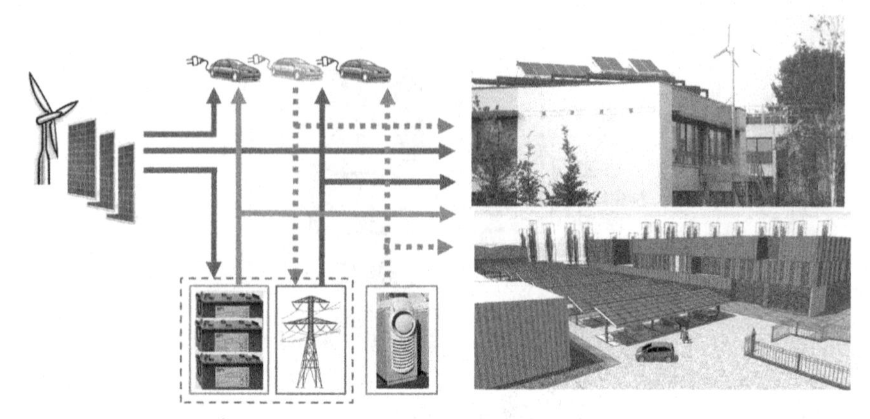

Figure 1.9 Urban energy management strategies using a DC microgrid.

injected into the grid. If available, the utility grid power is used only as a backup for the building and the electric vehicles or to charge the storage during the off-peak grid time period (usually during the night). The microturbine operates only if the utility grid power is not available.

Regarding the electric vehicles, if required for stringent situations, they can supply the building and/or provide energy to the grid. These operating modes are called vehicle-to-load and vehicle-to-grid modes.

In general, the messages received from the smart grid command the microgrid operating mode aiming at compliance the actual availability of the utility grid as well as energy total cost.

The control is ensured through an interface, which may be organized into several hierarchical layers such as a human-machine interface, prediction calculation, energy management and power optimization, and real-time operational power balancing. This supervisory system controls and optimizes the power system and interacts with the end-user interface, databases including weather forecasts and building energy consumption previsions, and smart grid communication.

Concrete applications in urban areas are:

- positive-energy buildings as consumers and producers, which may also consider self-consumption for energy self-sufficiency;
- infrastructure for charging electric vehicles that are expected to increase in the coming years; and
- a combination of positive-energy buildings and infrastructure for charging electric vehicles for ecodistricts and other forms of positive territory overall energy.

3.4 Experimental Platform for Direct Current Microgrids

The urban DC microgrid analyzed and described in this book is associated with a laboratory-scale experimental platform that has been installed in Centre Pierre Guillaumat of Univesité de Technologie de Compiègne, France. The general overview of the experimental platform is given in Fig. 1.10.

The experimental platform consists mainly of 16 PV panels and electrochemical storage based on a lead-acid battery. The small-scale wind turbine generator that can be seen in the picture is not yet included in the DC microgrid; it will be the subject of further research work. The grid connection is emulated by reversible AC voltage sources, and the DC load is emulated with programmable DC electronic load. The power converters are based on IGBTs with commutation frequency at 10 kHz. The different solutions proposed in this book are developed and implemented with controllers based on dSPACE hardware. More details on this experimental platform are given in the last chapter, in which the main tests concerning the experimental validation of the urban DC microgrid are described, but necessary parts to test and validate this study step by step are presented in almost every chapter.

Regarding the scientific objectives, the experimental platform allows mainly testing and validating numerical or experimental models, algorithms

Figure 1.10 General overview of the DC microgrid experimental platform.

able to extract maximum power or limited power from renewable energy sources, and complex control algorithms that enable the DC microgrid to operate autonomously and in grid-connected mode.

4. CONCLUSIONS

This chapter attempted to make a comprehensive general presentation of the definition of the smart grid and microgrids and their topology in the urban electrical network landscape as the future shape of distribution networks. Microgrids, based on the loads and generators connected to them, are presented and analyzed following two major categories of AC and DC types. A wide range of research works in different aspects of microgrids, including control, protection, and optimal structure for energy management, shows that microgrid systems will become more and more complex and selective according to the needed applications. Furthermore, the state of the art highlights the complexity of the overall problem of the smart grid, the realization of which requires interaction with microgrids. The difficulty of defining microgrid control leads to deeply analyzing the control of specific components of the utility grid and its power balancing.

Regarding protection, in DC systems the problem is related to the lack of a natural zero crossing of the current. Solutions based on semiconductor switches, or hybrid structures such as static and electromechanical, exist and have the advantage of speed and good reliability; however, at present, they remain expensive solutions.

As an element of the future smart grid, microgrids should be developed in control and energy management considering interactions with the smart grid. In literature, the aspects of control and energy management of the microgrid are treated separately, and smart grid interaction is merely proposed. Thus a further study of microgrid control combined with energy management and smart grid messages must be developed and proposed in real smart grid operation. Global energy management is often related to economic energy sources dispatching as day-ahead operation and online power flow optimization. Nevertheless, to build efficient energy management, adapted to microgrid operation integrated with the smart grid communication, it is necessary to study and propose appropriate models for each type of microgrid.

According to the increasing interest of the whole networking formed by the smart grid, smart cities and smart buildings, the urban DC microgrid may be seen as a microgrids class by itself. The urban DC microgrid may improve the penetration level of distributed small renewable energy sources with less impact in the utility grid while optimizing local power. Compared with the current research studies, the urban DC microgrid offers integration of the smart grid messages by adopting the suggested urban smart grid topology. In addition, for tertiary buildings with increased use of power electronic loads and new developments in DC microgrids, the redefinition of the internal electric distribution bus is required. Therefore the DC networks may come back. By removing multiple energy conversions and taking into account the converter efficiency characterization and analysis, the well-designed and carried-out urban DC microgrid may improve the overall system efficiency for local production and local consumption. Nevertheless, there are two main drawbacks preventing the simple implementation of the DC bus distribution: the protection devices and the adapted standard appliances for DC power supply.

Concerning the microgrid controller interface, the main requirements are to meet utility grid demands for limited injection or absorption power on one hand and to meet the DC load power demand ensuring the stability of the DC bus on the other hand. However, these requirements must consider many dynamic interactions and data exchanges between the microgrids and the smart grid, including public grid availability, dynamic pricing, injection and power demand predictions, forecasting, smart metering, monitoring, and energy cost optimization. Following the sources that make up the urban DC microgrid, several operating strategies may be developed with maximal energy efficiency.

The proposed urban building-integrated DC microgrid described and developed in this book is connected to the smart grid following the described manner. By applying the approach of multiphysics and a multi-scale system to obtain a global response, the DC microgrid applied in urban areas leads to studying the interaction between the microgrid and the utility grid at different time scales involving different types of physical parameters and with different operating modes that meet specific needs.

REFERENCES

[1] Blaabjerg F, Teodorescu R, Liserre M, Timbus AV. Overview of control and grid synchronization for distributed power generation systems. IEEE Trans Ind Electron 2006;53(5):1398–409.

[2] Rodriguez P, Timbus AV, Teodorescu R, Liserre M, Blaabjerg F. Flexible active power control of distributed power generation systems during grid faults. IEEE Trans Ind Electron 2007;54(5):2583–92.

[3] Krichen L, Francois B, Ouali A. A fuzzy logic supervisor for active and reactive power control of a fixed speed wind energy conversion system. Electr Power Syst Res 2008;78(3):418–24.
[4] Zhang L, Sun K, Xing Y, Feng L, Ge H. A modular grid-connected photovoltaic generation system based on DC bus. IEEE Trans Power Electron 2011;26(2):523–31.
[5] Hatziargyriou N. Microgrids: architectures and control. Wiley–IEEE; 2014.
[6] Guerrero J, Chandorkar M, Lee T, Loh P. Advanced control architectures for intelligent microgrids part I: decentralized and hierarchical control. IEEE Trans Ind Electron 2013;60(4):1254–62.
[7] Wang BC, Sechilariu M, Locment F. Intelligent DC microgrid with smart grid communications: control strategy consideration and design. IEEE Trans Smart Grid 2012;3(4):2148–56.
[8] Hadipaschalis I, Poullikkas A, Efthimiou V. Overview of current and future energy storage technologies for electric power applications. Renew Sustain Energy Rev 2009;13:1513–22.
[9] Papaefthymiou SV, Karamanou E, Papathanasiou S, Papadopoulos M. A wind-hydro-pumped storage station leading to high RES penetration in the autonomous island system of Ikaria. IEEE Trans Sustain Energy 2010;1(3):163–72.
[10] Chen C, Duan S, Cai T, Liu B, Hu G. Optimal allocation and economic analysis of energy storage system in microgrids. IEEE Trans Power Electron 2011;26(10):2762–73.
[11] Falcao DS, Oliveiraa VB, Rangelb CM, Pintoa AMFR. Review on micro-direct methanol fuel cells. Renew Sustain Energy Rev 2014;34:58–70.
[12] Bernal-Agustin JL, Dufo-Lopez R, Rivas-Ascaso DV. Design of isolated hybrid systems minimizing costs and pollutant emissions. Renew Energy 2006;31(14):2227–44.
[13] Benghanem M. Low cost management for photovoltaic system in isolated site with new IV characterization model proposed. Energy Convers Manag 2009;50(3):748–55.
[14] Vazquez MJV, Marquez JMA, Manzano FS. A methodology for optimizing stand-alone PV-system size using parallel-connected DC/DC converters. IEEE Trans Ind Electron 2008;55(7):2664–73.
[15] Liserre M, Sauter T, Hung JY. Future energy systems, integrating renewable energy sources into the smart power grid through industrial electronics. IEEE Ind Electron Mag 2010;4(1):18–37.
[16] Ding G, Gao F, Zhang S, Loh PC, Blaabjerg F. Control of hybrid AC/DC microgrid under islanding operational conditions. J Mod Power Syst Clean Energy 2014;2(3):223–32.
[17] Patterson BT. DC, come home: DC microgrids and the birth of the "Enernet". IEEE Power Energy Mag 2012;10(6):60–9.
[18] Kanchev H, Colas F, Lazarov V, Francois B. Emission reduction and economical optimization of an urban microgrid operation including dispatched PV-based active generators. IEEE Trans Sustain Energy 2014;5(4):1397–405.
[19] Kumar Nunna HSVS, Doola S. Multiagent-based distributed-energy-resource management for intelligent microgrids. IEEE Trans Ind Electron 2013;60(4):1678–87.
[20] Lu D, Fakham H, Zhou T, François B. Application of Petri nets for the energy management of a photovoltaic based power station including storage units. Renew Energy 2010;35(6):1117–24.
[21] Sechilariu M, Wang BC, Locment F, Jouglet A. DC microgrid power flow optimization by multi-layer supervision control. Design and experimental validation. Energy Convers Manag 2014;82:1–10.
[22] Sechilariu M, Locment F, Wang BC. Photovoltaic electricity for sustainable building. Efficiency and energy cost reduction for isolated DC microgrid. Energies 2015;8(8):7945–67. Special Issue on Solar Photovoltaics Trilemma: Efficiency, Stability and Cost.

[23] Wu H, Sechilariu M, Locment F. Impact of power converters' efficiency on building-integrated microgrid. In: European Conference on power electronics and applications (EPE'15 ECCE-Europe); 2015.
[24] Karlsson P, Svensson J. DC bus voltage control for a distributed power system. IEEE Trans Power Electron 2003;18(6):1405−12.
[25] Salomonsson D, Sannino A. Low-voltage DC distribution system for commercial power systems with sensitive electronic loads. IEEE Trans Power Deliv 2007;22(3):1620−7.
[26] www.sma.de. PV inverters SUNNY BOY 1100/1200/1700 installation Guide. SB11_12_17-IEN100132 | IMEN-SB11_17 | Version 3.2.
[27] Kim Y-H, Ji Y-H, Kim J-G, Jung Y-C, Won C-Y. A new control strategy for improving weighted efficiency in photovoltaic AC module-type interleaved flyback inverters. IEEE Trans Power Electron 2013;28(6):2688−99.

CHAPTER 2

Photovoltaic Source Modeling and Control

1. PHOTOVOLTAIC SOURCE MODELING

In recent years, environmental problems have become a worldwide issue. Although the energy demand is significantly increasing against the decreasing reserves of fossil and fissile resources, photovoltaic (PV) energy as an inexhaustible and clean source can reply to that demand. Either at large- or small-scale plants, PV sources are increasingly used to produce electricity, especially in distributed renewable energy systems, for grid-connected and off-grid systems.

1.1 Photovoltaic Cell

The PV effect, which is the conversion of solar irradiance into electricity, was discovered by the Becquerel family and presented to the French Academy of Sciences by the end of 1839. This conversion is done by the PV cell, the electrical characteristics of which resemble those of the PV source. A PV panel is composed of one or more PV cells, and a PV source consists of one or several PV panels, as illustrated in Fig. 2.1.

The PV electric energy comes from the direct conversion of the energy carried by photons contained in sunlight or artificial light. This effect involves the absorption of photons in a semiconductor material to create electron-hole pairs. This phenomenon depends on the semiconductor's

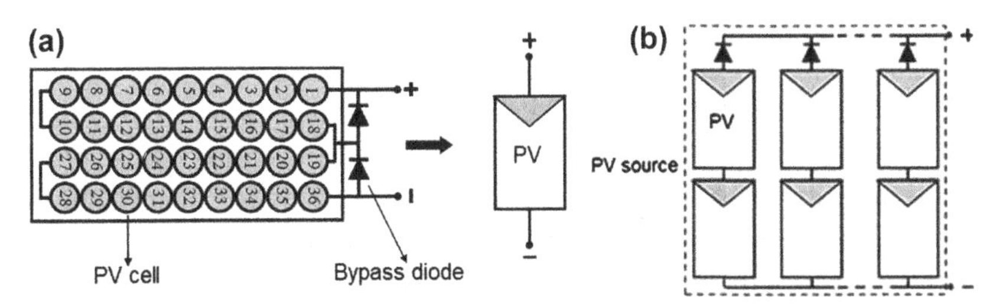

Figure 2.1 (a) Photovoltaic (PV) cell and PV panel; (b) PV source.

Urban DC Microgrid
ISBN 978-0-12-803736-2
http://dx.doi.org/10.1016/B978-0-12-803736-2.00002-5

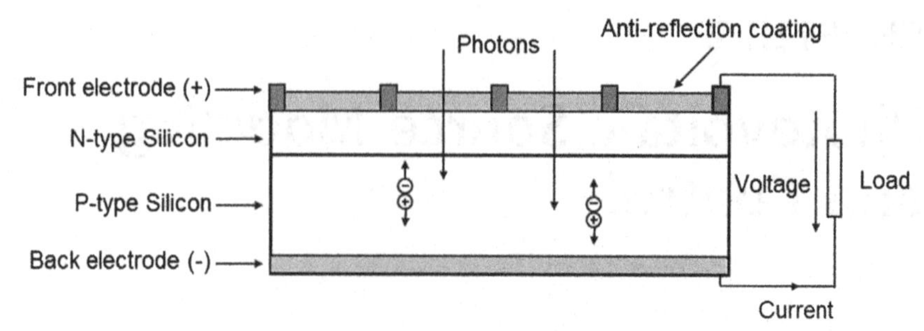

Figure 2.2 Photovoltaic cell schema.

ability to absorb photons and the reflection (and treatment of the illuminated surface) of the wavelength of the incident light [1].

The PV cell is a wafer formed mainly by two semiconductor materials, usually silicon, doped differently and connected by a P-N junction. The simplified diagram of a PV cell is shown in Fig. 2.2.

1.1.1 Operating Principle of a Photovoltaic Cell

Two electrodes are used on the wafer: (1) the front electrode, allowing the passage of solar irradiance, is deposited on the upper surface, and (2) the back electrode, which completely covers the rear face. To increase the amount of light absorbed, a thin layer of antireflection coating covers the PV cell. When the PV cell is subjected to irradiance, the PV effect occurs only if the photon energy is greater than or equal to the energy of the band gap of the semiconductor (1.12 eV for silicon [2]). The energy gained by the photons generates electron-hole pairs in the different areas of the cell (area N, area P, and junction). The photons having an energy less than 1.12 eV will not be absorbed. Many electron-hole pairs reach the P-N junction, and an electric field is created. The separation of these pairs then takes place:

- In an N- or P-type layer, minority carriers (holes in the N-type layer, electrons in P-type layer) will be disseminated by the concentration gradient effect. They will then be accelerated in opposite areas of their home region by the electric field prevailing on the junction. This movement gives rise to a photocurrent that is proportional to the light intensity (ie, solar irradiance), the cell temperature, and the manufacturing process. Under the effect of charge separation, the junction is polarized in the forward direction and the voltage barrier is lowered. The value of this voltage depends on the material and the joining structure; it is equal to 0.6 V for silicon.

- In P-N junctions, where photogenerated electron–hole pairs are separated by the electric field, which subsequently forces the electrons to move toward the N-type layer and the holes toward the P-type layer.

The electrodes collect these particles, and by connecting the electrodes to an external electrical load, a direct current (DC) current is established.

1.1.2 Electrical Characteristics of a Photovoltaic Cell

For a given solar irradiance and cell temperature, Fig. 2.3 presents the evolution of the current i_c function of the voltage v_c across the PV cell. The current-voltage $i_c = f(v_c)$ characteristic of a PV cell is nonlinear and depends on the solar irradiance level and its temperature, whence a maximum power point (MPP). Fig. 2.3 also shows the evolution of the power $p_c = i_c \cdot v_c$.

These two characteristics are nonlinear and can be decomposed into three parts:

- zone 1: operating like current source;
- zone 2: operating like voltage source; and
- zone 3: intermediate operating mode in which can be found the MPP.

From Fig. 2.3, the following parameters can be defined:

- $i_{c_{SC}}$: short-circuit current ($v_c = 0$);
- $v_{c_{OC}}$: open-circuit voltage ($i_c = 0$); and
- $p_{c_{MPP}} \left(p_{c_{MPP}} = i_{c_{MPP}} \cdot v_{c_{MPP}}\right)$: maximum power for which the MPP is in zone 3, typically $0.85 \leq \frac{i_{c_{MPP}}}{i_{c_{SC}}} \leq 0.99$ and $0.75 \leq \frac{v_{c_{MPP}}}{v_{c_{OC}}} \leq 0.9$ [3].

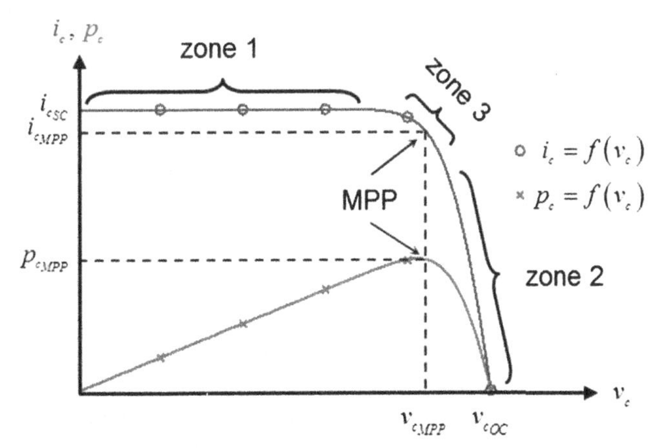

Figure 2.3 Characteristics $i_c = f(v_c)$ and $p_c = f(v_c)$ of a photovoltaic cell. *MPP*, maximum power point.

The PV cell efficiency η_c is defined as the ratio between the maximum power of the cell $p_{c_{MPP}}$ and the potential power that is the product of solar irradiance g (W/m^2) and the cell surface S_c (m^2). The PV cell efficiency is expressed by Eq. [2.1]:

$$\eta_c = \frac{p_{c_{MPP}}}{g \cdot S_c} = \frac{i_{c_{MPP}} \cdot v_{c_{MPP}}}{g \cdot S_c} \quad [2.1]$$

For a PV cell made of silicon, the manufacturing technology (monocrystalline, polycrystalline, amorphous, etc.) influences the efficiency. Performance also depends on the irradiance spectrum and the temperature of the PV cell. The use of an antireflection layer reduces the reflection, increases the absorption, and improves the efficiency.

Depending on the solar irradiance g at a constant PV cell temperature, the characteristics $i_c = f(v_c)$ and $p_c = f(v_c)$ are shown in Fig. 2.4(a) and (b), with $g_1 < g_2 < g_3 < g_4$. The characteristic $i_c = f(v_c)$ and $p_c = f(v_c)$ functions of the PV cell temperature θ (Kelvin) for a constant irradiance are shown in Fig. 2.4(c) and (d), with $\theta_1 < \theta_2 < \theta_3 < \theta_4$.

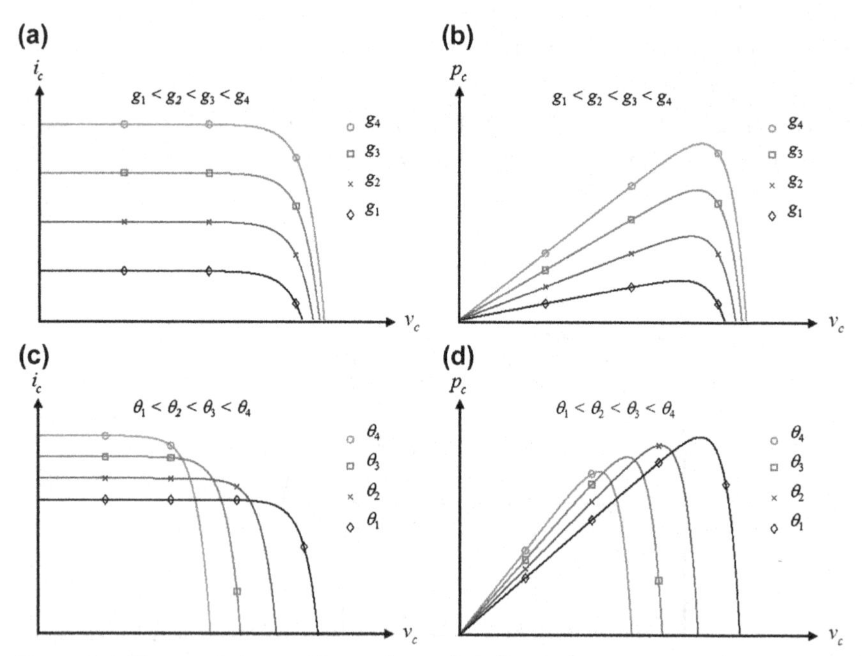

Figure 2.4 Characteristics $i_c = f(v_c)$ and $p_c = f(v_c)$ of a photovoltaic (PV) cell depending on irradiance (a and b) at a constant PV cell temperature and (c and d) depending on PV cell temperature for a constant irradiance.

These curves show that the output power of the PV cell is nonlinear and critically influenced by irradiance and cell temperature. The PV cell temperature effect is less than the solar irradiance one. Each curve has an MPP at which the PV cell most efficiently operates.

1.2 Photovoltaic Source Modeling

The manufacturers give just the electrical features of the PV panel under the standard test conditions (STCs; ie, irradiance of 1000 W/m^2, solar spectrum of air mass 1.5, and PV cell temperature at 25°C). However, the PV source modeling is essential to evaluate the PV behavior under all operating conditions. Thus many works are focused on the identification of the unknown parameters of the PV panel or the PV model. The classical PV models proposed in published literature are based on the electrical equivalent circuit of the PV cell. The parameter identification problem is very complex because the model will invariably have more than one set of parameters that generate the PV behavior under certain operating conditions. Therefore, under each different operating condition, the PV model will require a different set of parameters. In general, the models based on parameter identification correspond to a set of operating conditions; thus these models allow for the study of PV behavior under the same strict operating conditions taken into account. These models can be built using different analytical methods (simplified graphical method, robust linear regression methods, artificial neural network) combined with experimental methods (adjusting the current-voltage characteristic curve at the three points, open circuit, MPP, and short circuit, given by the manufacturer).

1.2.1 Photovoltaic Power Prediction

Regarding the PV power prediction, very often required by the PV source or microgrid energy management, the PV model involves a weather forecast service. Nowadays, worldwide solar irradiance prediction service can be found. Day-ahead solar irradiance forecasting can provide 36% root mean square error for regional forecasts, but 4-h ahead solar irradiance and air temperature forecasting gives the smallest error.

Given solar irradiance and air temperature prediction, an adequate PV model should be chosen to give PV power production. As described earlier, the PV source has nonlinear characteristics with an MPP. These characteristics vary depending on the operating conditions (ie, solar irradiance g and PV cell temperature θ).

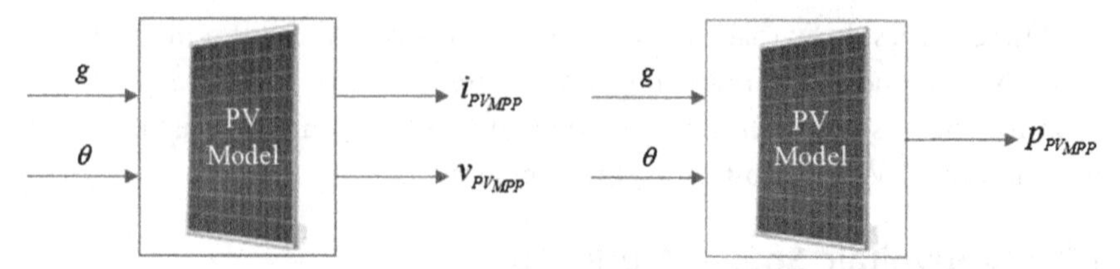

Figure 2.5 Photovoltaic (PV) source modeling generic representations.

As illustrated in Fig. 2.4, for a given PV cell temperature, the power varies proportionally with the solar irradiance, whereas a given solar irradiance power decreases while increasing the temperature of the PV cell. Hence, the PV model has to give enough accurate prediction while being simple to implement. Thus the model of the PV source must be able to estimate the MPP current $i_{PV_{MPP}}$ and MPP voltage $v_{PV_{MPP}}$, or directly the MPP power $p_{PV_{MPP}}$, within all possible operating conditions, as shown schematically given in Fig. 2.5.

This model involves requirements that could easily lead to a reliable prediction of the PV production. In the literature there are many PV source models. As already mentioned, PV models are often very useful and used in numerical simulations to understand the PV physical phenomenon, whereas it is much more difficult to find robust models for the PV power prediction. In this section three PV models—a single-diode PV model, linear model, and purely experimental model—are proposed and compared for PV power production calculation.

1.2.2 Equivalent Circuit Photovoltaic Model

The most common and most used model in the scientific community is the model based on the equivalent circuit schema of the PV cell. In this model it is assumed that a PV cell may be modeled by a current source (photocurrent source) that is proportional to the irradiance and that is linearly dependent on the PV cell temperature [2]. Nonlinear phenomena within the PV cell are modeled by a diode connected in parallel with this current source. Thus the ideal PV cell may be equivalent to a current source in parallel with a diode [4]. To refine this model, two resistors, which symbolize the losses, are often used: one parallel and the other one in series. A single-diode model provides enough precision with fewer components. It is adopted to study the nonlinear characteristics of the PV source for the whole power range. Note that there are also models based on several diodes for obtaining finer models, thus more accurate, for which the applications focus on PV phenomena studies. Regarding the PV power prediction, the

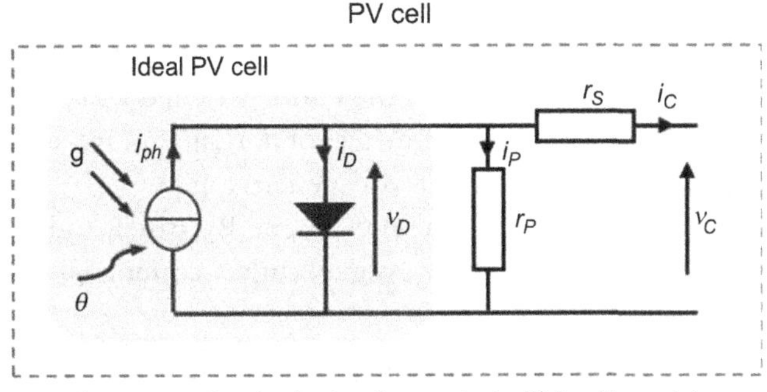

Figure 2.6 Single-diode photovoltaic (PV) cell model.

single-diode model is simple enough, yet accurate, to be considered in this study. Fig. 2.6 shows the equivalent circuit schema of a PV cell.

This PV cell equivalent circuit includes a series resistance r_S and a parallel resistance r_P, with i_{ph} the current through the photocurrent source, i_D the diode current, v_D the diode voltage, and i_P the current through the parallel resistance r_P. The PV cell output current is expressed by Eq. [2.2]:

$$i_c = i_{ph} - i_D - i_P \quad [2.2]$$

In this model the photocurrent source i_{ph} changes according to the solar irradiance and the PV cell temperature as

$$i_{ph} = i_{cSC}\left(\frac{g}{g^*}\right)\left(1 + K_I\left(\theta - \theta^*\right)\right) \quad [2.3]$$

where g^* is the solar irradiance STC reference (1000 W/m^2), K_I is the temperature coefficient for current, and θ^* is the STC cell temperature reference ($\theta^* = 298.15$K).

The diode saturation current i_{sat} is a function of PV cell temperature:

$$i_{sat} = \frac{i_{cSC}\left(\frac{\theta}{\theta^*}\right)^{\left(\frac{3}{n}\right)} \exp\left(\frac{-qE_G}{nK}\left(\frac{1}{\theta} - \frac{1}{\theta^*}\right)\right)}{\left(\exp\left(\frac{qv_{cOC}}{nK\theta^*}\right) - 1\right)} \quad [2.4]$$

where E_G is the band gap energy of the semiconductor ($E_G = 1.12$ eV for silicon); n is the diode ideality factor, which typically takes a value between 1 and 2; and K and q are the Boltzmann constant and electron charge, respectively. The final current of the PV cell i_c is given in Eq. [2.5]:

$$i_c = i_{ph} - i_{sat}\left(\exp\left(\frac{v_c + i_c r_S}{nV_T}\right) - 1\right) - \frac{v_c + i_c r_S}{r_P} \quad [2.5]$$

where $V_T = K\theta/q$ is the thermal voltage. Eq. [2.5] has five unknown parameters: i_{ph}, i_{sat}, n, r_S, and r_P.

According cases, the variations of the currents i_{ph} and i_{sat} could be taken into account whereas the other parameters of n, r_S, and r_P are kept constant or adjusted to better fit the current-voltage curve [5].

Taking into account N_S cells in series in one PV panel, and N_{PV} panels in series in the PV source, the PV source output current i_{PV} is given by Eq. [2.6]:

$$i_{PV} = i_{ph} - i_{sat}\left(\exp\left(\frac{v_{PV} + i_{PV}Nr_S}{nNV_T}\right) - 1\right) - \frac{v_{PV} + i_{PV}Nr_S}{Nr_P} \qquad [2.6]$$

where v_{PV} is the voltage across the PV source and $N = N_S \cdot N_{PV}$.

Thus the $i_{PV} = f(v_{PV})$ curve is an implicit and nonlinear equation with four variables—g, θ, i_{PV}, and v_{PV}—and three unknown parameters, which are n, r_S, and r_P. This function can be expressed as $i_{PV} = f(g, \theta, i_{PV}, v_{PV}, n, r_S, r_P)$. Hence, the maximum power of the PV source corresponding to the MPP on the $i_{PV} = f(v_{PV})$ curve is given in Eq. [2.7]:

$$p_{PV_{MPP}} = i_{PV_{MPP}} \cdot v_{PV_{MPP}} = f(g, \theta, i_{PV_{MPP}}, v_{PV_{MPP}}, n, r_S, r_P) \qquad [2.7]$$

The classical single-diode model, with three unknown parameters for which the values depend on the weather operating conditions, involves an important computation effort. Thus, for each set of weather operating conditions, the model parameters must be computed using the measurement of weather conditions as well as the current and voltage measurement.

The parameter identification method uses experimental test results obtained at the MPP of the PV source, for days having different solar irradiance profiles, and focuses on minimizing the error between the measured current and the calculated current from the mathematical model [5]. This minimization provides optimal values of the desired parameters. It is noted that the parameter values vary greatly depending on weather conditions. This model can be applied with a reasonable margin of error only for a posteriori applications.

The model presented here shows that its use requires measures of current and voltage, or power, in parameter identification. The use of parameter values shows that for every given pair of g and θ, the modeling of the PV source implies knowledge of current and voltage measurements of the PV source to recalculate the parameters. To obtain reliable and valid results for any operating condition, this model can lead to significant errors. Thus, knowing that the goal is to find a reliable model whatever the meteorological conditions of the operation, this model cannot be adopted for power prediction.

1.2.3 Linear Power Photovoltaic Model

To simplify the previous model, the maximum power of a PV panel can be estimated considering the PV panel as linear power source according to the PV cell temperature and solar irradiance level. Thus the PV source output power at MPP, $p_{PV_{MPP}}$, is obtained from Eq. [2.8] [6–8]:

$$p_{PV_{MPP}} = P_{MPP_{STC}} \frac{g}{1000}[1 + \gamma(\theta - 298.15)]N_{PV} \qquad [2.8]$$

where $P_{MPP_{STC}}$ is the PV panel maximum power at STC, which is available from the manufacturer data sheet, and γ is the power temperature coefficient at the MPP.

The estimation of $p_{PV_{MPP}}$ requires knowledge of the γ parameter value. Because the value of γ is highly dependent on the solar irradiance and PV cell temperature, g and θ, respectively, it should be estimated in experimental reference tests following a real weather conditions day. Once this parameter is estimated and validated, the power $p_{PV_{MPP}}$ could be calculated according to the weather data for any day.

The results obtained for the parameter γ are highly dependent on the operating conditions, especially on the cell temperature. When the cell temperature is predominantly greater than 25°C, the value is negative, reflecting the negative effect of increasing the cell temperature on the PV cell power. Conversely, the coefficient γ takes a positive value. Although the errors between the PV power calculation following this model and the experimental measurements are not considered important, this model cannot be used. Indeed, the variability of the γ sign depending on the weather still remains an issue for the PV power.

To overcome this problem, the difference between the cell temperature and the temperature of 25°C may not be taken into account, as shown in Eq. [2.9]:

$$p_{PV_{MPP}} = P_{MPP_{STC}} \frac{g}{1000}[1 + \gamma'\theta]N_{PV} \qquad [2.9]$$

The results obtained with Eq. [2.9] are very interesting because the coefficient γ' takes values near a constant value. It is possible to go even further with Eq. [2.10]:

$$p_{PV_{MPP}} = P_{MPP_{STC}} \frac{g}{1000}\gamma'''[1 + \gamma''\theta]N_{PV} \qquad [2.10]$$

Eq. [2.10] takes into account the proportionality of the PV power at the MPP with solar irradiance and PV cell temperature, and two coefficients have to be identified. This equation leads to fairly consistent values of the γ'' coefficient and values of the γ''' coefficient with changes in cell temperature and solar irradiance, but nevertheless, it highlights the fact that these parameters are strongly dependent on weather conditions.

The linear power model seems simple and easy to use for power prediction once the temperature coefficient γ is determined. In addition, it does not need special computation effort. However, the value of γ varies according to the meteorological conditions of the operating conditions, and the γ parameter calculation involves, at given weather conditions, current and voltage measurements corresponding to maximum power of the PV source. Using a fixed value of γ corresponding to 1 day for any given day involves errors that can range from acceptable to very large.

Therefore this model is simple enough to use, but it remains highly dependent on weather conditions. It does not take into account the various existing correlations with the PV cell temperature. The linear model of power could allow the PV power prediction only with a set of γ values and assuming an error for which the value must be compared with the errors of the weather forecast. Therefore it should be noted that correct use of this model requires a mapping of values of γ for different irradiance and PV cell temperature values.

1.2.4 Purely Experimental Photovoltaic Model

The PV model based on the PV cell equivalent circuit and the linear power model showed their strong dependence on operating conditions and the fact that the parameter identification implies knowledge of current measurements and voltage measurements at the output of the PV source. Therefore, with a reasonable margin of error, both models can only be applied a posteriori. Given the complexity inherent phenomena of the PV cell, a purely experimental model is proposed. It is based on measurements of current and voltage, which are performed for different pairs of irradiance and cell temperatures. The goal is to obtain a simple model and validates the PV power source, whatever the operating conditions.

To achieve this goal (ie, forecasting PV source production), a purely experimental model based on measurements of current $i_{PV_{MPP}}$ and voltage $v_{PV_{MPP}}$ measurements done for different couples (g, θ) at the MPP is proposed as in [9]. This experimental prediction model is based on the PV panel indoor experimental tests. Depending on different PV technology and manufacturers, PV panel characteristics greatly vary. General PV models could be limited in reflecting these differences. One exemplary PV panel is tested under different levels of constant irradiance and different cell temperatures. The model is drawn from a series of MPP voltages and currents ($v_{PV_{MPP}}$ and $i_{PV_{MPP}}$) of a PV panel, measured for several levels of irradiance at given cell temperatures. The data are obtained from the measurement of maximum output voltage and current variation depending on PV cell temperature increasing under a constant irradiance. Then, as shown in Fig. 2.7, they are combined to form two 2-dimension look-up tables (LUTs), the inputs of which are g and θ, and the outputs are $v_{PV_{MPP}}$ and $i_{PV_{MPP}}$. The LUTs can then be used in prediction of PV source power.

Given the difficulty of making such measurements with the desired solar irradiance (irradiance rapid change, or unwanted cell temperature), for this model a bench test and measurement based on an artificial irradiance are proposed and realized.

Test Bench Description

The indoor experimental tests have been done on a single PV panel, Solar-Fabrik SF-130/2−125, the electrical characteristics of which are given in Table 2.1.

The PV panel is excited by an irradiance emulator composed of 42 halogen lamps and 21 kW total power, as shown in Fig. 2.8. Although this electric power seems impressive compared with 125 W at STC for the used PV panel, it is necessary to obtain a homogeneous irradiance on all of the cells constituting the PV panel. Halogen is used because it is a light source that has the best relationship between approximation of the solar spectrum and acquisition cost. Depending on the projection distance, this test bench provides an irradiance range from 200 to 1200 W/m^2. The irradiance is measured with two pyranometers for which the spectral sensitivity is between 400 and 1100 nm.

Two PT100 temperature sensors are also used to measure the cell temperature. Pyranometers and the PT100 sensors are shown in Fig. 2.8a and the halogen lamp excitation is shown in Fig. 2.8b. Pyranometers and temperature sensors are placed so as to verify the homogeneity of the total area of the PV irradiance and the cell temperature. If the PV panel is shined homogeneously, then the pyranometer measurements are the same; the same consideration is taken for temperature sensors. Tests were conducted only within these operating conditions.

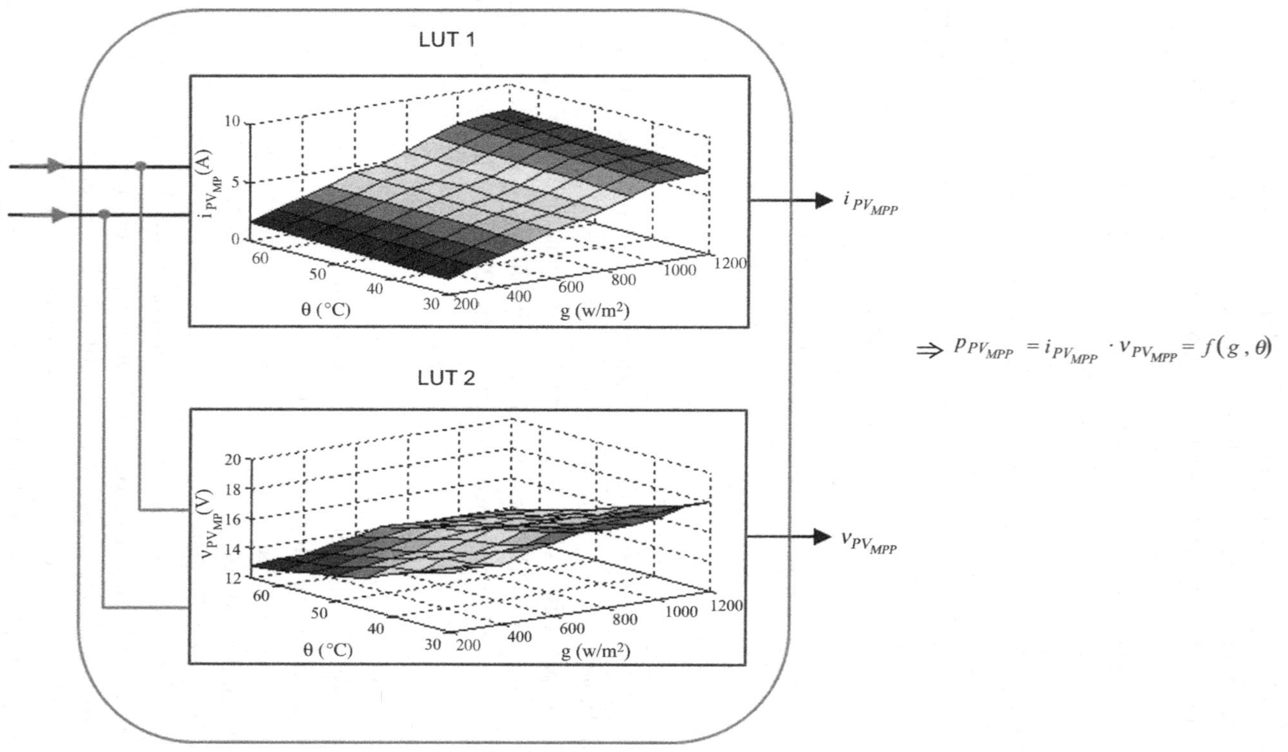

Figure 2.7 Experimental prediction model generic schema. *LUT*, look-up table.

Table 2.1 Electrical STC Specifications of Photovoltaic Panel **Solar-Fabrik SF-130/2-125**

N_S	Number of cells in series	36
I_{SC}	Short-circuit current	7.84 A
V_{OC}	Open-circuit voltage	21.53 V
I_{MPP}	Maximum power point current	7.14 A
V_{MPP}	Maximum power point voltage	17.50 V
P_{MPP}	Power maximum	124.95 W
K_I	Temperature coefficient for current	0.00545 A/K
θ^*	Cell temperature reference	298.15 K
g^*	Irradiance reference	1000 W/m^2

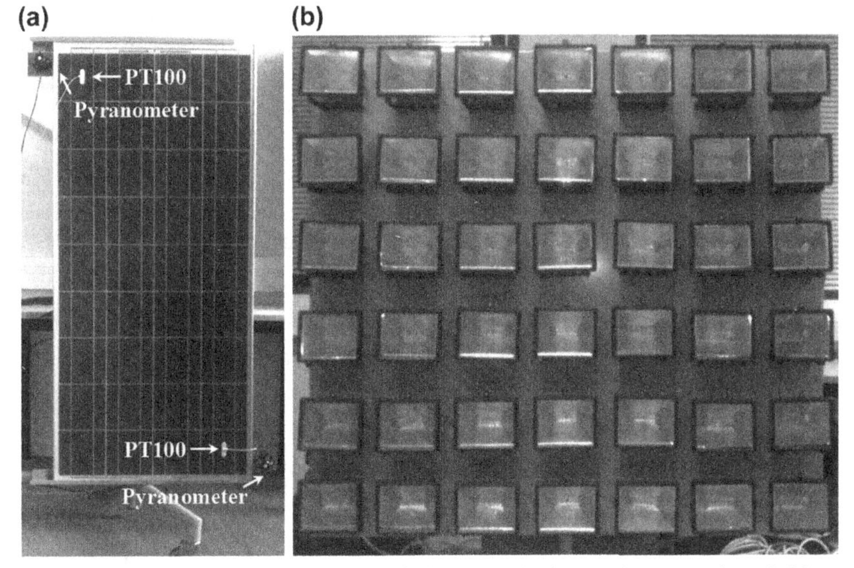

Figure 2.8 Images of the experimental devices (a) photovoltaic panel and (b) its excitation.

Process Measurements and Data Acquisition

To impose the desired current, a DC-DC converter, the equivalent circuit of which is shown in Fig. 2.9, was developed.

To absorb and dissipate energy extracted from the PV, a programmable electronic load (PEL) is used. To impose to PV panel, a current ramp and to simplify the command, a hysteresis control is implemented. A real-time system (dSPACE 1104) performs control and monitoring of

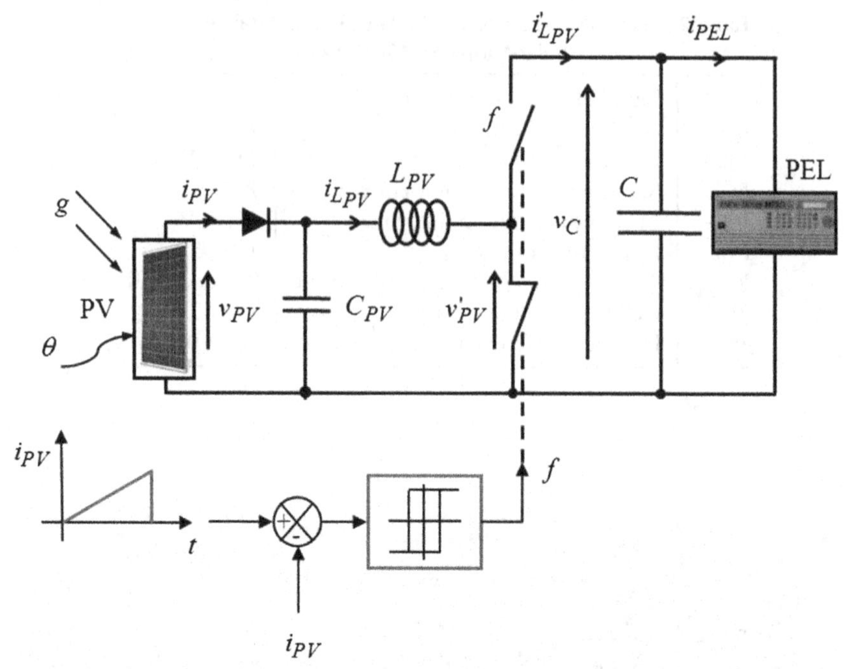

Figure 2.9 Direct current (DC)-DC converter equivalent circuit used on photovoltaic (PV) panel process measurements. *PEL*, programmable electronic load.

the process (electrical measurements, converter switching function, as well as measurements of g and θ). The other device values are $L_{PV} = 20$ mH, the internal resistance of which is neglected, and $C_{PV} = 1000\ \mu$F.

Fig. 2.10 shows the temporal evolution of the voltage measured for an imposed current ramp according to increasing temperature and under constant irradiance of 1000 W/m^2.

The experimental evolution of the power-voltage characteristic depending on the PV cell temperature measured for a constant irradiance of 1000 W/m^2 is given in Fig. 2.11. The negative effect of PV cell temperature can be noted.

Concerning the maximum irradiance, 1200 W/m^2, and cell temperature, 65°C, for which the experimental tests were performed, these are normal limits according to findings on the actual local weather conditions measured in Compiegne, France. However, the lower limits of irradiance and temperature that could be ensured during testing in our laboratory are 200 W/m^2 and 30°C respectively. This is due to the technical limitations of the experimental setup.

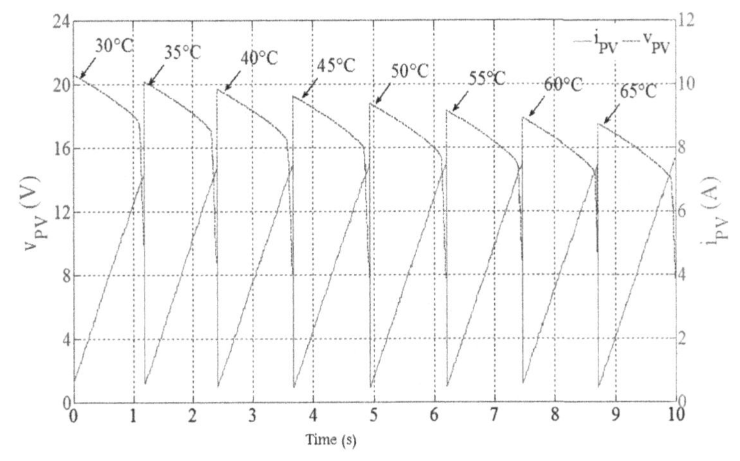

Figure 2.10 Measured current and voltage ramp according to increasing cell temperature and for a constant irradiance of 1000 W/m^2.

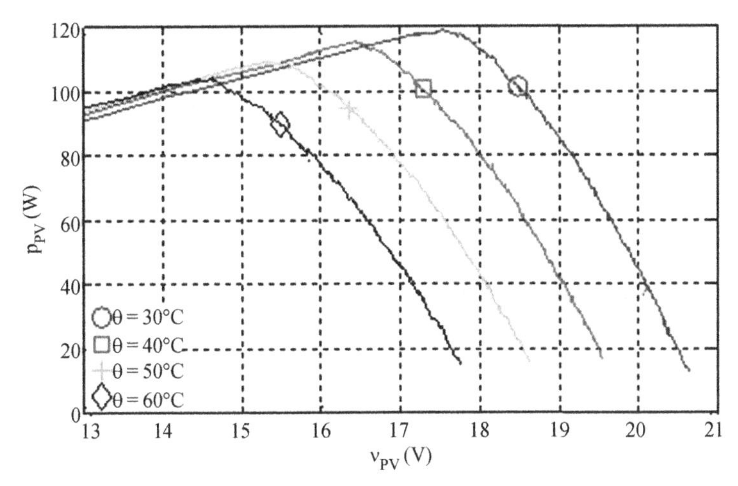

Figure 2.11 Evolution of power-voltage characteristics according to increasing cell temperature and for a constant irradiance of 1000 W/m^2.

Figs. 2.12 and 2.13 show the experimental evolution of the current and the voltage at the MPP according to the irradiance and the cell temperature. These figures illustrate that the current increases as a function of irradiance and the cell temperature whereas the voltage slightly varies depending on the irradiance and substantially decreases with the cell temperature growth.

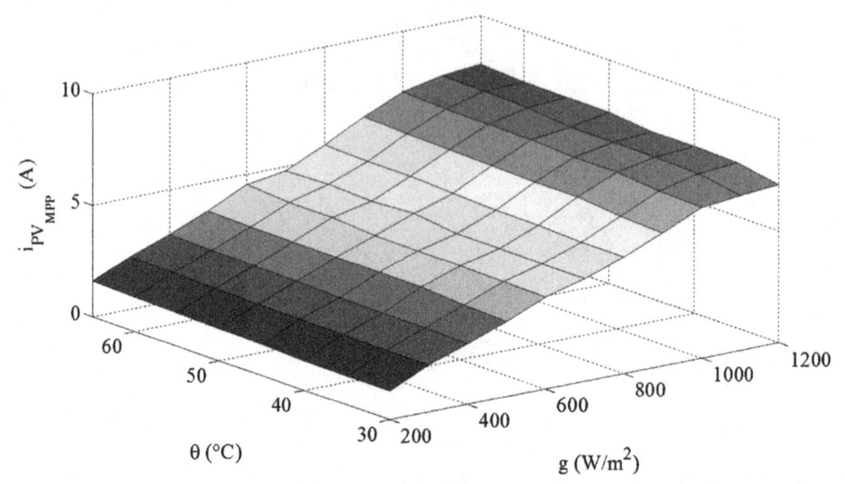

Figure 2.12 Experimental data of current model, measured according to g and θ evolution.

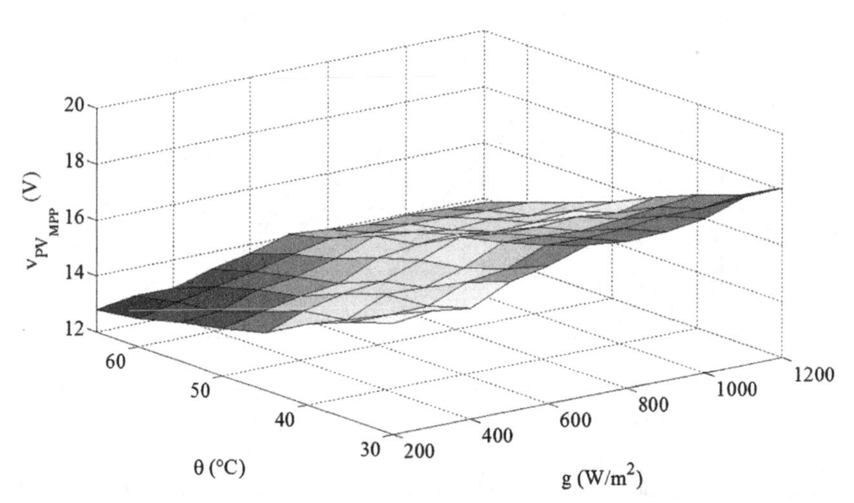

Figure 2.13 Experimental data of voltage model, measured according to g and θ evolution.

The obtained experimental data are stored in two LUTs, one for current and the other for voltage; hence, knowing the irradiance and cell temperature measurement, the values $v_{PV_{MPP}}$ and $i_{PV_{MPP}}$ can be found and lead to calculating the power at the MPP. For any irradiance or cell temperature value included in the LUT but not experimentally determined, each LUT applies linear approximation.

The purely experimental model is simplified and easy to use for numerical simulation of the overall system and PV power prediction. However, this model is not perfect; it involves

- achieving a more or less costly experimental device (which can, however, be used to validate the maximum power point tracking (MPPT) algorithm or other...),
- neglecting the PV panel aging,
- repeating all tests when using different PV panel, and
- accepting uncertainties induced by the adopted linear approximation.

Designed for a single PV panel, this purely experimental model can be extended to any PV source composed by several identical PV panels and identical to the modeling used PV panel.

1.3 Experimental Comparison of Photovoltaic Power Models

One of the purposes of this chapter is to develop a model of PV sources for power prediction, simple to implement, which from known operating conditions (solar irradiance and PV cell temperature), calculate or estimate the potentially recoverable power. For this, three different models were studied and need to be experimentally compared. To eliminate the disadvantages of the first two models, a purely experimental model is used.

For the experimental comparison of these three models, various days of solar irradiance and PV cell temperature completely different profiles are used. There are outdoor experimental tests based on a test bench similar to the test bench presented in Fig. 2.9, but they use a PV source built on the roof of our university and are composed of 16 PV panels (SF-130/2–125), as shown in Fig. 2.14.

Figure 2.14 Photovoltaic source considered for the experimental tests.

The differences between the calculated powers and the measured powers are observed and analyzed. To summarize this analyze, one notes that

- Regarding the purely experimental model, the models based on the equivalent circuit of the PV cell and linear power offer the nearest power curves of the power curve measured.
- Furthermore, the model based on the equivalent circuit of the PV cell provides the smallest difference between calculation and measurement.

Finally, the first two models can be successfully applied in applications only a *posteriori*, but their use in numerical simulation of the overall system and PV power prediction would provide randomly a good result as well as a bad result. On the other hand, it is interesting to note that the linear power model is very simple and requires no computational effort. For a purely experimental proposed model, a significant error can be observed for very sunny days. Indeed, a purely experimental model was performed on a clean PV, almost new and without taking into account the wiring resistance (a high solar irradiance means a solar conversion high current, which leads to high Joule losses). Despite this deficit, unlike the other two models, the purely experimental model can be useful in many applications a priori because it does not require parameter identification.

To illustrate this comparison, Fig. 2.15 shows the operating conditions for the day of October 20, 2011 (solar irradiance and PV cell temperature). Fig. 2.16 gives the PV source powers calculated by the three models and the measured PV source power.

In conclusion, the purely experimental model is the only one that can be used in PV power prediction. This model could be improved by at least

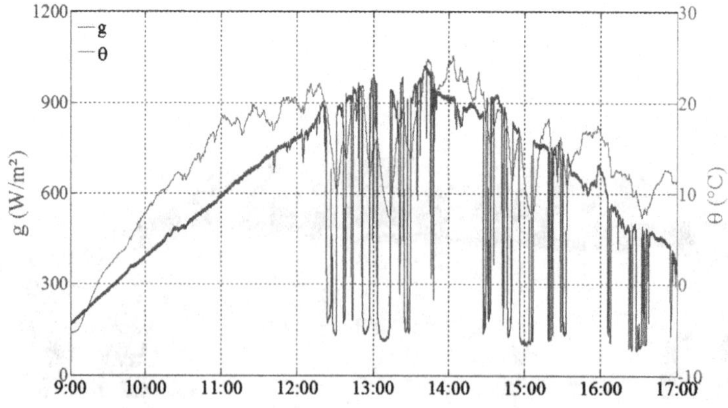

Figure 2.15 Operating conditions of the experimental test.

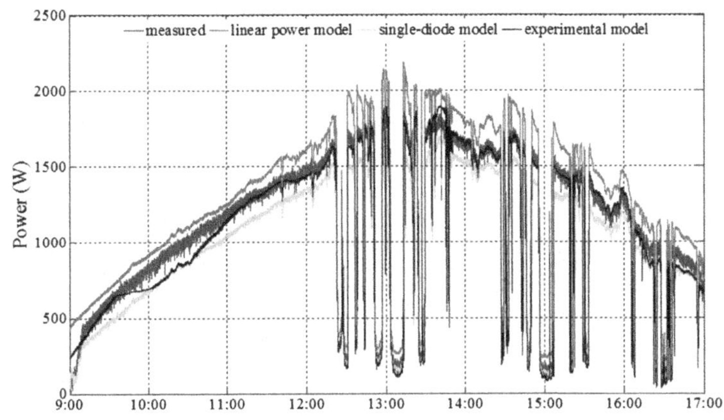

Figure 2.16 Comparison of photovoltaic source output power measured and calculated.

adding more data, calculating the PV cell temperature from ambient temperature and solar irradiance, and taking into account the PV aging phenomena. Adapted for each PV panel manufacturer, this model could be successfully used to predict the power.

1.4 Photovoltaic System Efficiency and Optimal Operating Points

Regardless of new PV technologies at the stage of research and development and focusing on the PV silicon-based first generation and second generation called thin film, PV sources do not have high efficiency; therefore it is necessary to optimize power extraction. To operate a PV source within its MPP, whatever the solar irradiance and cell temperature variations, a MPPT method is needed to find and maintain the peak power.

Fig. 2.17 shows some curves, $i = f(v)$ and $p = f(v)$, of a single PV panel, SF-130/2−125, measured at 30°C and under the solar irradiance variation of $200\ W/m^2 \leq g \leq 600\ W/m^2$.

The numerical modeling under MATLAB-Simulink presented here shows the importance of the use of the MPPT algorithm.

2. MAXIMUM POWER POINT TRACKING

During recent years, for PV systems, many MPPT algorithms have been proposed and developed to maximize the produced energy. Regarding the design manner, these methods vary in many aspects, such as

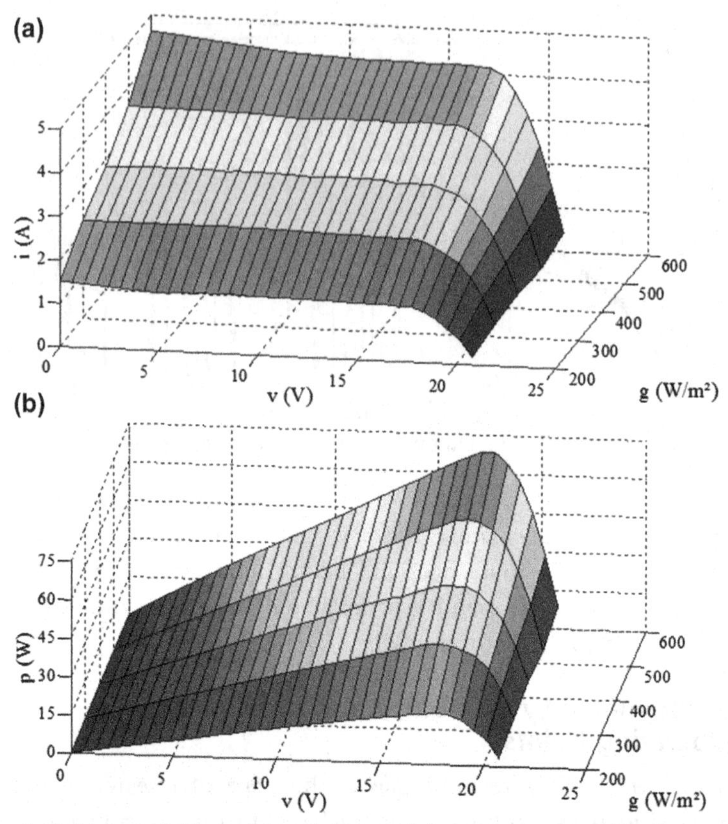

Figure 2.17 Measured (a) $i = f(v)$ and (b) $p = f(v)$ characteristics of the SF-130/2-125 photovoltaic (PV) panel under variable solar irradiance and constant PV cell temperature (30°C).

implementation simplicity, power or energy efficiency, convergence speed, sensors required, and cost effectiveness. Some comparative studies, based on widely adopted MPPT algorithms, presented in the literature give results obtained either from a simulation tool, which provides simultaneous operating systems, or using a real PV test bench under a solar simulator to reproduce the same operating solar conditions. This section presents an experimental comparison, under real solar irradiance, of four most used MPPT methods for PV power systems: perturb and observe (P&O) and incremental conductance (InC) as a constant tracking step and improved P&O (ImP&O) and fuzzy logic (FL)-based MPPT as a variable tracking step. Using four identical PV systems, under strictly the same set of technical and meteorological conditions, an experimental comparison of these four algorithms is done. Following two criteria, energy efficiency and cost effectiveness, this comparison shows the advantage of using an MPPT with a variable tracking step. The extracted energies by all four methods are almost identical with a slight advantage for the ImP&O algorithm.

2.1 Maximum Power Point Tracking Method Overview

An efficient MPPT algorithm, as its name suggests, is able to search and find the MPP of a nonlinear electrical generator regardless of the operating meteorological conditions. In our case, nonlinear electric generators are PV sources (Figs. 2.14 and 2.17). The MPPT algorithm must operate in every moment all PV panels composing a source to its MPP and that despite highly variable meteorological conditions.

In recent years there has been great interest in the study, improvement, and implementation of MPPT algorithms. Fig. 2.18 shows the evolution of the number of articles in international journals, based on years of publication, for two of the largest publishers in electrical engineering—the Institute of Electrical and Electronics Engineers and Science Direct.

From 2000 to 2010, a cyclical increase in the number of articles can be observed. Since 2011, there is a very strong growth that attempts to prove the great current interest in the study of these algorithms. In this literature it is possible to separate more than 15 different algorithms. Furthermore, there

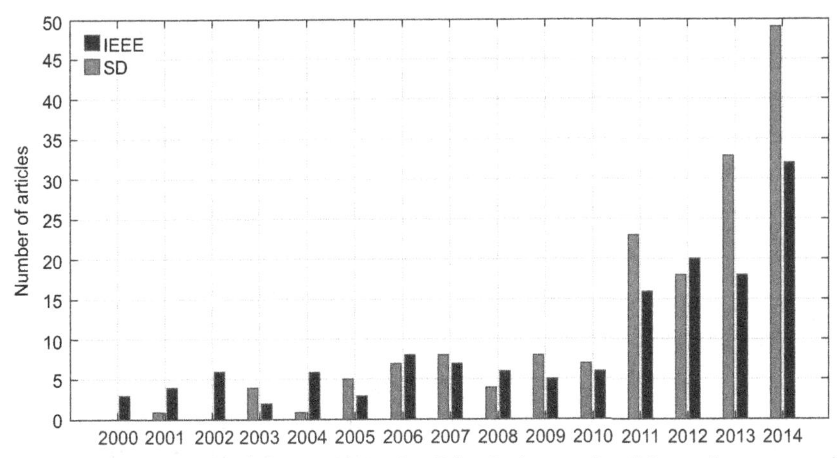

Figure 2.18 Evolution of the number of articles in International journals concerned with MPPT algorithms.

are so many extensions and/or variations of these algorithms that the choice of an algorithm for a particular application becomes very difficult. The most adequate MPPT method involves the best response under the particular PV system constraints. Thus what are the most important criteria to consider? These include the rapidity, efficiency, complexity, implementation, etc. Moreover, these criteria are often related to each other, which makes the choice even more complicated. This is why an analysis of the MPPT method is proposed, based on the most commonly used algorithms, in research and industrial applications, to identify selection criteria and provide support for a decision.

The MPPT strategy aims to find the voltage or current at which the PV system provides the maximum output power. The MPPT methods can be classified, following MPP process seeking, into indirect and direct method. The indirect methods, such as short-circuit and open-circuit methods, need a prior evaluation of the PV panel, or they are based on mathematical relationships or a database not valid for all operating meteorological conditions. Therefore they cannot obtain exactly the maximum power of a PV panel at any irradiance and cell temperature. Conversely, the direct methods operate at any meteorological condition and act in real time on the voltage reference variable, corresponding to the maximum power provided by the PV system. The most used direct methods are P&O, InC, and FL-based MPPT. Some direct MPPT algorithms can also be classified according to the method by which the command variable is changed. Thus fixed-step MPPT algorithms and variable-step MPPT algorithms can be differentiated.

Another classification can be built based on the implementation simplicity, power or energy efficiency, convergence speed, sensors required, cost effectiveness, etc. According to the solar irradiance, homogeneous or not, there are some works that deal with the partial shading of the PV source; however, this case is not studied in the current book.

The following two sections present the fixed-step MPPT algorithms and variable-step MPPT algorithms.

2.2 Fixed-Step Size Maximum Power Point Tracking Algorithms

The most common fixed-step MPPT algorithms in research works and industrial applications are P&O and InC. These algorithms operate in real time on the voltage reference variable corresponding to the MPP of the PV power system.

2.2.1 Perturb and Observe

The P&O algorithm acts periodically by giving a perturbation to the operating voltage v and observing the power variation $p = v \cdot i$ to deduct the direction of evolution to give to the voltage reference v^*. Taking into account the power-voltage characteristic curve $p = f(v)$ obtained under given conditions g and θ, the goal is to track the operating point at the MPP as shown in Fig. 2.19.

This algorithm measures at each z instant the variables $i(z)$ and $v(z)$, calculates $p(z)$, and then compares with the power calculated at the $z - 1$ instant $p(z - 1)$. For all of the operating points where the power and current variations are positive, the algorithm continues to perturb the system in the same direction in increasing the voltage reference v^*; otherwise, if these variations are negative, then the direction of perturbation is reversed. The increasing or decreasing of the reference v^* is done by the tracking step ΔV. The flow chart of the P&O algorithm is presented in Fig. 2.20.

Theoretically, the algorithm is simple to implement in its basic form. However, it produces some oscillations around the MPP in steady-state operating and induces power losses. Its functioning depends on the tracking step size applied to the voltage reference v^* and on the adopted computation step. For the same computation step, the oscillations, and consequently the power losses, could be minimized if the tracking step is small. Nevertheless, the response of the algorithm becomes slower, then, under rapid solar irradiance changes, this algorithm may deviate [10].

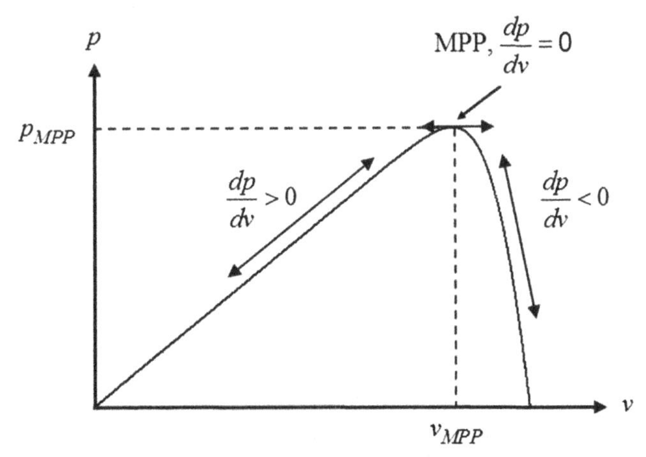

Figure 2.19 Maximum power point (MPP) highlighted on the power-voltage characteristic of the photovoltaic panel.

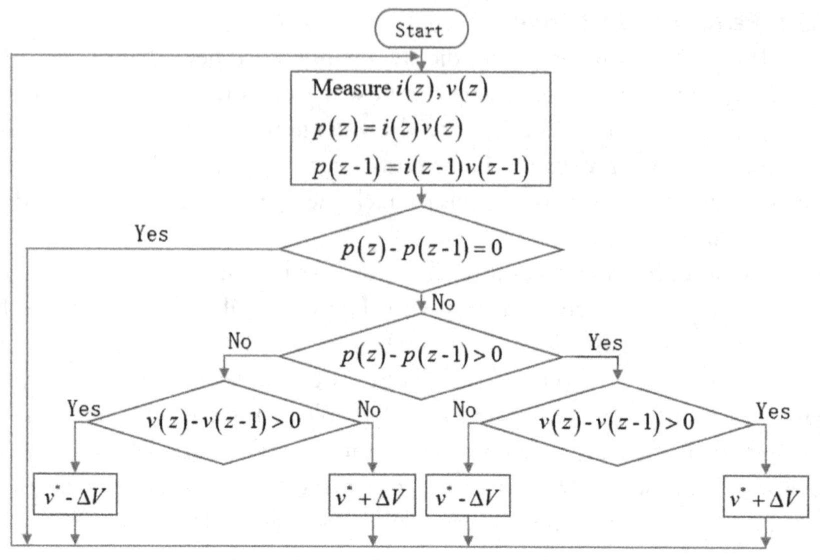

Figure 2.20 Flow chart of the perturb and observe algorithm.

2.2.2 Incremental Conductance

To find out the position of the actual operating point in relation to the MPP, this algorithm uses the derivate of the conductance di/dv. It is based on the fact that the slope tangent of the characteristic $p = f(v)$ is zero in the MPP, positive on the MPP left side, and negative on the MPP right side, as illustrated in Fig. 2.19. Because the power is equal to the product of current and tension, the calculation of this slope is given by Eq. [2.11].

$$\frac{dp}{dv} = \frac{d(vi)}{dv} = i + v\frac{di}{dv} \qquad [2.11]$$

The development of this derivate in MPP is given by [2.12].

$$\frac{i}{v} + \frac{di}{dv} = 0 \qquad [2.12]$$

When the operating point is located in the left side of the MPP, then $i/v + di/dv > 0$, whereas $i/v + di/dv < 0$ when the operating point is on the other side. From the instantaneous measurement of i and v, the following approximations can be done:

$$\begin{cases} di \approx \Delta i = i(z) - i(z-1) \\ dv \approx \Delta v = v(z) - v(z-1) \end{cases} \qquad [2.13]$$

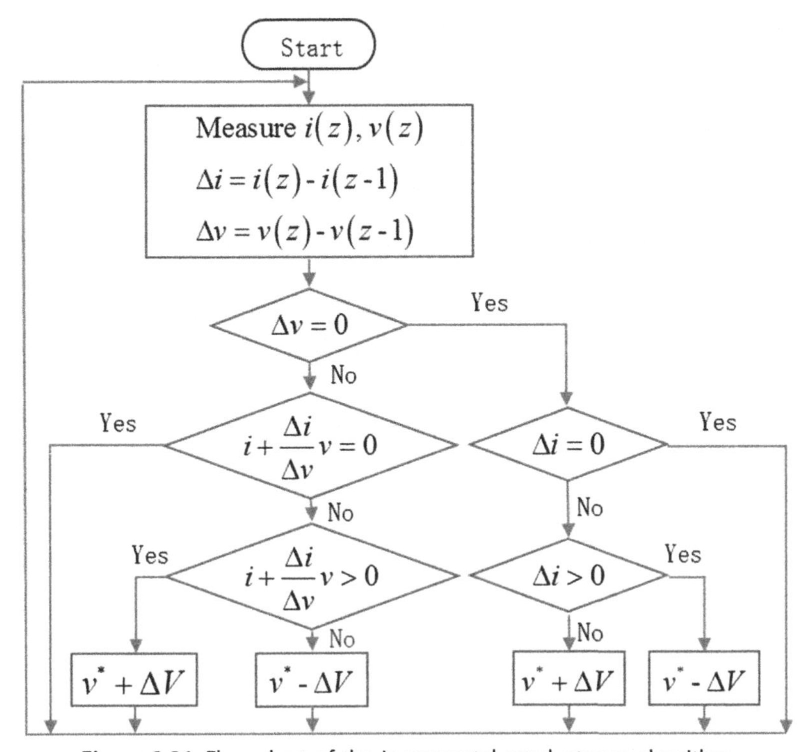

Figure 2.21 Flow chart of the incremental conductance algorithm.

Therefore the algorithm can instantly calculate i/v and di/dv to deduct the direction of the perturbation leading to the MPP. This is done by acting on v^*. Fig. 2.21 shows the flow chart of the InC algorithm.

Concerning power efficiency, this method could theoretically provide a better tracking of the MPP than the P&O algorithm. However, because of the noise and error measurements, it seems interesting to note that experimentally Eq. [2.12] is never satisfied. It produces oscillations around the MPP and power loss. Furthermore, the complexity of the InC algorithm, compared with the P&O, increases the calculation time.

2.2.3 Comparison and Discussion on Perturb and Observe and Incremental Conductance

Maximum energy efficiency, calculation time, and simple implementation are the principal criteria taken into account in this chapter for choosing the most suitable MPPT algorithm for a PV power system.

The P&O method has been highlighted in numerous studies in the literature to be less efficient than other methods but on the basis of one criterion to extract maximum power. However, this method, analyzed according to a criterion to extract maximum energy, has proved as effective as the InC. In the literature, comparisons between these two MPPT methods are not operated in technical and meteorological conditions that are strictly identical under real solar irradiance. On the basis of the experimental platform installed on the roof of our university, tests were performed in accordance with strictly identical real conditions; thus the performance of these two fixed-step tracking MPPT algorithms was quantified by energy efficiency [11]. The comparison, based on three tests made at the same sample time, and three different fixed-step trackings shows the influence of the tracking step value on the operation of MPPT algorithms. According to this work, it is concluded that P&O, when the tracking step value is correctly chosen, can have an energy efficiency equivalent to that obtained with InC, whereas P&O is easier to implement compared with other MPPT methods.

2.3 Variable-Step Size Maximum Power Point Tracking Algorithms

This section proposes an improved P&O algorithm, as noted earlier as ImP&O, for which the tracking step is variable according to the operating meteorological conditions. As one of the most common MPPT methods with variable-step tracking, the FL method is also studied in this section.

2.3.1 Improvement of the Perturb and Observe Method

An evolution of tracking step ΔV according to characteristic slope dp/dv is supposed as shown in Fig. 2.22 (solid line). If the PV power system is near the MPP, $|dp/dv| \approx 0$, then the tracking step oscillates between two values ΔV_{S2} and $-\Delta V_{S2}$. This is why there are oscillations around the MPP that cause some power losses as previously mentioned. The performance of P&O can be improved by making the tracking step variable following the variations of operating conditions. The tracking step can vary according to the logic: if the PV power system is far away from the MPP $|dp/dv| > 0$, then the tracking step must be large, otherwise it must be smaller. The same logic is used in [12], in which the time variation of solar irradiance dg/dt is taken into account. The operating principle of the proposed ImP&O method is illustrated by the dashed line in Fig. 2.22.

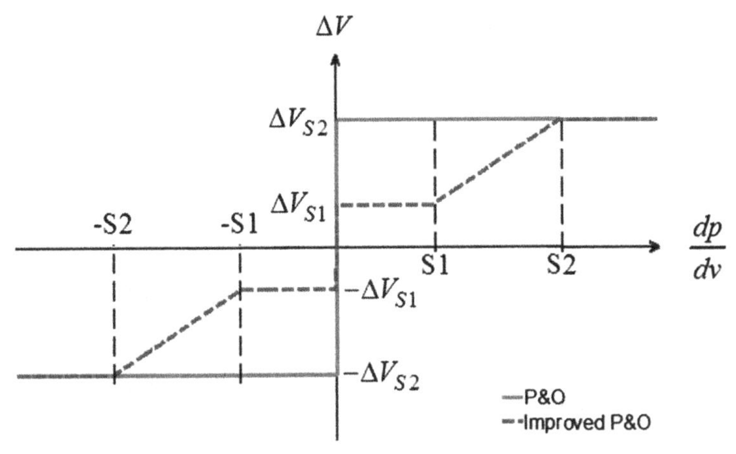

Figure 2.22 Principal operation of perturb and observe (P&O) and improved P&O (ImP&O) algorithms.

Therefore if $0 \leq |dp/dv| \leq |S1|$, then the tracking step value is settled at ΔV_{S1}, whereas for $|S1| \leq |dp/dv| \leq |S2|$ it varies linearly (is an arbitrary choice that is proposed in this work). For $|dp/dv| > |S2|$, the tracking step is settled at ΔV_{S2}, with $\Delta V_{S2} > \Delta V_{S1}$.

As said at the beginning, to analyze the energy performance of the most used MPPT algorithms, a comparison according experimental tests is proposed. Thus for consistency reasons, the value of ΔV_{S2} must be identical to the tracking step value used in fixed-step P&O and InC algorithms. Moreover, the value of ΔV_{S1} and the thresholds $S1$ and $S2$ must be well determined (empirical methods). Finally, based on the assumption that the variation of tracking step value between the thresholds $S1$ and $S2$ is linear, the implementation of the ImP&O algorithm requires an iterative methodology and certainly some experience.

2.3.2 Fuzzy Logic Maximum Power Point Tracking Approach

During recent years, the FL MPPT has been used increasingly for the PV systems [13]. The FL MPPT can handle imprecise inputs, works with the nonlinear systems, and offers a robust control; however, the designer must have much knowledge and experience with PV systems. The FL consists of mapping the input space and the output space through logical operations. The inputs/outputs are expressed as linguistic variables, called fuzzy sets, overlapped between each other. Using logical operations, the relationship is done between the input fuzzy sets and the outputs sets. Values are then assigned to the outputs. This is why the design of FL MPPT requires experience and knowledge on PV power system operation. The FL MPPT generally consists of three stages: fuzzification, fuzzy reasoning, and defuzzification.

Fuzzification

During the fuzzification, the numerical values of the inputs/outputs are transformed into fuzzy sets. Therefore they are expressed in linguistic variables characterized by a membership called the subset, which represents each point of input space, called the universe of discourse. For a membership function, the estimation is usually given more in degree rather than in value. Comparing with the Boolean logic, in which the response is either 0 or 1, the degree of membership function takes many values between 0 and 1. A well-designed FL MPPT satisfies every instant the equality $|dp/dv| = 0$, whatever the operating conditions. Therefore two inputs are taken into account, $e_1 = dp/dv$ and $e_2 = d(dp/dv)/dt$, which gives information about the direction and the rate of algorithm convergence toward the MPP, and the output is the tracking step ΔV as shown in Fig. 2.23.

The FL MPPT is performed by iterative methodology; therefore the two inputs are numerically defined by Eq. [2.14].

$$\begin{cases} e_1 = \dfrac{dp}{dv} = \dfrac{p(z) - p(z-1)}{v(z) - v(z-1)} \\ e_2 = e_1(z) - e_1(z-1) \end{cases} \quad [2.14]$$

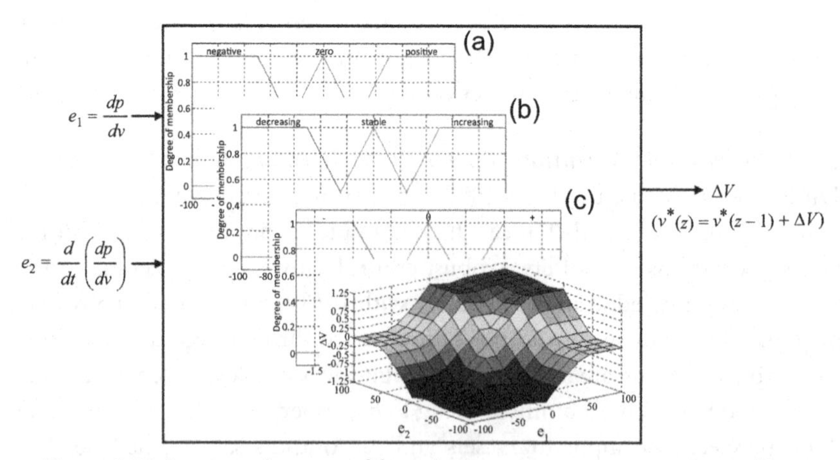

Figure 2.23 Inputs and output of fuzzy logic maximum power point tracking.

The three fuzzy subsets that define the inputs and the output are illustrated in Fig. 2.24 as follows: $e_1 \in$ {negative; zero; positive}; $e_2 \in$ {decreasing; stable; increasing}; $\Delta V \in \{-;\ 0;\ +\}$, where the universe of discourse of the inputs and the output are [−100, 100] and [−1.75, 1.75], respectively. To simplify this study, triangular and trapezoidal membership functions were chosen instead of Gaussian or sigmoidal shapes.

The proposed FL MPPT design is considered accurate enough, but for more accuracy considerations some studies used five subsets as demonstrated in [14].

Fuzzy Reasoning

Because the inputs and the output are fuzzified, mapping all of the inputs with the output is known as fuzzy inference of fuzzy reasoning. This is accomplished using fuzzy "If-Then" rules. The "If-Then" statement consists of two parts: the "If" part called the antecedent and the "Then" part called the consequent. The antecedent defines a fuzzy region in the input space whereas the consequent specifies the output in the fuzzy region. Table 2.2 summarizes the "If-Then" rules used in the proposed FL MPPT.

The fuzzy reasoning consists of two phases. The first one includes inputs fuzzifying and determination of the activated rules. The interpretation of the antecedent defines a region in the universe of discourse and returns values between 0 and 1, which are the values of the membership functions in that region.

The second phase of fuzzy reasoning seeks to determine the consequent of all activated rules and combine them. Firstly, a logical operator is applied to the membership function values to evaluate the degree of each activated rule, which expresses how the antecedent part of each rule is satisfied. Then, the resulting number is used to determine the fuzzy sets in the consequent of each rule; this is called the implication. Finally, the fuzzy sets or the output of each rule are combined or aggregated to form one fuzzy set. In this case study, one of the most common systems supported in MATLAB, a fuzzy toolbox called the Mamdani fuzzy inference system, is used. It applies "And" ("min") operation for evaluation of the activated rule degrees and "Or" ("max") for aggregating the fuzzy sets.

Defuzzification

The defuzzification main goal is to interpret the fuzzy set resulting from the aggregation into a numerical value to be used by the designer (ie, the value of the tracking step ΔV). At least seven defuzzification operators can be

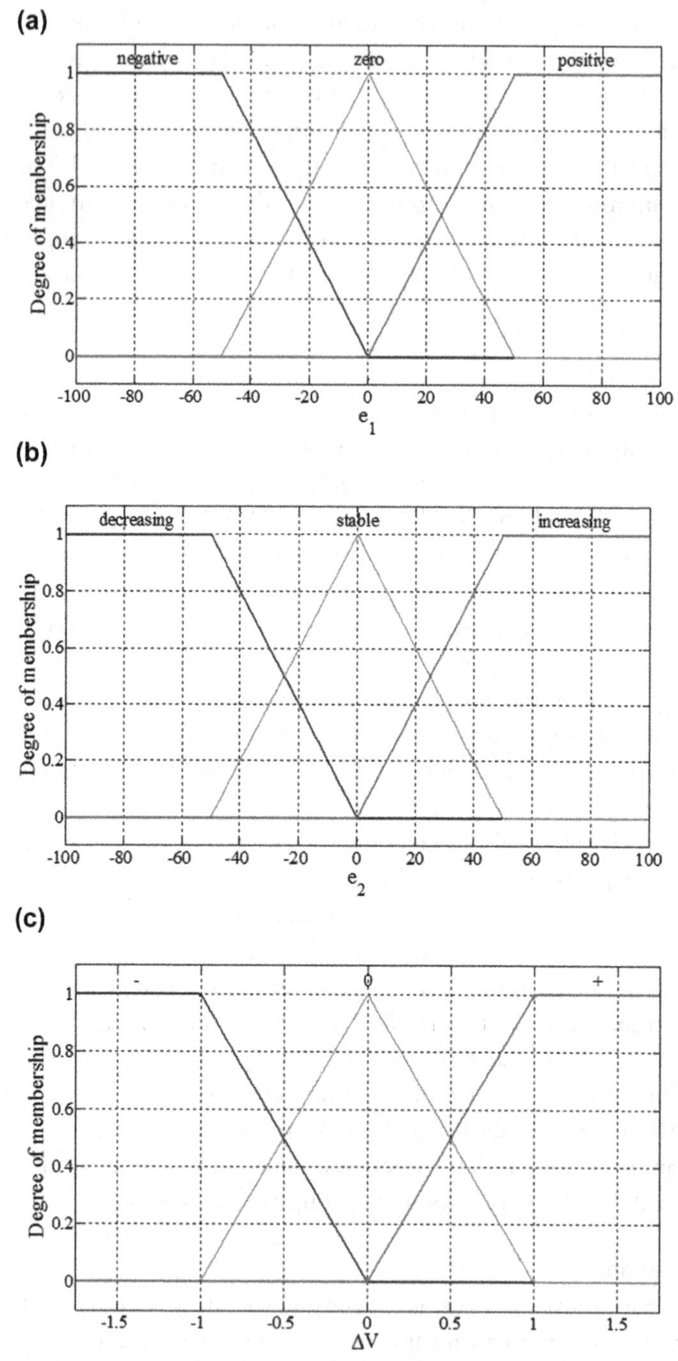

Figure 2.24 Membership functions of (a and b) inputs and (c) of output.

Table 2.2 Fuzzy Rules Table

	ΔV	e_2 Decreasing	Stable	Increasing
e_1	Negative	–	–	0
	Zero	–	0	+
	Positive	0	+	+

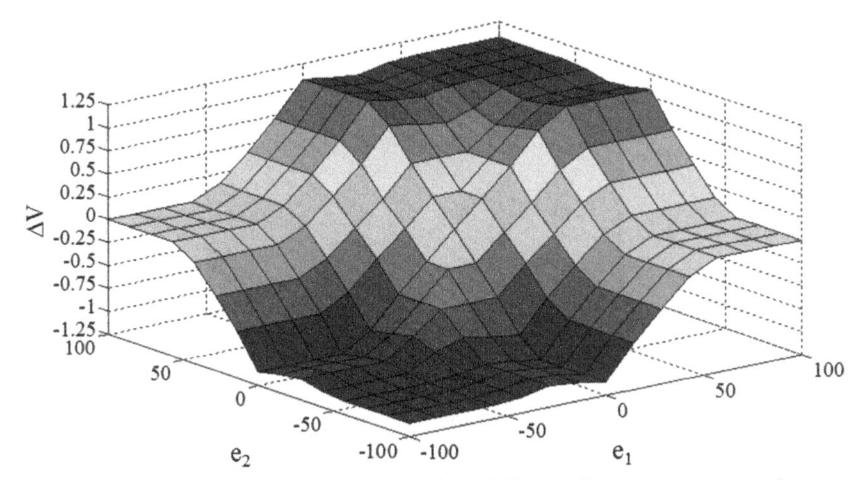

Figure 2.25 Evolution of the output values ΔV according to inputs e_1 and e_2.

used, but the most used one is the centroid, or center of gravity, generally computed by $\int \mu(\Delta V)\Delta V d(\Delta V) / \int \mu(\Delta V) d(\Delta V)$, where $\mu(\Delta V)$ is the degree of membership of the aggregated fuzzy set for the output ΔV. If all of the membership functions and If-Then rules are built, the MATLAB fuzzy toolbox computes it for any input. The all output values ΔV corresponding to the variation of the inputs e_1 and e_2 over the universe of discourse is given in Fig. 2.25.

The flow chart shown in Fig. 2.26 summarizes the proposed FL MPPT.

2.4 Experimental Comparison Between Different Maximum Power Point Tracking Algorithms

Concerning the widely adopted MPPT algorithms for PV systems, some comparative studies presented in the literature give results obtained from simulation tools, which provide simultaneous operating systems. On the

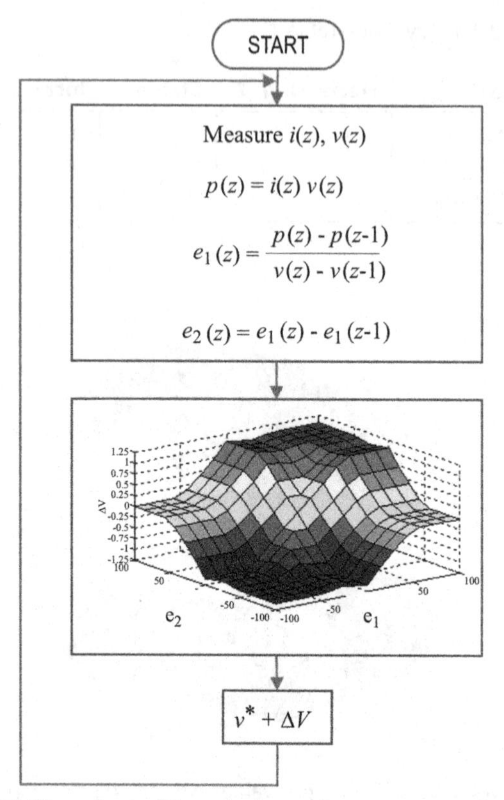

Figure 2.26 Flow chart of fuzzy logic maximum power point tracking.

other hand, for a real PV test bench, to reproduce the same operating meteorological conditions, comparative studies were done under a solar irradiance simulator and often only for one PV panel. Therefore this section aims to meet the gap between simulation systems and experimental tests under a solar irradiance simulator to compare the four most used MPPT methods and to offer support for a decision. On the other hand, the originality of this study lies also in the experimental tests based on four identical PV power systems under strictly the same set of technical and meteorological conditions. Finally, energy efficiency is emphasized by tests performed in situ and based on operation during the daylight (8—10 h) [15] whereas other studies use as a main criterion the maximum power during a few tens of seconds.

The principal criteria, which are taken into account for choosing the most suitable MPPT algorithm for a PV power system, are maximum energy efficiency, calculation time, and simple implementation. For different weather conditions an experimental validation of each algorithm has been done to determine its energy performance.

2.4.1 Maximum Power Point Tracking Experimental System Description

The experimental system is shown in Fig. 2.27. It is composed of four identical PV arrays (PVAs) as PVA_N ($N = 1, 2, 3, 4$); each one consists of four PV panels, SF-130/2-125, connected in series. All of the PVA_N send the power to a DC voltage bus ($v_c = 400$V) through four legs (B_N) of an insulated gate bipolar transistor converter (SKM100GB063D) and a $L_{PV_N}C_{PV_N}$ filter. The value of the DC bus capacitor is $C = 1100$ μF, and the filter elements have the following values: inductance $L_{PV_N} = 10$ mH with internal resistance (not shown in Fig. 2.27) $R_{PV_N} = 22.5$ mΩ and the capacitor $C_{PV_N} = 1000$ μF. At the STCs, filter element values allow for obtaining the admissible voltage and current ripples. These values are also chosen within a reasonable cost. A DC PEL (Chroma 63,202, 2.6 kW) is used for dissipating the power produced by all of the PVA_N.

As previously mentioned, a linear control is implemented to each leg power converter B_N. There are two storage energy elements in each converter leg (L_{PV_N} and C_{PV_N}). Thus there are two state variables: the PVA

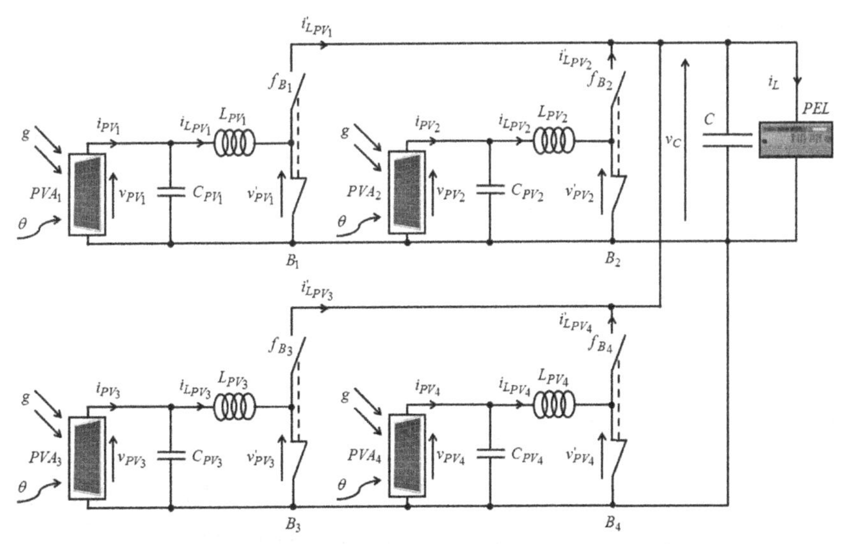

Figure 2.27 Maximum power point tracking experimental system.

current $i_{L_{PV_N}}$ and the PVA voltage v_{PV_N}. According to the first and second Kirchhoff laws, Eq. [2.15] is given:

$$\frac{di_{L_{PV_N}}}{dt} = \frac{1}{L_{PV_N}}\left(v_{PV_N} - v'_{PV_N} - R_{PV_N} i_{L_{PV_N}}\right)$$
$$\frac{dv_{PV_N}}{dt} = \frac{1}{C_{PV_N}}\left(i_{PV_N} - i_{L_{PV_N}}\right) \qquad [2.15]$$

These equations show that $i_{L_{PV_N}}$ is a state variable and command variable, v_{PV_N} is a state variable and it is a disturbance at the same time, v'_{PV_N} is a command variable, and i_{PV_N} is a disturbance. On the basis of these observations, the references $i^*_{L_{PV_N}}$ and $v'^*_{PV_N}$ are obtained as in Eq. [2.16]:

$$i^*_{L_{PV_N}} = -C_V\left(v^*_{PV_N} - v_{PV_N}\right) + i_{PV_N}$$
$$v'^*_{PV_N} = -C_I\left(i^*_{L_{PV_N}} - i_{L_{PV_N}}\right) + v_{PV_N} \qquad [2.16]$$

where C_V and C_I are, respectively, the voltage and current loop controllers. The modulated current and voltage are defined as follows:

$$\begin{bmatrix} i'_{L_{PV_N}} \\ v'_{PV_N} \end{bmatrix} = \alpha_N \begin{bmatrix} i_{L_{PV_N}} \\ v_C \end{bmatrix} \qquad [2.17]$$

where α_N represents the average values of switching functions f_{B_N} over one operating period T ($\alpha_N = \frac{1}{T}\int f_{B_N} dt,\ \ \alpha_N \in [0,\ 1]$).

From Eqs. [2.16] and [2.17], α^*_N can be determined:

$$\alpha^*_N = \frac{1}{v_C}\left(-C_I\left(-C_V\left(v^*_{PV_N} - v_{PV_N}\right) + i_{PV_N} - i_{L_{PV_N}}\right) + v_{PV_N}\right) \qquad [2.18]$$

Eq. [2.18] expresses a nested loop control, but the matter is to determine the controllers' values C_V and C_I for which the two loops will be uncoupled.

Fig. 2.28 shows a synoptic block diagram of the used control, where α_N is compared with a voltage carrier reference v_{REF} (triangle repeating sequence of 20 kHz frequency) to achieve pulse width modulation (PWM), which determines the switching function $f^*_{B_N}$.

Regarding the correctors C_V and C_I, their structures must be chosen and their values consequently determined. It is supposed that the command variables ($v^*_{PV_N}$ and $i^*_{L_{PV_N}}$) are slowly variable, capacitors C_{PV_N} are pure integrators, and inductances L_{PV_N} are first-order systems. Controller C_V has

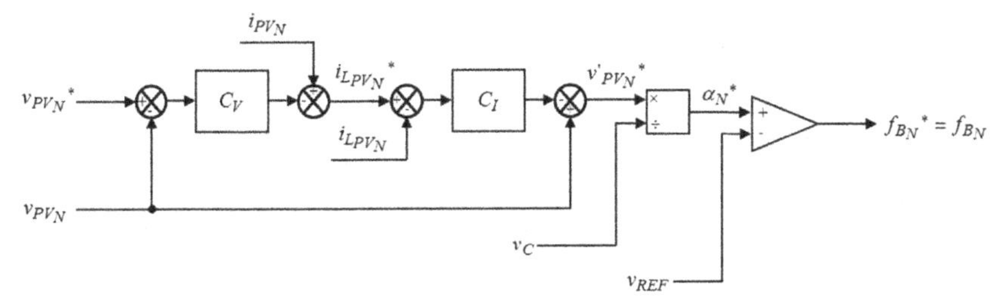

Figure 2.28 Synoptic block diagram of converter control.

to be at least a proportional corrector; however, because there is no pure real system, an integral controller may be added. Therefore a proportional integral (PI) or integral proportional (IP) controller can be used. If the disturbance i_{PV_N} is correctly compensated, and if the converter is modeled by a unitary static gain, the voltage closed loop using an IP controller is illustrated in Fig. 2.29, where K_{1_V} and K_{2_V} are controller parameters to be determined according to the desired dynamic.

From the closed-loop transfer function given by Eq. [2.19], the parameters K_{1_V} and K_{2_V} are defined as in Eq. [2.20].

$$\frac{v_{PV_N}}{v^*_{PV_N}} = \frac{1}{1 + \frac{1}{K_{1_V}} s + \frac{C_{PV_N}}{K_{1_V} K_{2_V}} s^2} = \frac{1}{1 + 2\zeta_V \tau_V s + \tau_V^2 s^2} \quad [2.19]$$

$$K_{1_V} = \frac{1}{2\zeta_V \tau_V} \quad \text{and} \quad K_{2_V} = \frac{2\zeta_V C_{PV_N}}{\tau_V} \quad [2.20]$$

Depending on the damping coefficient ζ_V, the time constant of closed loop τ_V, and the value of capacitor C_{PV_N}, the parameters K_{1_V} and K_{2_V} are easily calculated. With a PI controller, the closed loop transfer function includes two poles (real or complex conjugate) and one zero to be taken into account. Therefore the parameter tuning becomes more difficult. This is why an IP controller is used.

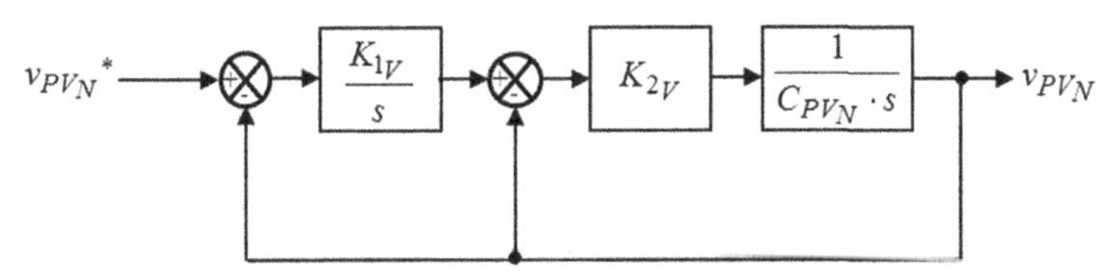

Figure 2.29 Block diagram of voltage closed loop.

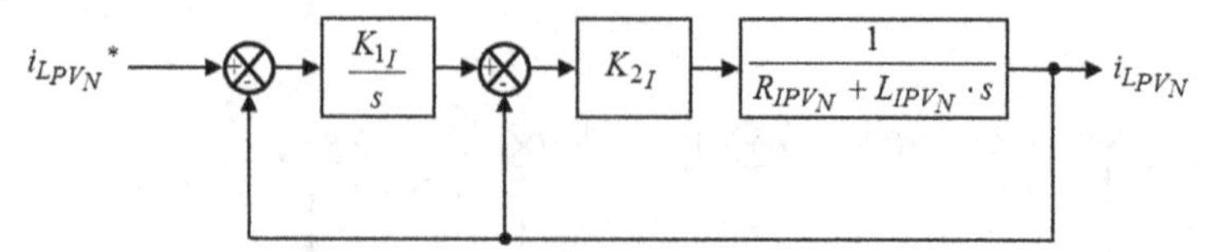

Figure 2.30 Block diagram of current closed loop.

According to a similar reasoning, the proposed structure of the controller C_I is shown in Fig. 2.30.

From the block diagram of current closed loop, the transfer function and controller parameters are given by Eq. [2.21]:

$$\frac{i_{L_{PV_N}}}{i^*_{L_{PV_N}}} = \frac{1}{1 + \frac{K_{2_I} + R_{PV_N}}{K_{1_I} K_{2_I}} s + \frac{L_{PV_N}}{K_{1_I} K_{2_I}} s^2} = \frac{1}{1 + 2\zeta_I \tau_I s + \tau_I^2 s^2}$$

$$K_{1_I} = \frac{L_{PV_N}}{(2\zeta_I L_{PV_N} - \tau_I R_{PV_N})\tau_I} \quad \text{and} \quad K_{2_I} = \frac{2\zeta_I L_{PV_N}}{\tau_I} - R_{PV_N} \qquad [2.21]$$

The values of parameters K_{1_I} and K_{2_I} are easily calculated depending on the desired values of the damping coefficient ζ_I, the time constant closed loop τ_I, and the values of L_{PV_N} and R_{PV_N}.

Similar to voltage loop, the PI controller brings a zero, but it may be easily eliminated in current loop by using a dominant pole compensation method. Nevertheless, this method is not robust because the value of R_{PV_N} varies following the temperature, and the value of L_{PV_N} varies according to the saturation degree. For these reasons, an IP controller in the current loop is used.

The detailed synoptic block diagram of the control proposed and implemented for MPPT comparison is shown in Fig. 2.31, where A_N refers to one of four tested MPPT algorithms.

Proper operation involves correct choice of the gains (K_{1_V}, K_{2_V}, K_{1_I} and K_{2_I}), which requires satisfying the following inequalities:

$$\tau_V >> \tau_I \leftrightarrow \frac{1}{\omega_V} >> \frac{1}{\omega_I} \leftrightarrow \frac{1}{2\pi f_V} >> \frac{1}{2\pi f_I} \Rightarrow f_I >> f_V \qquad [2.22]$$

with f_V and f_I the frequencies of the voltage and current closed loops, respectively. Then, the dynamic of algorithm A_N, which imposes $v^*_{PV_N}$, must be lower than the other one of the voltage loop.

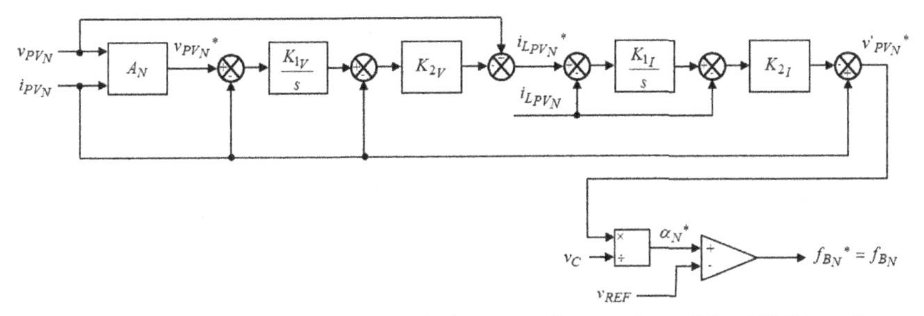

Figure 2.31 Detailed synoptic block diagram of control used for MPPT study.

2.4.2 Maximum Power Point Tracking Experimental Results Analysis and Discussions

Experimental 9-h continuous tests were operated at Compiegne in 2012 on the following days: April 3–5, 9, 11, and 12. The meteorological conditions on these days were different in solar irradiance and air temperature. Each MPPT algorithm is associated with a PVA_N. The goal is to compare the described MPPT algorithms under the strictly same conditions ensured by the system structure described in Fig. 2.27.

The imposed sample time is 0.1 s for all algorithms. The damping coefficient for voltage and current loops is 0.707 ($\zeta_V = \zeta_I = 0.707$) that leads to the best compromise between rapidity and precision. To satisfy the inequalities given by Eq. [2.22], the frequencies values of f_V and f_I are 5 and 500 Hz, respectively.

The MPPT algorithms and the dedicated control are implemented in MATLAB-Simulink. The experimental system is controlled in real time by dSPACE 1103 with a sample time equal to 100 μs (synchronized to the PWM at 20 kHz). The irradiance and PV cell temperature are respectively measured by a solar pyranometer CT-RM and temperature sensor PT100. During each test, the solar irradiance g, air temperature θ_{AIR}, PV cell temperature θ, PVA current i_{PV}, and PVA voltage v_{PV} are measured. Data acquisition is made by SL1000 (YOKOGAWA), which is also synchronized to the control of the dSPACE 1103 (100 μs).

PVA_1 and PVA_2 are respectively associated with the P&O and InC algorithms with the following setting: $\Delta V = 1$ V (which corresponds to a variation of ±10 V/s). This value is estimated to be a good value because it seems to work in almost all cases.

The ImP&O algorithm is implemented for PVA_3, and the following settings are made: $\Delta V_{S1} = \Delta V_{S2}/10 = 1/10 = 0.1$ V and $S1 = S2 = 2$W/V (values performed from some tests).

The FL MPPT is associated with PVA_4, and the same parameters and settings shown in Section 2.3.2 are kept; however, two scales factors are used (10 with the input e_1 and 1 with the input e_2) to modify the inputs instead of the general profile.

Figs. 2.32–2.37 show the meteorological operation conditions (solar irradiance, air temperature, and PV cell temperature) as well as the measured electrical power ($p_{PV_N} = v_{PV_N} i_{PV_N}$).

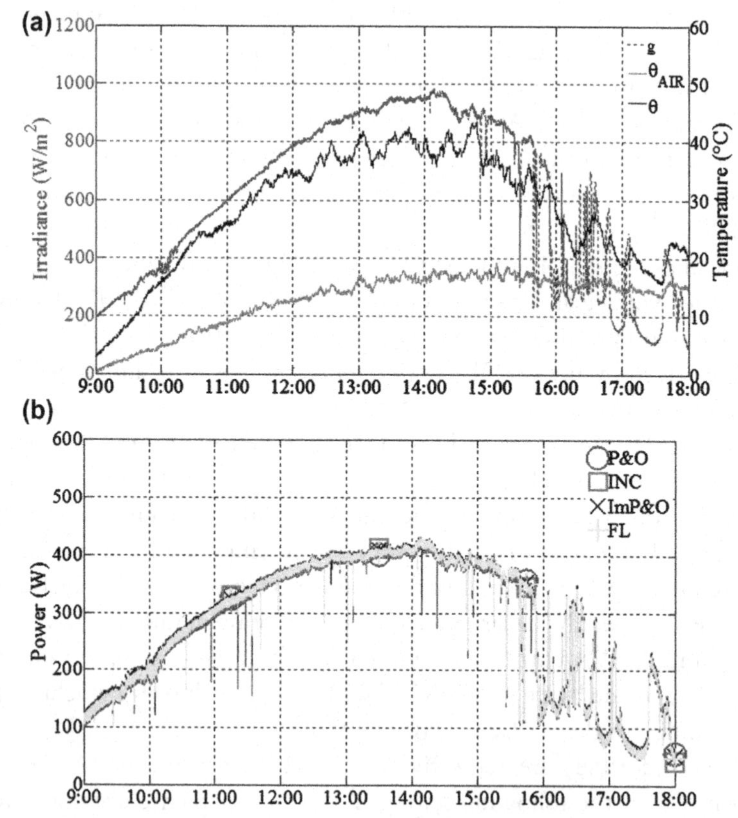

Figure 2.32 (a) Meteorological conditions and (b) electrical powers extracted on April 3, 2012. *P&O*, perturb and observe; *INC*, incremental conductance; *ImP&O*, improved P&O; *FL*, fuzzy logic.

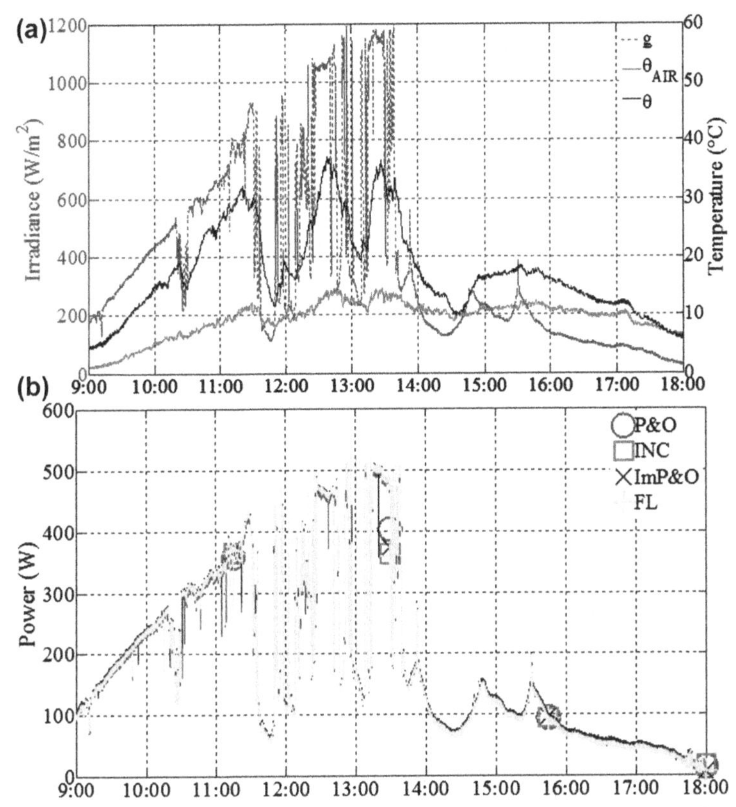

Figure 2.33 (a) Meteorological conditions and (b) electrical powers extracted on April 4, 2012. *P&O*, perturb and observe; *INC*, incremental conductance; *ImP&O*, improved P&O; *FL*, fuzzy logic.

The extracted energies are calculated using the trapeze method; Table 2.3 presents the obtained results.

For all tests days the four algorithms operate in satisfactory manner. Given the matching with the measured power curve, Figs. 2.32–2.37 show that four MPPT algorithms enable the extraction of almost the same power.

Table 2.3 shows that four MPPT algorithms extract almost the same energy; nevertheless, the maximum of energy is extracted by the ImP&O algorithm and the minimum of energy is extracted by the P&O algorithm. The relative errors between ImP&O and P&O vary from 1.56% (April 3) to 7% (April 5). It is important to know that meteorological conditions as measured on April 5 often occur north of France. Getting 7% more of energy is not negligible in the measurement of energy performance.

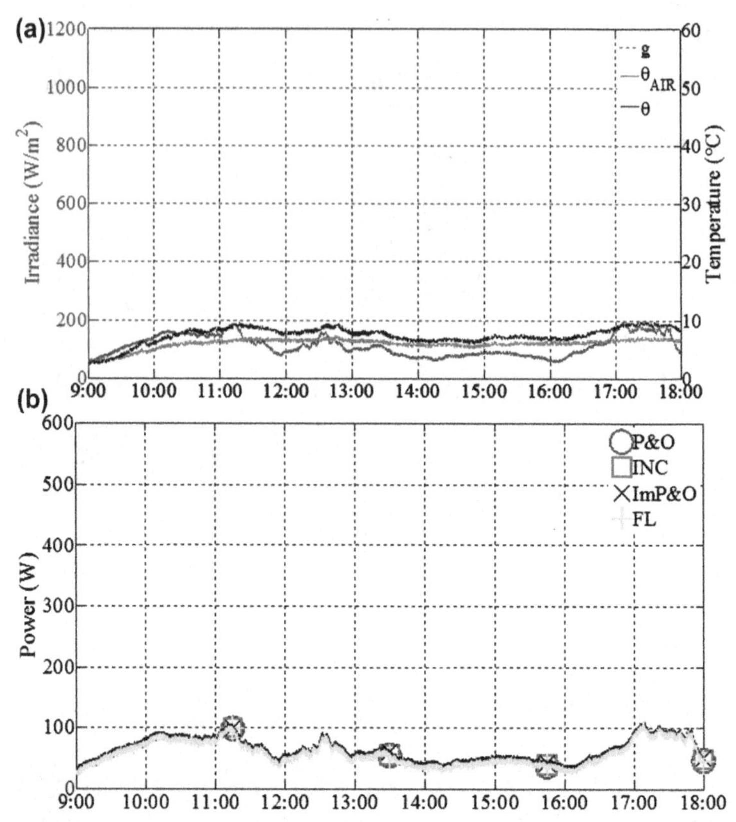

Figure 2.34 (a) Meteorological conditions and (b) electrical powers extracted on April 5, 2012. *P&O,* perturb and observe; *INC,* incremental conductance; *ImP&O,* improved P&O; *FL,* fuzzy logic.

Otherwise, for very sunny days, as April 3–12, which correspond to the highest extracted energy, the relative errors are small.

Regarding FL MPPT, it could not extract as much energy as the InC algorithm or ImP&O algorithm. This poor performance may be due to improper adjustment of the scale factor of the input $e_2 = d(dp/dv)/dt$. Indeed, the derivate adjustment of another derivate is not so easy.

The extracted electrical power evolutions show that some algorithms induce power envelopes larger than others. To further analyze, the total oscillation rate, on each of the four electrical powers signals, was performed. Having many points to be displayed, the graphical representation is not practical. Therefore the total oscillation rate is calculated for each power,

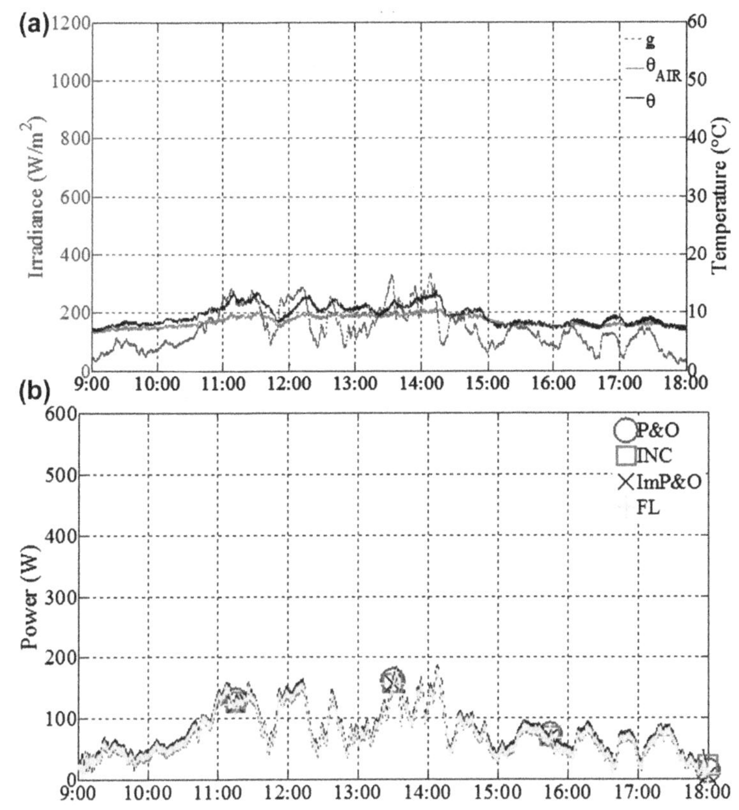

Figure 2.35 (a) Meteorological conditions and (b) electrical powers extracted on April 9, 2012. *P&O*, perturb and observe; *INC*, incremental conductance; *ImP&O*, improved P&O; *FL*, fuzzy logic.

and the carried out results are given in Table 2.4. Except for April 5, Table 2.4 shows that the ImP&O algorithm causes less noise than the other three methods.

Aiming to compare strictly under the same meteorological and technical conditions, the four MPPT algorithms, having consistency adjustment, are associated with four identical PVA coupled on a DC common bus through a four-leg power converter. The energy performance comparison, based on the measurement of four extracted powers, shows that the extracted energies are almost identical with a slight advantage for the ImP&O algorithm. Therefore the ImP&O algorithm optimizes the energy recovery, especially when solar irradiance is low. It also enables minimization of the power oscillations. Consequently, ImP&O operates better

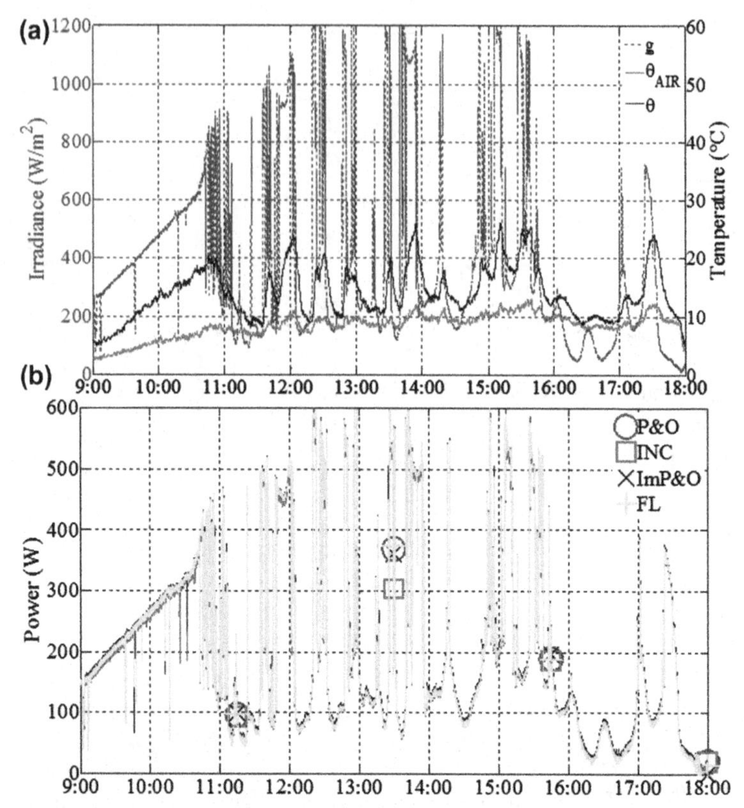

Figure 2.36 (a) Meteorological conditions and (b) electrical powers extracted on April 11, 2012. *P&O*, perturb and observe; *INC*, incremental conductance; *ImP&O*, improved P&O; *FL*, fuzzy logic.

than the other algorithms; in contrast, it requires know-how on PV power system operating.

Furthermore, following both criteria, energy efficiency and cost effectiveness, P&O has low cost effectiveness with very similar energy efficiency for a long time period. Although the ImP&O method has the greatest energy efficiency, this does not justify the higher implementation cost, unless if required specifically for research works. For PV industrial applications, where the economic criterion is the most important one, this study proves the effectiveness of P&O, but it is claimed in the literature to be inferior to other MPPT methods. In addition, this study justifies the most widely used MPPT method in industrial applications regarding other algorithms.

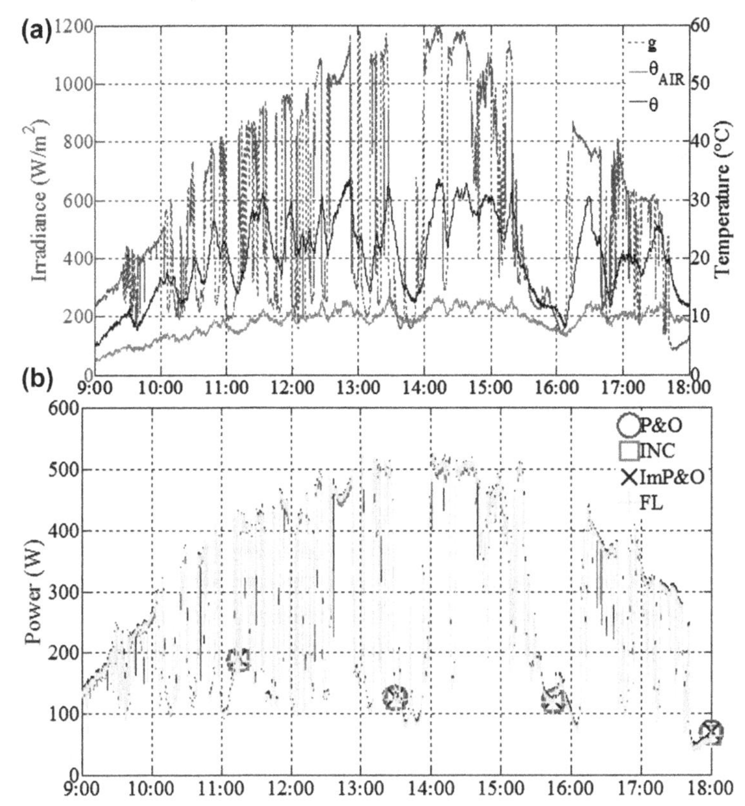

Figure 2.37 (a) Meteorological conditions and (b) electrical powers extracted on April 12, 2012. *P&O*, perturb and observe; *INC*, incremental conductance; *ImP&O*, improved P&O; *FL*, fuzzy logic.

Table 2.3 Extracted Energies During Experimental Tests (kWh)

	April 3	April 4	April 5	April 9	April 11	April 12
P&O	2.53	1.59	0.53	0.68	1.73	2.36
InC	2.56	1.61	0.55	0.69	1.76	2.39
ImP&O	**2.57**	**1.63**	**0.57**	**0.71**	**1.79**	**2.41**
FL	2.55	1.6	0.53	0.68	1.76	2.38

The bold formatting highlights the best results.

Table 2.4 Total Oscillation Rate of Four Extracted Powers

	April 3	April 4	April 5	April 9	April 11	April 12
P&O	19.67	21.14	**24.05**	43.17	40.53	23.99
InC	19.67	21.32	24.19	45.77	40.69	24.12
ImP&O	**10.07**	**21.07**	27.28	**19.16**	**38.71**	**23.86**
FL	10.69	22.1	32.01	22.31	40.53	24.67

The bold formatting highlights the best results.

3. PHOTOVOLTAIC-CONSTRAINED PRODUCTION CONTROL

PV systems usually work with an MPPT algorithm to produce power as much as possible. The PV output power changes mainly according to the solar irradiance; therefore PV systems are unpredictable power sources. For grid-connected PV systems, the PV MPPT production could be strongly fluctuating due to possible significant variations of solar irradiance, which could be an issue for grid power quality with consideration of future PV penetration level. On the other hand, for stand-alone PV systems, which usually work with energy storage, the MPPT production would cause extra problems in power balancing when the storage reaches its upper state of charge limit. In these cases the PV production should be constrained. This section proposes a PV-constrained production control strategy, which leads to control the PV power at any level within the MPPT production ability [16]. If the solar irradiance decreases and PV MPPT power is not able to put out the desired constrained power, then the control strategy continues to operate the PV system with the MPPT algorithm. The transition between MPPT and the constrained mode is continuous and seamless; the two modes separately operate. Experimental results validate the proposed control strategy.

3.1 Photovoltaic Power-Constrained Production Strategy

As energy and environmental problems become an issue, more and more PV sources are being installed. According to grid connection, the PV systems can be divided into two groups: a grid-connected system and a stand-alone PV system. PV sources usually work with MPPT strategy to extract power as much as possible. As already explained in the previous section, the PV MPPT production varies according to solar irradiance and PV cell temperature. Thus, because of possible significant variations of solar irradiance, the PV MPPT production is intermittent, random, and could be strongly fluctuating.

For a grid-connected PV system, the PV production fluctuations could affect the grid voltage and have a potential impact on the power quality [17]. Regarding the increase of PV production, there would be a potential risk for grid power quality by fluctuating PV MPPT productions. For a relatively small proportion of PV production injected into the grid, the PV intermittent production can be absorbed by the grid, but the high-level PV penetration influence on the grid still remains unknown. Meanwhile, the traditional grid regulation time measured in minutes is much longer than the PV fluctuation time, which is in seconds. It could be hard for a grid to regulate fluctuations with high PV penetration. In Ref. [18] the authors highlighted that the large-scale grid-connected PV stations could induce fluctuations in transmission line power and reduce the power qualified rate. Thus PV production should be better integrated into the grid by smoothing or downscaling fluctuations. Energy storage could be used to smooth PV MPPT power fluctuations for grid injection [19], but the cost would increase. If the PV production can be constrained as wished during fluctuations, then grid power fluctuations can be downscaled and the negative impact on the grid can be decreased. PV-constrained production could be a more cost-effective solution.

A stand-alone PV system usually works with energy storage to maintain sustainable power supply to the load. In the case of PV production excess, when the storage reaches its upper state of charge limit, the PV production has to be limited. In Ref. [20] the authors use constrained voltage to constrain PV production, which is an open-loop control and could be limited for precise PV-constrained production control. In Ref. [21] a closed-loop PV-constrained control algorithm is proposed, which can precisely control PV power. By the proposed control strategy in Ref. [21], MPPT function and constrained function are coupled and could affect each other during operation. Regarding a better PV integration for grid-connected and stand-alone PV systems, this section proposes a simple PV-constrained production control.

3.2 Photovoltaic-Constrained Power Control

Fig. 2.38 shows that the MPP power-voltage characteristic is nonlinear but monotonous for the whole voltage range. Hence, by splitting a curve by the MPP, each part of the curve, on the left side or on the right side of the MPP, is monotonous and can have an approximate linearization, which provide the possibility of power closed-loop control by voltage to constrain the PV system power.

On the basis of the system monotonicity, by choosing the power-voltage curve on the right side of the MPP for a constrained power closed-loop control, a PV system constrained production, following constrained power reference $p^*_{PV\_CONS}$, may be proposed. Let us consider that the PV system operates under the P&O MPPT algorithm unless a power-limiting instruction is transmitted. Fig. 2.39 illustrates the PV system operating strategy combining MPPT and constrained control.

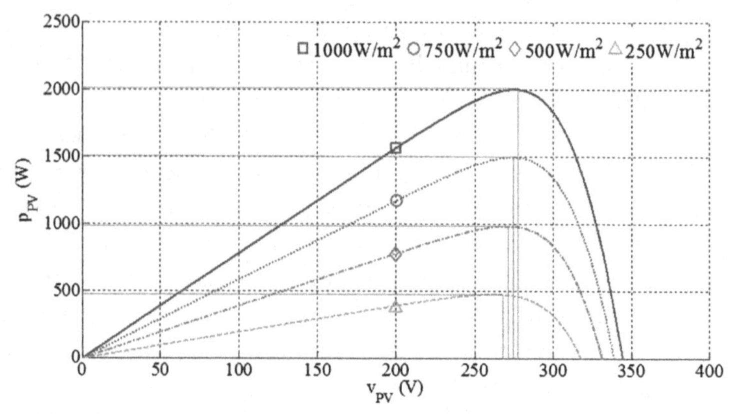

Figure 2.38 Maximum power point power-voltage characteristics modeling at different solar irradiances.

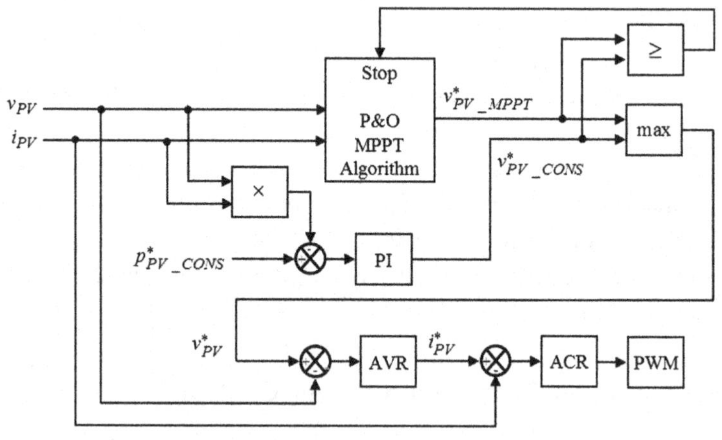

Figure 2.39 Photovoltaic (PV) system operating strategy combining maximum power point tracking (MPPT) and constrained control. *P&O*, perturb and observe; *PI*, proportional integral; *AVR*, automatic voltage regulator; *ACR*, automatic current regulator; *PWM*, pulse width modulation.

The P&O MPPT algorithm and the constrained production algorithm simultaneously provide corresponding voltage references $v^*_{PV\_MPPT}$ and $v^*_{PV\_CONS}$ to operate the PV system (ie, MPPT voltage reference and output of constrained power closed-loop control, respectively). The maximum of these two references is taken as the PV system voltage control reference v^*_{PV}, which represents the minimum power. Following v^*_{PV}, the PV system is operated by voltage and current double closed-loop control via an automatic voltage regulator and automatic current regulator.

During the MPPT operation, if a constrained power within the MPPT ability is given, then the PI controller would increase the $v^*_{PV\_CONS}$. When $v^*_{PV\_CONS}$ is greater than $v^*_{PV\_MPPT}$, the constrained reference is taken and the MPPT algorithm is stopped. By constrained power closed-loop control, the PI controller could control the PV power at the constrained level. In the case of low solar irradiance, the PV system output power ability is less than the constrained power reference; therefore the PI controller will decrease $v^*_{PV\_CONS}$ until the lower limit and $v^*_{PV\_MPPT}$ is taken to control the PV system. Thus the constrained power control does not affect the MPPT algorithm and MPPT power is produced. To avoid continuous decreasing of PI controller output toward infinite in this case, an anti-windup PI controller is used as the constrained power PI controller to avoid the integrator output saturation, as in Fig. 2.40.

An anti-windup PI controller is a typical PI controller with an extra feedback with the gain K_c to deal with the integrator saturation. When the output is within the saturation limits, the extra K_c branch does not work. If the integrator saturates, then the extra K_c branch interacts to keep the integrator output at the saturation lower or upper limit instead of integrating toward infinite, which could eliminate the control dead time caused by the delay for integrator output returning to the normal control range. Concerning the PI controller, the main goal is not to develop synthesis of the controller but to perform a control in a simple and robust manner.

3.3 Experimental Results

The considered PV system is represented by a PVA composed by 16 PV panels. The schema of the experimental circuit is shown in Fig. 2.41. The PVA is in parallel with a capacitor C_{PV}. A half-bridge converter working in boost mode is used to connect the PVA with a DC PEL set to constant voltage mode.

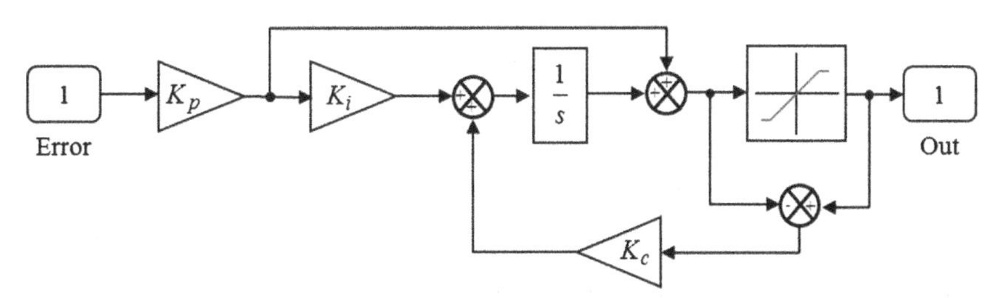

Figure 2.40 Proportional integral controller with anti-windup.

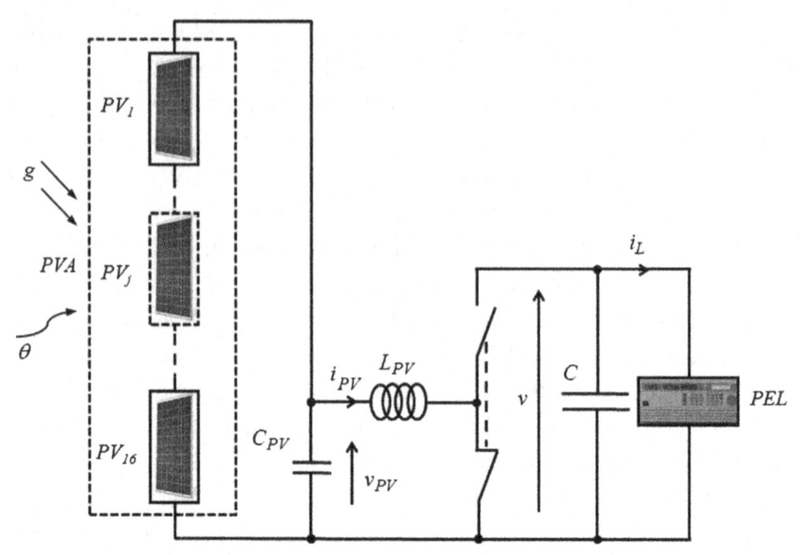

Figure 2.41 Test bench of a photovoltaic (PV) system operating strategy combining maximum power point tracking (MPPT) and constrained control. *PVA*, PV array; *PEL*, programmable electronic load.

Experimental tuning of controller parameters is obtained from few tests. In this application, K_p is given as a negative value to decrease the voltage reference when the controller input error is positive and vice versa. For experimental validation the following values are considered: $K_p = -0.2$, $K_i = 100$, and $K_c = 200$. The other elements of the test bench are detailed in Table 2.5.

Table 2.5 Element Detail for Photovoltaic-Constrained Power Control Experimental Test

Element	Parameter	Device
Photovoltaic system (16 panels)	$I_{MPP} = 7.14$ A, STC $V_{MPP} = 17.5$ V, STC	Solar-Fabrik SF-130/2-125
Direct current programmable electronic load	2.6 kW	Chroma 63202
Controller board		dSPACE 1103
Power electronic converter	600 V/100 A	SEMIKRON SKM100GB063D
C_{PV}	1000 μF	
L_{PV}	40 mH	
C	1100 μF	

The experimental tests are first operated for a very short time and then for a daylight time. Figs. 2.42–2.45 present the short-time experimental test results.

Following the solar irradiance and the PV cell temperature given in Fig. 2.42a and b shows the evolution of PVA power output following the constrained production control strategy, the values of which are arbitrarily chosen. Before the time of 9.6 s, the constrained power reference $p^*_{PV\_CONS}$ is greater than the PVA MPPT power, thus the PVA works with the MPPT algorithm and the power is not limited. Between 9.6 and 20.2 s, $p^*_{PV\_CONS}$ is given within the ability to constrain the PVA production at 500 W. After 20.2 s, $p^*_{PV\_CONS}$ is given beyond the MPPT ability and the PVA again returns to MPPT production.

Fig. 2.42c shows the voltage reference evolution, which illustrates how the proposed control strategy works. First, before 9.6 s, the constrained power reference $p^*_{PV\_CONS}$ is greater than the PV MPPT power, and the PI controller would continuously decrease $v^*_{PV\_CONS}$ so to maintain it at its lower limit. Because the maximum voltage reference is picked to control the PVA, the PVA works with the MPPT algorithm reference, therefore $p^*_{PV\_CONS}$ does not affect the MPPT operation. Then, between 9.6 and 20.2 s, $p^*_{PV\_CONS}$ is given within the MPPT ability. At the beginning of the reference change, as $p_{PV} > p^*_{PV\_CONS}$, the negative input error makes the PI controller increase the voltage reference $v^*_{PV\_CONS}$ until p_{PV} is controlled at $p^*_{PV\_CONS}$. During this process, after $v^*_{PV\_CONS} > v^*_{PV\_MPPT}$, the PVA no longer works with the MPPT algorithm, and p_{PV} is under power closed-loop control following the reference of $p^*_{PV\_CONS}$. Finally, after 20.2 s, the constrained power reference is given beyond the MPPT ability; the positive error makes the PI controller to decrease $v^*_{PV\_CONS}$. For $v^*_{PV\_CONS} < v^*_{PV\_MPPT}$, the PV again returns to MPPT production. The transition between the MPPT algorithm and constrained control strategy is continuous and seamless.

Fig. 2.43a and b shows the performance of voltage and current control of the PV system. In the whole period, the PVA voltage v_{PV} and current i_{PV} are able to follow corresponding references v^*_{PV} and i^*_{PV}. However, pulses in current control, both i^*_{PV} and i_{PV}, can be observed in the case of a sudden change in the voltage reference.

The pulse around 9.6 s shows that the voltage controller decreases i^*_{PV} to increase the voltage for constraining PV production. The pulse around 20.2 s represents the discharge of the capacitor to decrease the PVA voltage following the step change of v^*_{PV}. To satisfy a quick voltage response, the

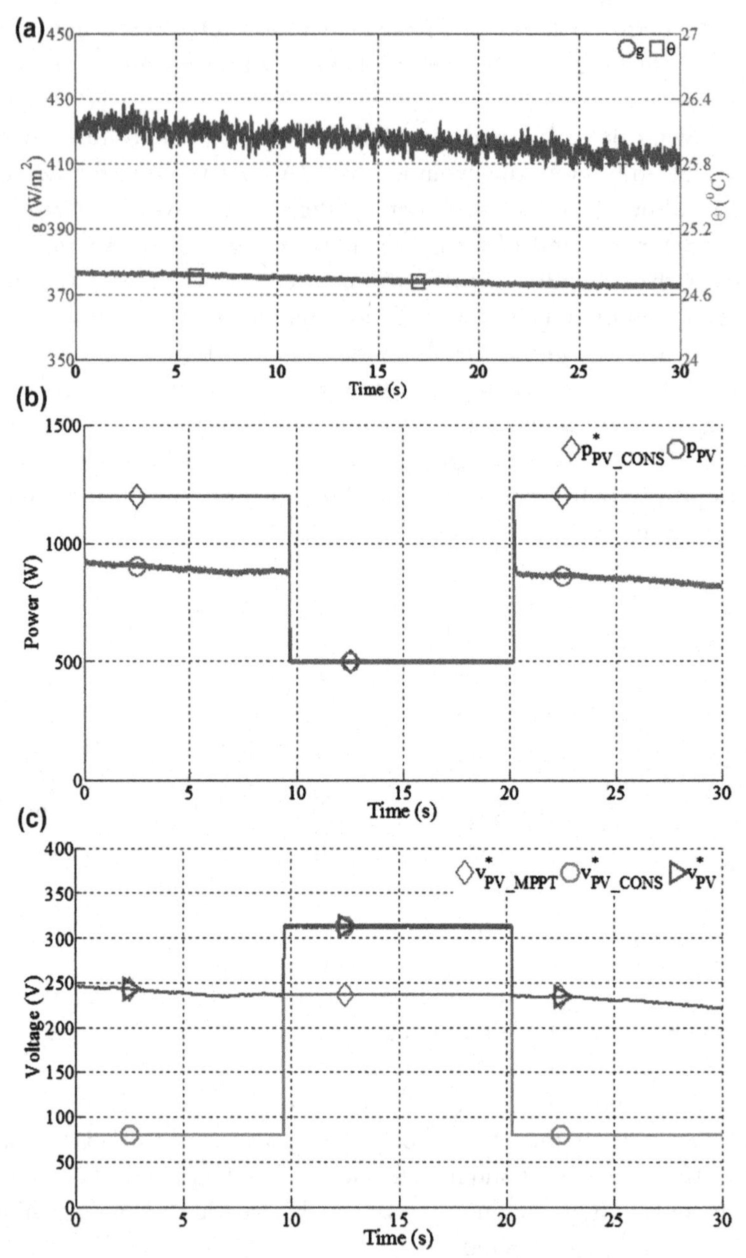

Figure 2.42 (a) Meteorological conditions, (b) photovoltaic array (PVA) power evolution, and (c) PVA control voltage reference evolution.

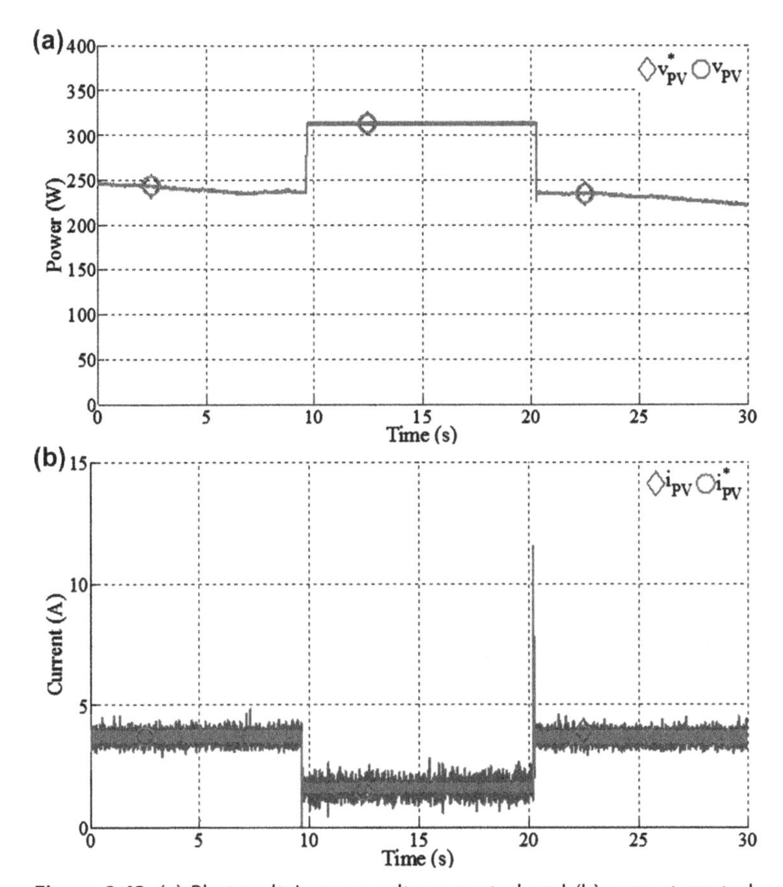

Figure 2.43 (a) Photovoltaic array voltage control and (b) current control.

two current pulses occurred. By decreasing the voltage controller gain or adding output limit to the voltage controller, the pulse can be reduced at the cost of responding time. According to different application conditions, the controller could be optimized by trade-off between the response time and pulse levels.

Fig. 2.44 shows the power and voltage transient performance during the transition from MPPT production to constrained PVA production. It can be seen after the power reference $p^*_{PV\_CONS}$ change, the *PVA* power change with a delay of approximately 10 ms. This is due to the transition time for $v^*_{PV\_CONS}$ to exceed $v^*_{PV\_MPPT}$ by the PI controller regulation. Within 70 ms, which is related to the control transient by the power, voltage, and current regulator, the PV power is well controlled to follow

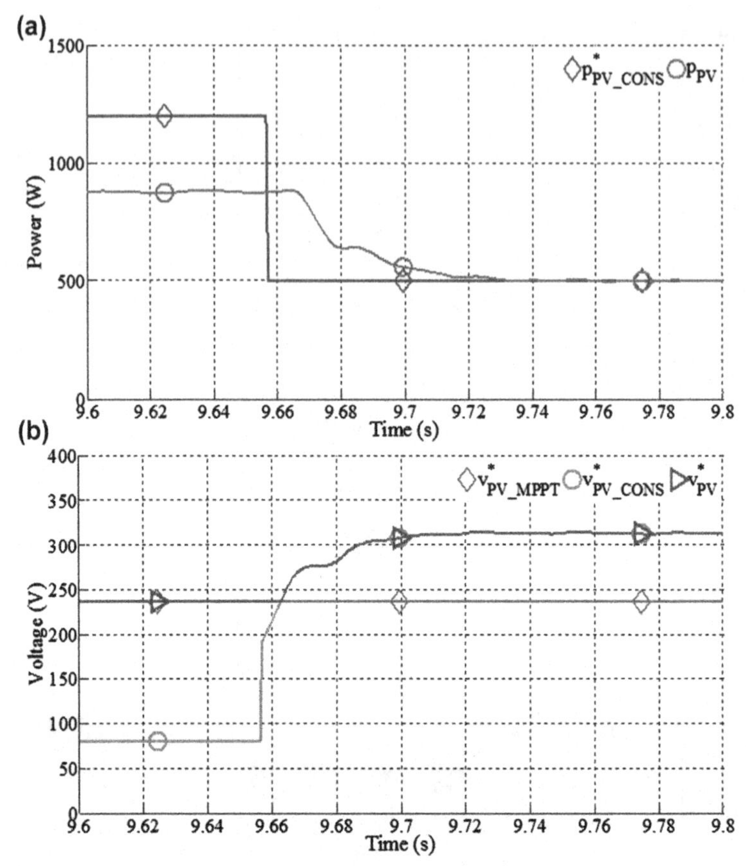

Figure 2.44 (a) Power and (b) voltage reference transient evolution during transition from maximum power point tracking to constrained production.

$p^*_{PV\_CONS}$. The transient performance is without overshoot. By tuning the PI controller parameters, the delay time and establishing time can be shortened, but power oscillations could be introduced.

Fig. 2.45a and b, shows the power and voltage reference transient evolution during the transition from constrained production to MPPT production. The transition lasts approximately 70 ms with a brief surge in PVA power at approximately 20 ms. This is explained by the fact that before arriving at MPPT mode, the PI controller tries to follow $p^*_{PV\_CONS}$ at 1500 W, which is greater than the MPPT power.

Therefore it controls the voltage reference to suddenly decrease, and the quick discharge of the capacitor C_{PV}, the capacitor stored energy, causes the overshoot in PVA current and therefore causes the PVA power overshoot

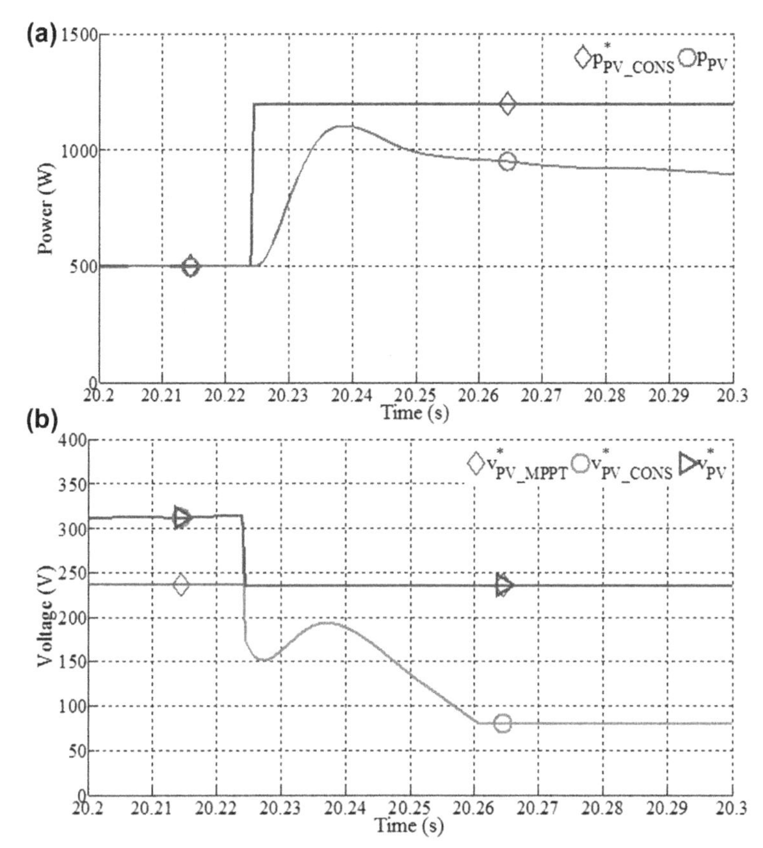

Figure 2.45 (a) Power and (b) voltage reference transient evolution during transition from constrained production to maximum power point tracking.

before stabilizing in MPPT mode. The delay between the reference step change and overshoot is from the controller response time for a step change. The overshoot could also be reduced by limiting the $p^*_{PV\_CONS}$ change rate, tuning power controller parameter values, or by optimizing capacitor value. The optimization should be performed according to applications and is out of the scope of this chapter.

For a better experimental validation, a day test was performed on December 2, 2011. An arbitrary PVA power limit is given as 1000 W. The solar irradiance is given in Fig. 2.46a. Fig. 2.46b shows that the PVA power is well controlled to not exceed the power limit.

However, pulses in PVA power evolution can be observed when transiting from constrained production mode to MPPT mode, as analyzed in transient performance. It can be seen that the PVA production

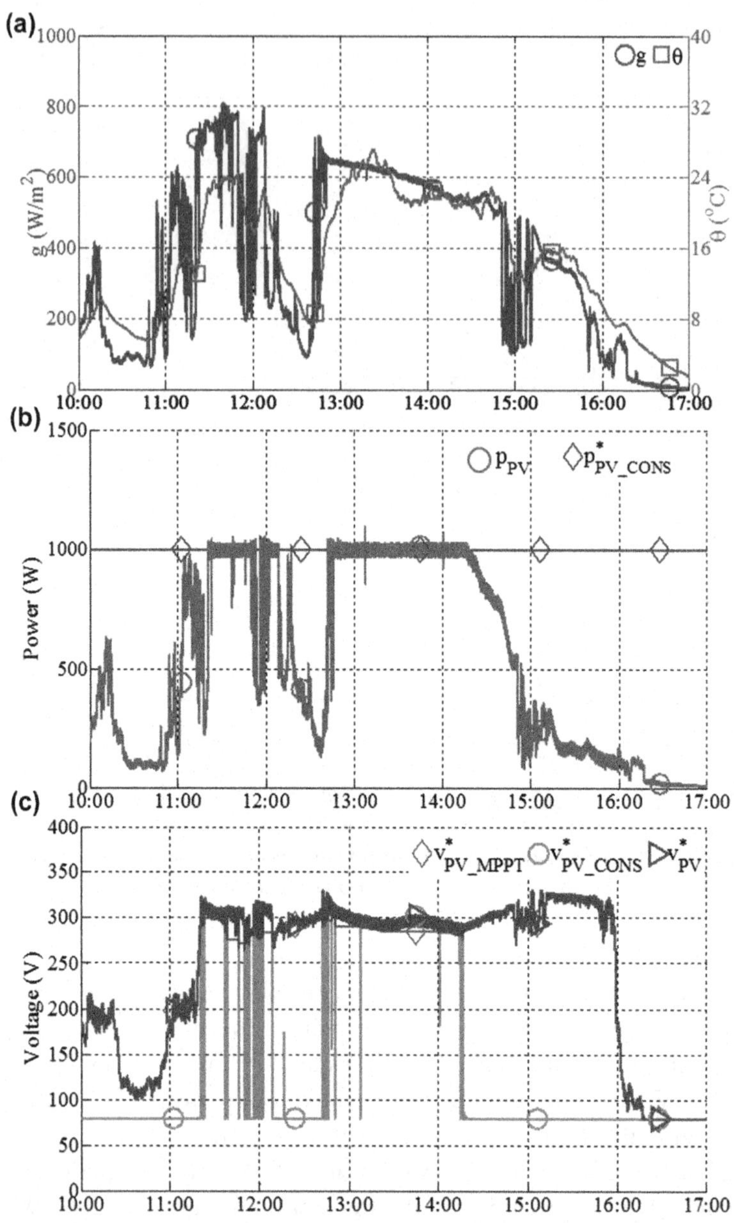

Figure 2.46 (a) Meteorological conditions, (b) photovoltaic array power evolution, and (c) voltage reference evolution during a day test.

fluctuations could be downscaled by a power constraint limit. Fig. 2.46c shows the voltage reference evolution. During operating transition, the voltage reference v_{PV}^* smoothly changes. At each moment only one algorithm voltage reference is taken to control the PVA, and the two algorithms separately work.

Regarding a better PV grid integration, this study proposed a simple PV-constrained production control algorithm. The proposed control strategy can control PV panels to output power at any level within its MPPT ability. If the PV MPPT ability does not match constrained power reference, then the PV is operated with the MPPT algorithm. The two modes operate separately and the transition between the two operation modes is continuous and seamless. Experimental results validated the proposed control strategy. No additional sensor is needed on the basis of the commonly used MPPT algorithms.

4. CONCLUSIONS

Aiming at energy modeling concerning microgrid components, this chapter focuses on the study of the PV source, modeling, and control. The goal is to choose a reliable and independent model of operating conditions on the one hand and implementing an energy-efficient control at reasonable cost on the other.

For this the two most common models of the PV source in the literature have been highlighted for ease of use and accuracy. The study shows that they can be used only for a posteriori applications because of their dependence on actual measurements. To meet the requirement of the PV power prediction, which is essential for microgrid control, a purely experimental model has been proposed and developed; it is reliable and easy to use, allowing for simulation and PV power prediction knowing only meteorological data. An experimental comparison on the performance of these three models emphasizes the use of the proposed experimental model.

The PV source power control is studied in the second part of this chapter. Three MPPT algorithms, most often used in PV systems, are analyzed, and an improvement of one of the three algorithms was proposed. The studied MPPT methods optimally control the PV source, but they may have deficits because of certain factors, such as the wrong choice concerning the increment-step calculation, sample time calculation, or incorrect setting of the algorithm. On the basis of one of the most common MPPT algorithms, a simple improved algorithm that is easy to implement was developed. It allows for varying the step tracking-based power variations relative to voltage variations. Experimental comparison shows better energy performance of this improved algorithm and lower power rate oscillations. However, for variable-step tracking, it is always possible to optimize the operation of an MPPT method, but this requires a particular expertise based on many observations during different meteorological profile days. Therefore, knowing that the energy performances of these four MPPT algorithms are very close, it is preferable for industrial applications to use the one at the lowest cost.

PV-constrained production control could be used to improve PV source control operating in any situations. The proposed control strategy can control PV sources to output power at any level. The MPPT control operates separately of constrained control but it can be switched from one to the other without discontinuities. Experimental results illustrate this continuous and seamless transition and validated the proposed control strategy. Note also that the constrained PV power control does not require any additional sensor. PV-constrained production control could be used to improve PV integration. For a grid-connected PV system, it helps to improve grid power quality by reducing fluctuation during significant variations of solar irradiance. For stand-alone PV systems, it helps to protect storage from overcharge and to reduce storage usage. For a microgrid-integrated PV system, it provides more feasibility to inject controlled power into the grid or to supply a load with respect to energy storage limits.

REFERENCES

[1] Luque A, Hegedus S. Handbook of photovoltaic science and engineering. Wiley; 2003.
[2] Villalva MG, Gazoli JR, Filho ER. Comprehensive approach to modeling and simulation of photovoltaic arrays. IEEE Trans Power Electron 2009;24(5):1198–208.
[3] Reddy PJ. Science and technology of photovoltaics II. BS Publications India; 2010.
[4] Tan YT, Kirschen DS. A model of PV generation suitable for stability analysis. IEEE Trans Energy Convers 2004;19(4):748–55.
[5] Houssamo I, Sechilariu M, Locment F, Friedrich G. Identification of photovoltaic array model parameters modelling and experimental verification. In: International conference on renewable energies and power quality ICREPQ; 2010.
[6] Riffonneau Y, Bacha S, Barruel F. Optimal power flow management for grid connected PV systems with batteries. IEEE Trans Sust Energy 2011;2(3):309–20.
[7] Skoplaki E, Palyvos JA. On the temperature dependence of photovoltaic module electrical performance: a review of efficiency/power correlations. Sol Energy 2009;83(5):614–24.

[8] Fuentes M, Nofuentes G, Aguilera J, Talavera DL, Castro M. Application and validation of algebraic methods to predict the behaviour of crystalline silicon PV modules in Mediterranean climates. Sol Energy 2007;81(11):1396–408.
[9] Houssamo I, Wang B, Sechilariu M, Locment F, Friedrich G. A simple experimental prediction model of photovoltaic power for DC microgrid. In: IEEE international symposium on industrial electronics ISIE; 2012.
[10] Esram T, Chapman PL. Comparison of photovoltaic array maximum power point tracking techniques. IEEE Trans Energy Convers 2007;22(2):439–49.
[11] Houssamo I, Locment F, Sechilariu M. Maximum power tracking for photovoltaic power system: development and experimental comparison of two algorithms. Renew Energy 2010;35(10):2381–7.
[12] Locment F, Sechilariu M, Houssamo I. Energy efficiency experimental tests comparison of P&O algorithm for PV power system. In: International power electronics and motion control conference EPE/PEMC; 2010.
[13] Gounden NA, Peter SA, Nallandula H, Krithiga S. Fuzzy logic controller with MPPT using line-commutated inverter for three-phase grid-connected photovoltaic systems. Renew Energy 2009;34(3):909–15.
[14] Bounechba H, Bouzid A, Nabti K, Benalla H. Comparison of perturb & observe and fuzzy logic in maximum power point tracker for PV systems. Energy Procedia 2014;50:677–84.
[15] Houssamo I, Locment F, Sechilariu M. Experimental analysis of impact of MPPT methods on energy efficiency for photovoltaic power systems. Int J Electr Power Energy Syst 2013;46(1):98–107.
[16] Wang BC, Houssamo I, Sechilariu M, Locment F. A simple PV constrained production control strategy. In: IEEE international symposium on industrial electronics ISIE; 2012.
[17] Woyte A, Van Thong V, Belmans R, Nijs J. Voltage fluctuations on distribution level introduced by photovoltaic systems. IEEE Trans Energy Convers 2006;21(1):202–9.
[18] Wang Y, Liao H, Wu C, Xu H. Large-scale grid-connected photovoltaic power station's capacity limit analysis under chance-constraints. In: International conference on sustainable power generation and supply SUPERGEN; 2009.
[19] Yonghua C. Impact of large scale integration of photovoltaic energy source and optimization in smart grid with minimal energy storage. In: International symposium on industrial electronics ISIE; 2010.
[20] Lin M, Kai S, Teodorescu R, Guerrero JM, Xinmin J. An integrated multifunction DC/DC converter for PV generation systems. In: IEEE international symposium on industrial electronics ISIE; 2010.
[21] Chiang SJ, Hsin-Jang S, Ming-Chieh C. Modeling and control of PV charger system with SEPIC converter. IEEE Trans Ind Electron 2009;56(11):4344–53.

CHAPTER 3

Backup Power Resources for Microgrid

1. DIFFERENT BACKUP RESOURCES FOR DIFFERENT OPERATING MODES

A backup resource is a key element in a microgrid. In general, there are two kinds of backup resources: (1) bidirectional sources characterized by energy absorption and injection and (2) unidirectional sources characterized by only energy injection. The bidirectional backup resources are used as an energy buffer; when the renewable energy sources produce an energy surplus the backup resource serves as an energy reservoir (with respect to its physical limits) whereas it supplies energy when the renewable energy sources do not produce enough power. In contrast, the unidirectional backup resources, very often based on fossil fuel or biofuel, can only supply the microgrid, but with the advantage that they can be used at any time and, depending on the fuel tank level, for very long time periods. For microgrids, the most common bidirectional backup resources are electrochemical storage, including fuel cells, and utility grid connections. Regarding the unidirectional backup resources, microturbines and diesel generators are typically used in microgrids.

This section presents some general notions on the electrochemical battery and capacitor, fuel cell, and microturbine, whereas important characteristics of a lead-acid battery, diesel generator, and utility grid connection are highlighted in the other sections of this chapter.

1.1 Electrochemical Battery and Capacitor

There are several technologies of electrochemical storage, including the conventional electrochemical battery, the electrochemical capacitor (also known as a supercapacitor or ultracapacitor), and the fuel cell. Depending on the operating mode, all of these technologies can be used in a microgrid. To evaluate performances and differentiate these electrochemical storages, the well-known Ragone chart, plotting their respective energy and power densities, shows that supercapacitors feature very large power densities but

Urban DC Microgrid
ISBN 978-0-12-803736-2
http://dx.doi.org/10.1016/B978-0-12-803736-2.00003-7

low energy densities, whereas batteries and fuel cells have large energy densities but low power densities because of their slow reaction kinetics. On the other hand, the Ragone chart also compares the charging and discharging time duration of energy storage devices. It highlights that the battery and fuel cell provide very large storage capacity for a longer time than a supercapacitor (respectively hours and seconds).

Otherwise, a battery and fuel cell are able to provide some degree of backup power whereas a supercapacitor requires an oversizing of energy density to become backup power. Different technologies of the electrochemical battery are presented and discussed in the next section, aiming to present one of the most used in building-integrated microgrids, the lead-acid battery.

Concerning the supercapacitor, because of the high power density and short duration for charging-discharging combined with a high number of charge and discharge cycles, it is not used as backup power, but mostly in association with a main source, as complementary power, to optimize energy management according to their features. Supercapacitors can be used to reduce power fluctuations due to the renewable power intermittency or to compensate the slow dynamic of a diesel generator. On the basis of different energy storage mechanisms, there are typically two categories of supercapacitors: (1) electric double-layer capacitors, which store charges physically in electric double layers forming near the electrode/electrolyte interfaces (this process is considered highly reversible and therefore the cycle life becomes almost infinite) and (2) pseudocapacitors, the charge storage of which is operated not only in an electric double layer but also because of reactions as fast surface oxidation-reduction, also called redox, and possible ion intercalation in the electrode. To summarize, supercapacitors are more interesting for power storage than served as energy storage; therefore they more used in applications linked to power regulation to increase the energy quality.

1.2 Fuel Cell

A fuel cell is an electrochemical converter that transforms the chemical energy of a gas or a liquid fuel into electrical energy. A fuel cell can be based on simple and derived hydrocarbons such as alcohols, but the most commonly used fuel is hydrogen; nevertheless, a fuel cell also requires oxygen. The use of hydrocarbons leads to high fuel-cell efficiency, but the CO_2 emissions are significant whereas hydrogen does not generate byproducts such as CO and CO_2. A fuel-cell system running on hydrogen converts the chemical energy into electrical energy, with simultaneous production of water and heat, by redox chemical reaction. The basic principle of the technology of the fuel cell is based on the separation of the oxidation reaction of hydrogen and the reduction of oxygen. Therefore the basic cell of a fuel cell consists of two electrodes (anode and cathode) separated by an electrolyte. Several cells are assembled to form a stack able to generate a significant amount of direct current (DC) electricity.

There are several types of fuel cells that are based mainly on the used electrolyte that differentiates them from each other and giving names to the large fuel-cell families. The fuel-cell operating temperature is also a differentiator that permits classifying low-temperature fuel cells and high-temperature fuel cells. The fuel-cell main technologies are alkaline fuel cells, proton-exchange membrane fuel cells, direct methanol fuel cells, phosphoric acid fuel cells, molten carbonate fuel cells, and solid oxide fuel cells.

The fuel-cell conversion efficiency is generally between 40% and 60%, but it can be increased up to 85% in the case of cogeneration (ie, combined heat and power) when the wasted heat is used.

Concerning greenhouse gases and pollutants, depending on the fuel source, fuel cells may produce very small amounts of nitrogen dioxide, but they are generally considered to be low emissions, and hydrogen-based fuel cells have a near-zero pollutant profile.

From an economic standpoint, the development of hydrogen production technology will allow storing the renewable energies as hydrogen when it is convenient; it can then be used by fuel cells to produce power without pollutant emissions. Therefore fuel cells become a clean energy source than can be used in grid-connected and off-grid microgrids. Taking into account these aspects and its bidirectional character, hydrogen-based fuel cells, seen as backup sources, are considered a very promising future solution for the development of flexible and adaptable energy systems with enough capacity to meet demand safely and with a reduced environmental impact. Furthermore, for building-integrated microgrids and those based on photovoltaic (PV) sources, fuel cells may very efficiently complement PV generators in terms of the high efficiency of electricity supply considering the cogeneration operating mode of fuel cells. Indeed, the fuel-cell total energy production is efficiently used in wintertime whereas PV high electricity production is mostly during the summertime.

1.3 Microturbines

Based on multiple fuels and biofuels, microturbines are a relatively new technology for electric power generation that are able to produce heat and power, very often combined, with high efficiencies. A microturbine is a form of gas turbine, but compared with traditional gas turbines it provides high electrical efficiency because of the recovered exhaust energy to preheat compressed inlet air, thereby increasing the electrical efficiency compared with a simple-cycle machine. In general, a microturbine is composed of the following main subassemblies, often using just one stage of each: a radial gas compressor, a combustion chamber, a turbine, an air-to-air heat exchanger (heat "recuperator"), and a high-speed generator. The rotating components are mounted on a single shaft supported by air bearings; therefore this assembly is also called a turbogenerator (ie, turbine components plus the generator).

The operating principle may be summarized as follows. Air enters through the intake located beside the generator and then flows into the compressor, where it is pressurized and then directed into the cold side of the air-to-air heat exchanger. The exhaust heat is used to preheat the air before it enters in the combustion chamber; thus it significantly reduces fuel consumption. The combustion chamber then mixes the heated air with fuel and burns it. This mixture expands through the turbine, which drives the compressor and generator at high speed. The combusted air is then exhausted through the "recuperator" before being discharged at the exhaust outlet.

Regarding gas turbines in the same size class, microturbines provide high energy efficiency because of the double thermal energy recuperation: a portion of the exhaust energy is back into the energy conversion process, and the other part of the exhaust energy is used to produce heat, cooling, or combined. However, the electrical efficiency range is still low to provide an attractive economic return on investment. The efficiency advantage of the microturbine lies in applications in which the clean exhaust heat can be recovered and productively used (ie, where polygeneration is totally useful (combined heat and power or combined cooling, heat, and power)).

For most generator-based internal combustion engines, because of the slow dynamics, the main inconvenience of microturbine use in a microgrid is that it requires several tens of seconds to reach the rated output power and several seconds for a 50% change in power output. Therefore in many cases it could be necessary to use short-term storage, such as supercapacitors, to reduce the impact of the microturbine slow dynamic to the fast power balancing required by the microgrid.

2. LEAD-ACID STORAGE RESOURCE

2.1 Characteristics of Electrochemical Storage

The electrochemical cell is the basic element of a battery unit; several battery units constitute electrochemical storage. The electrochemical cell is composed of two metal electrodes separated by a conductive material; it serves to transform the chemical energy stored in the active materials of the electrodes into electrical energy. This conversion is performed using the electrochemical redox reaction [1], in which the electrons move from one electrode to the other through an external electrical circuit. The redox reaction may be represented by the following equation, where e^- is the number of transferred electrons:

$$\text{reducer} \underset{\text{reduction}}{\overset{\text{oxydation}}{\rightleftharpoons}} \text{oxydant} + n \cdot e^- \qquad \text{[i]}$$

At the same time, the electrolyte participates in the redox reaction by separating the charges and leading them to the two electrodes. The reversibility of the redox reaction can store electricity during charging or return it during discharge. In general, the electrochemical cell is composed of three major elements:

- One cathode, which is the positive electrode; it is the place of the redox. The cathode is an oxidant, which consumes electrons.
- One anode, which is the negative electrode; it is the place where the redox occurs. The anode consists of a reducer material, which can donate electrons to the external circuit.
- An electrolyte, which is a substance, liquid or solid, that allows the passage of electric current by ion displacement. The electrolyte, a conductive substance, is the medium in which the anode and cathode are immersed [1].

Inside the accumulator, the two electrodes are physically isolated by a separator to prevent short-circuiting. However, the separator is permeable to the electrolyte to maintain the ionic conductivity between the electrodes.

To obtain a large amount of energy stored, it is necessary to have many electrons exchanged per mole and a reaction between a highly oxidizing element and another good reducer. The electrodes are made of conducting material, which defines the type of accumulator. Among the most traded accumulators there are batteries based on lead acid, nickel cadmium (NiCd) or nickel metal hydride (NiMH), and lithium ion (Li-ion). Compared with other types of batteries, because of their low cost and high rate of recycling of approximately 97%, lead-acid batteries remain the most used for microgrids.

The batteries are characterized by specific variables and parameters as well as some necessary terms for the description and quantification of the phenomena associated with the operation. Variables and used terms are as follows:

- Nominal voltage: the nominal voltage of a battery measured in volts (V) is determined by the potential difference of the electrochemical reaction in each electrode.
- Nominal capacity (C_{NOM}): the nominal capacity of the battery, expressed in ampere hours (Ah), is the amount of electricity it can provide under certain conditions. It depends on the discharge rate, the voltage minimum allowable value during discharge, and the electrolyte temperature. Thus, depending on the application, a battery may have several capabilities. Therefore it is often useful to define the nominal capacity (C_{NOM}) as the value given by the manufacturer indicating the battery capacity at the beginning of its life cycle and for a given discharge rate.
- Charging regime (C_r): the charging or discharging current can be expressed following the charge or discharge mode, which is subjected during a determined time period r expressed in hours (h) $C_r = C/r$. For example, C_{10} means a regime that implies a battery with a nominal capacity of 130 Ah that can deliver a current of 13 A during 10 h.
- State of charge (SOC): it is the ratio between the capacity at a certain moment and the nominal capacity of the battery. It indicates the amount of power available in the battery. It varies between 0 and 1. $SOC = 0$ means that the battery is fully discharged, and then it is full if $SOC = 1$.
- Depth of discharge (DOD): this is the opposite of the state of charge. The discharge depth indicates the quantity of electricity supplied by the battery.
- Mass energy: this is the amount of electrical energy stored in a particular amount of mass; it is expressed in watt-hours per kilogram (Wh/kg).

- Self-discharge: the phenomenon of self-discharge occurs in most batteries even if they are not used. It results in a loss of capacity. For lead-acid batteries, it comes from side reactions occurring at the electrodes corresponding to the redox reactions of the water molecules.
- Internal resistance: the internal resistance of the battery (R_{Ω}) results in opposition to the passage of a current through the battery. It is the sum of all of the ionic resistance of the electrolyte with the electronic resistance. Ionic resistance represents the factors affecting the ions' movement in the battery, such as electrolyte conductivity, porosity of the electrode, area of the electrode, side reactions, etc. [1]. The electronic resistance is seen as the set of resistances of active materials, electrodes, and any possible connection.
- Total life cycle: the number of charge-discharge cycles that the battery can endure before losing 20% of its nominal capacity.
- Effect of temperature: the temperature influences the internal resistance, capacity, and self-discharge rate. For very low temperature (eg, $-15°C$) for a lead-acid storage battery the viscosity of the electrolyte increases; thus the internal resistance of a lead storage battery will grow. This generates a voltage drop. In contrast, the increase in temperature of $10°C$ doubles the reaction kinetics. Thus the capacity and self-discharge rate increase.

Table 3.1 presents the synthetic comparison of four types of accumulators: lead acid, NiCd, NiMH, and Li-ion [2]. The characteristics presented in this table allow a first choice of batteries in a given application. However, the economic criteria, such as cost of production and operation, and recycling opportunities are the assessment criteria for the final decision.

Table 3.1 Comparison of four types of accumulators

	Lead acid	Nickel cadmium	Nickel metal hydride	Lithium ion
Nominal voltage (V)	2	1.2	1.2	3.5
Mass energy (Wh/kg)	From 20 to 35	From 40 to 45	~65	90
Self-discharge (% per day)	~2	0.5	5	0.33
Internal resistance (mΩ) per accumulator of 1 Ah	~22	~6	~6	Very low
Operating temperature (°C)	Ambient	From -40 to $+80$	Ambient	Ambient
Total life cycle	<800	<1200	<1000	>1000

2.2 Operating Principle of a Lead-Acid Battery

The lead-acid battery has the following components:

- a positive electrode (ie, cathode or oxidant), which is the lead dioxide (PbO_2);
- a negative electrode (ie, anode or reducer), which is the lead (Pb); and
- an electrolyte, which is an aqueous solution of sulfuric acid (H_2SO_4).

2.2.1 Discharge-Charge

The overall reaction taking place during a charge-discharge cycle is given by the following equation:

$$PbO_2 + Pb + 2H_2SO_4 \underset{\text{charge}}{\overset{\text{discharge}}{\rightleftharpoons}} 2PbSO_4 + 2H_2O \qquad [ii]$$

Eq. [ii] shows that the sulfuric acid combines with the lead and lead dioxide to produce lead sulfate and water. Therefore the acid becomes increasingly diluted during discharging so that it concentrates the electrodes and returns them to their initial states during charging. The variation of the concentration of sulfuric acid during the discharge or the charge implies the *SOC* variation because an electric energy is being released during the reactions. To better understand the operating mechanism, the overall reaction of Eq. [ii] can be reduced to discharge reactions occurring at each electrode:

- Negative electrode:

$$Pb + HSO_4^- + H_2O \xrightarrow{\text{oxydation}} PbSO_4 + H_3O^+ + 2e^- \qquad [iii]$$

- Positive electrode:

$$PbO_2 + HSO_4^- + 3H_3O^+ + 2e^- \xrightarrow{\text{reduction}} PbSO_4 + 5H_2O \qquad [iv]$$

During the discharge, the Pb anode, as shown in Eq. [iii], oxidizes. The resulting electrons then move through an external electrical circuit and are consumed during the reduction of the lead(IV) oxide (PbO_2) cathode described in Eq. [iv]. The movement of electrons gives rise to a current for which the direction is reversed; from the PbO_2 cathode toward the Pb anode. By convention, the above equations are reversed during charging. The nominal equilibrium potential of the accumulator is $E_0 = 2$ V, equal to the difference between the equilibrium potential of the two electrodes (ie, $E_{0,\, PbSO_4/PbO_2} = 1.7$ V and $E_{0,\, Pb/PbSO_4} = -0.3$ V, respectively) [3]. For a desired current and/or desired voltage, more accumulators can be assembled in series or more rarely in parallel. This gives rise to a battery. For example, a 12-V battery is made up of six accumulators in series. Thereafter, v_{BAT} and i_{BAT} symbolize the voltage and current of a battery.

2.2.2 Others Electrochemical Reactions

In addition to the above reactions, other reactions, called secondary reactions, occur continuously, such as the electrolysis of water. This is reflected by the oxygen evolution on the positive electrode Eq. [v] and by the evolution of hydrogen on the negative electrode Eq. [vi].

$$3H_2O \rightarrow \frac{1}{2}O_2 + 2H_3O^+ + 2e^- \qquad [v]$$

$$2H_3O^+ + 2e^- \rightarrow 2H_2O + H_2 \qquad [vi]$$

The overall reaction of the water electrolysis is as follows:

$$H_2O \rightarrow \frac{1}{2}O_2 + H_2 \qquad [vii]$$

2.2.3 Manufacturing Technologies

According to the battery manufacturing technology, two families were formed: (1) conventional batteries, or "open" (vented batteries), and (2) sealed batteries, or "closed" (valve-regulated lead acid (VRLA) batteries). Regardless of the manufacturing technology, the electrolyte is typically composed of 65% water and 35% sulfuric acid. An open battery type is equipped with a valve to allow gas emissions from escaping. Therefore the amount of electrolyte will decrease. This requires an intervention from time to time to readjust the level of the electrolyte with distilled water to the proper functioning of batteries. In addition, local exhaust ventilation is required with this type of battery to avoid possible explosion due to hydrogen mixing with ambient air. As for the newer type of sealed battery, the electrolyte is immobilized or gelled by adding silicon dioxide (SiO_2); ie, retained by a high capillary surface fiberglass separator (absorbent glass mat) [3]. The immobilization of the electrolyte allows oxygen (O_2), produced on the positive electrode, to be circulated to the negative electrode where it is reduced as in Eq. [viii].

$$\frac{1}{2}O_2 + 2H^+ + 2e^- \rightarrow H_2O \qquad [viii]$$

Therefore recombinant O_2 reduces the emission of hydrogen H_2 in the water production. Unlike the vented batteries, water consumption and gas emissions are extremely low in VRLA batteries. They do not require maintenance, and they may be positioned in any premises without specification of ventilation.

2.3 Dynamic Phenomena of a Lead-Acid Battery

The main dynamic phenomena occurring in an electrochemical system are the charge transfer, mass transport, and the double-layer capacitance.

The charge transfer phenomenon shows the kinetics of the electrochemical reaction and results in the noted transfer resistance R_{tc}.

Mass transport is the movement of ions from the electrolyte on the surface of the electrode and vice versa. This displacement is due to three processes:

- diffusional transport under the influence of concentration gradient,
- transport by migrating under the effect of the electric field, and
- transport by convection as a result of pressure temperature gradient or mechanical agitation.

The charges are separated when a metal is in contact with an electrolyte. The corresponding area to this separation is called the double layer, the capacitance of which is denoted C_{dl}. Charging separation depends on many factors such as the electrochemical properties of the solid, adsorption of water molecules, or hydrated cations. This phenomenon can be represented by a plane capacitor according to the Helmholtz model.

2.4 Modeling of Lead-Acid Battery

The modeling of lead-acid batteries is the subject of several research works. The interest of a battery model lies in the ability to simulate its performance and estimate its tension and its state of charge, taking into account factors such as temperature, aging, and the number of charge-discharge cycles. According to the literature, the modeling of a battery is performed either from a chemical point of view or an electrical point of view.

The chemical modeling is to consider the dynamic phenomena such as reaction kinetics, diffusion, and polarization (voltage change during charging or discharging) [4–6]. These phenomena are represented by differential algebraic equations for which the solution becomes complex by the initial conditions and the imposed limits. In addition, some specific measurements of the electrolyte are not possible for the batteries of the sealed type. In this book, given the specificity of the overall modeling of a microgrid, these phenomena are not considered.

Among the best known electric models available in the literature, there are two families of models based on the equivalent electrical circuit: static and dynamic. These models are presented in the following paragraph.

2.4.1 Battery Static Modeling

The simplest model of the battery is shown in Fig. 3.1. It consists of an ideal voltage source E_{eq} modeling the voltage across an ideal battery in series with a constant internal resistance R_0 [7,8]. The voltage v_{BAT} at the terminals of the battery is given by Eq. [3.1].

$$v_{BAT}(t) = E_{eq} + R_0\, i_{BAT}(t) \qquad [3.1]$$

The current i_{BAT} is negative during discharge and positive during charge. Voltage E_{eq} is measured when the battery is empty. The resistance R_0 can be experimentally identified, as shown in Fig. 3.2. By applying a discharge current to the accumulator output, the voltage drop Δv_{BAT} is due to the resistance R_0. This resistance R_0 is determined by the following equation:

$$R_0 = \frac{\Delta v_{BAT}}{\Delta i_{BAT}} \qquad [3.2]$$

This very simplified model does not take into account the variation of the internal resistance as a function of SOC and temperature. It is valid only if one considers the effect of SOC as negligible.

2.4.2 Battery Dynamic Modeling

The dynamic model aims to address the dynamic phenomena in the battery, the phenomena described in Section 2.3. It consists of a voltage source modeling the equilibrium voltage of the accumulator E_{eq}, internal resistance R_Ω, and a parallel connection of two branches, which models the dynamic phenomena addressed. The first branch is the double-layer capacitance C_{dl} whereas the second branch contains the transfer resistance R_{tc} in series with the impedance expressing the diffusion (ie, the Warburg impedance

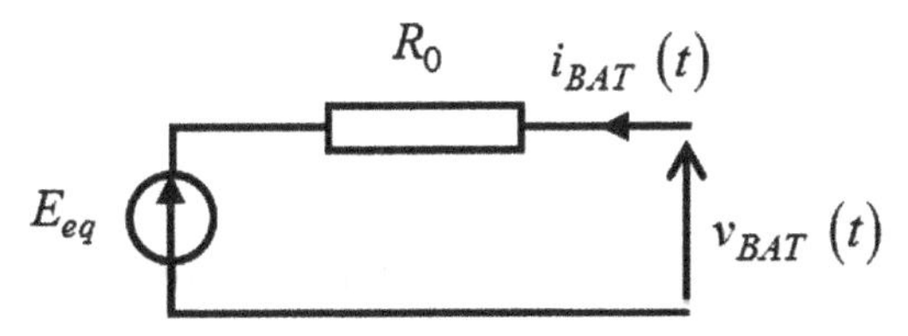

Figure 3.1 Static model of the battery.

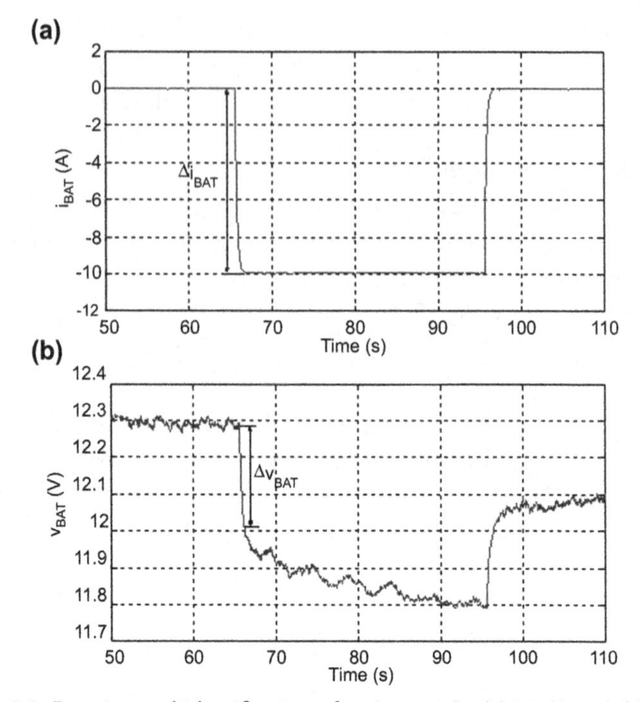

Figure 3.2 Experimental identification of resistance R_0: (a) $i_{BAT}(t)$ and (b) $v_{BAT}(t)$.

denoted by Z_w). This dynamic model, which is based on Randles schema [9], is given in Fig. 3.3.

The dynamic model of the battery based on Randles schema can be simplified by reducing it to the circuit diagram shown in Fig. 3.4.

In this case, semifinished diffusion conditions are taken into account. Thus the Warburg impedance is expressed by a nonintegral power transfer function of the Laplace variable as

$$Z_w(s) = \frac{(1+\tau_2 s)^{n_2}}{(\tau_1 s)^{n_1}} \quad [3.3]$$

with τ_1 and τ_2 as two time constants and n_1, n_2 as two real numbers between 0 and 1.

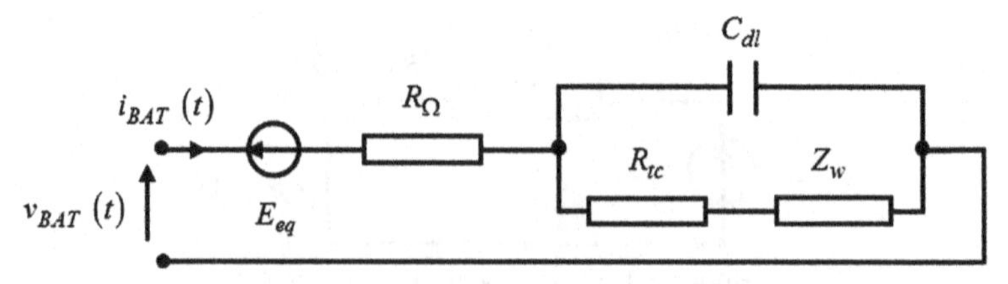

Figure 3.3 Dynamic model of the battery based on Randles schema.

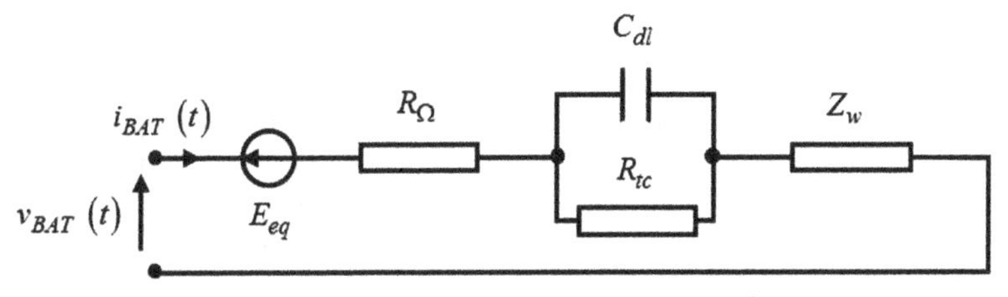

Figure 3.4 Dynamic model of the battery based on simplified Randles schema.

According to the presented work in [9], Eq. [3.3] can be approximated in a form of hyperbolic tangent

$$Z_w(s) = \frac{k_2}{\sqrt{s}} \tanh\left(\frac{k_1}{k_2}\sqrt{s}\right) \quad [3.4]$$

with k_1 and k_2 two parameters to identify. The inverse transform of $Z_w(s)$ may be written in the following form:

$$\begin{aligned} Z_w(t) &= \mathrm{L}^{-1}(Z_w(s)) = \mathrm{L}^{-1}\left(\frac{k_2}{\sqrt{s}}\tanh\left(\frac{k_1}{k_2}\sqrt{s}\right)\right) \\ &= 2\frac{k_2^2}{k_1}\sum_{k_w=1}^{n_w}\exp\left(\frac{-(2k_w-1)^2\pi^2 k_2^2 t}{4k_1^2}\right) \end{aligned} \quad [3.5]$$

with n_w the number of parallel branches in the Warburg impedance. Subsequently, this expression can be developed by

$$\begin{aligned} Z_w(t) = {} & \frac{1}{C_1}\exp\left(-\frac{t}{R_1 C_1}\right) + \cdots + \frac{1}{C_{k_w}}\exp\left(-\frac{t}{R_{k_w} C_{k_w}}\right) + \cdots \\ & + \frac{1}{C_{n_w}}\exp\left(-\frac{t}{R_{n_w} C_{n_w}}\right) \end{aligned} \quad [3.6]$$

Eq. [3.6] leads to a recursive structure of a parallel resistive-capacitive circuit, said Foster structure [10]. Thus the equivalent circuit diagram of the dynamic model is presented in Fig. 3.5.

The value of the elements of the Warburg impedance (Fig. 3.5) are calculated based on k_1 and k_2 with

$$\begin{cases} C_{k_w} = \dfrac{k_1}{2k_2^2} \\ R_{k_w} = \dfrac{8k_1}{(2k_w-1)^2\pi^2} \end{cases} \quad [3.7]$$

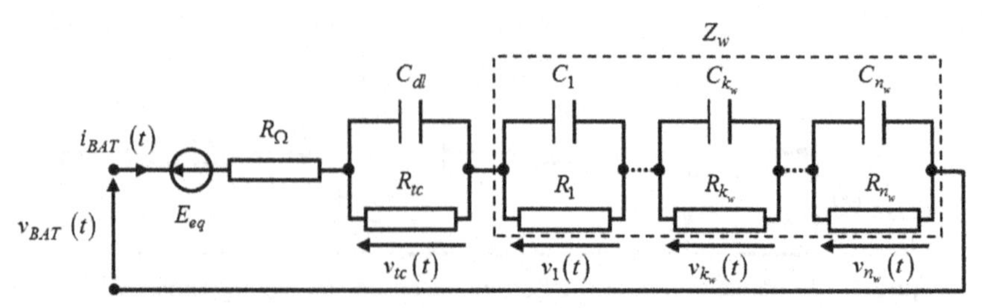

Figure 3.5 Dynamic model of the lead-acid battery.

The use of the dynamic model involves identification of parameters R_Ω, R_{tc}, C_{dl}, R_{k_w}, and C_{k_w}. Among the most used methods to identify the parameters of the dynamic model, there is the method by impedance meter and identification from the state model.

Whatever the number of recursive structures in the Warburg impedance, the parameters of the dynamic model are R_Ω, R_{tc}, C_{dl}, k_1, and k_2. Knowing that R_{tc} and C_{dl} can be expressed according to k_1 and k_2, the number of parameters can be reduced to three: R_Ω, k_1, and k_2.

A finite number n_w of recursive structures of the Warburg impedance would allow for a moderate approximation for a reasonable calculation time. Thus identification of parameters is performed from the state system.

The system state is described by first-order differential equations in terms of the state variables that summarize the past and help predict the future of the system. These equations are written as follows:

$$\frac{dx(t)}{dt} = Ax(t) + Bu(t)$$
$$y(t) = Cx(t) + Du(t) \qquad [3.8]$$

where x is the state vector, u is the input vector, and y is the output vector. A, B, C, and D are matrices. From Fig. 3.5, the state vector is identified as

$$x(t) = \begin{pmatrix} v_{tc}(t) \\ v_1(t) \\ \vdots \\ v_{k_w}(t) \\ \vdots \\ v_{n_w}(t) \end{pmatrix} \qquad [3.9]$$

the input vector as

$$u(t) = i_{BAT}(t) \qquad [3.10]$$

and the output vector as

$$y(t) = v_{BAT}(t) - E_{eq} = R_{\Omega}\, i_{BAT}(t) + v_{tc}(t) + \sum_{k_w=1}^{n_w} v_{k_w}(t) \qquad [3.11]$$

From Fig. 3.5 and Eqs. [3.8–3.11], the matrix A, B, C, and D values are

$$A = \begin{pmatrix} -\frac{1}{R_{tc}C_{dl}} & 0 & \cdots & 0 & \cdots & 0 \\ 0 & -\frac{1}{R_1 C_1} & \cdots & 0 & \cdots & 0 \\ \vdots & \vdots & \cdots & \vdots & \cdots & \vdots \\ 0 & 0 & \cdots & -\frac{1}{R_{k_w} C_{k_w}} & \cdots & 0 \\ \vdots & \vdots & \cdots & \vdots & \cdots & \vdots \\ 0 & 0 & \cdots & 0 & \cdots & -\frac{1}{R_{n_w} C_{n_w}} \end{pmatrix},$$

$$B = \begin{pmatrix} \frac{1}{C_{dl}} \\ \frac{1}{C_1} \\ \vdots \\ \frac{1}{C_{k_w}} \\ \vdots \\ \frac{1}{C_{n_w}} \end{pmatrix}, \; C = (1 \;\; 1 \cdots 1 \cdots 1) \text{ and } D = R_{\Omega} \qquad [3.12]$$

Having thus defined the matrices of the dynamic model, for a given n_w, the parameters R_{Ω}, k_1, and k_2 are estimated by minimizing the least square according to the procedure next described.

Iteratively and from the initial values of parameters, the model calculates the output voltage for an input current. The voltage from the model is compared thereafter with the measured voltage on the basis of the sum of the least square errors. A function of minimizing (ie, "lsqcurvefit") under the MATLAB environment can minimize errors by converging the parameter values to those that satisfy a local minimum. The resulting parameters of the minimizing step are introduced to calculate voltage again and so on. The results of the identification and model validation are shown in Section 2.5.

This model is a model estimating the battery terminal voltage; that is, it can provide this voltage for a charging or discharging given current. However, it makes no estimate or calculation of *SOC*.

An experimental approach is proposed to have a simple model, which for a charging or discharging given current is able to estimate an output voltage and the *SOC*.

2.4.3 Purely Experimental Battery Model

The purely experimental model that is proposed here is a static model type. This model, based on a battery, is derived from a process for measuring the current and voltage and estimating the state of charge of the battery during several charge-discharge cycles. However, the charge state is difficult to estimate. This could be determined using measurements of certain specifications of the electrolyte as the gravity or specific density [11]. The linear relationship between the gravity of the electrolyte and the charge state can estimate the state of charge, but this method is unsuitable for sealed batteries for which the modeling must be done in this study. Therefore in this case it is possible to establish a linear relationship between the charge state and the open circuit voltage [11]. The open circuit voltage measurement must be made in a period of rest (ie, after disconnecting the battery and letting it sit a while). Because of the difficulty of the establishment of the open-circuit voltage measurement procedure and the associated time durations to complete the required number of tests, in this study the state of charge is estimated by integrating the current (ie, measuring the amount of electricity entering or leaving the battery [11]):

$$SOC = SOC(0) - \frac{1}{C_{NOM}} \int_0^t i_{BAT}(t)dt \qquad [3.13]$$

with $SOC(0)$ and C_{NOM} being the initial state of charge and the nominal capacity expressed in ampere seconds, respectively.

The experimental tests performed include five charge–discharge cycles corresponding to the discharge conditions C_{20}, C_{15}, C_{10}, C_7, and C_5 with the sealed 12-V/130-Ah Solar Sonnenschein lead-acid battery. Assigned

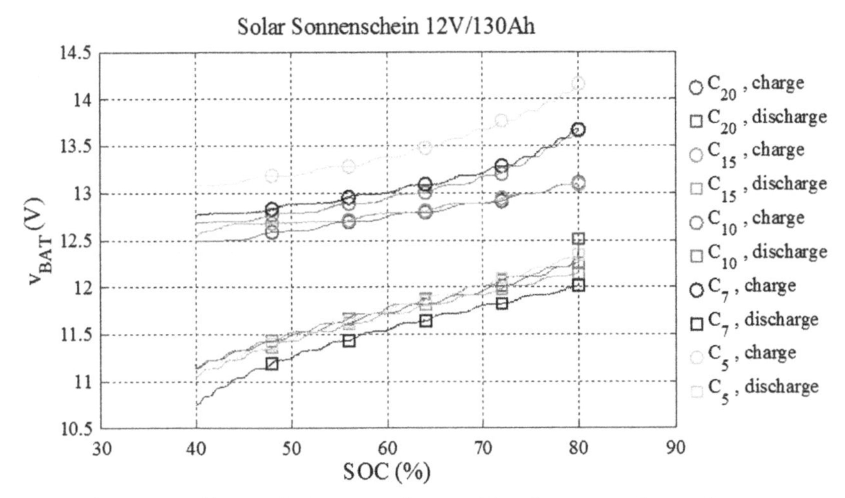

Figure 3.6 Charge-discharge cycles resulting from experimental tests.

discharge conditions are provided by the manufacturer. The test consists of imposing on the battery, through a DC-DC converter driven by a simple linear order, a current corresponding to the given discharge rate. During discharge, the battery delivers to a programmable electronic load. A voltage-controlled DC source is used to charge the battery. The charge-discharge cycles obtained are shown in Fig. 3.6.

The purely experimental model described is achieved by exploiting the collected experimental data and corresponding to *SOC* from 40% to 80%, as illustrated in Fig. 3.7. The effective implementation of a purely experimental model is feasible using the look-up table-type data tables under the MATLAB environment. Under this model, for a given current, *SOC* is estimated according to Eq. [3.13]. In addition to the value of *SOC*, the model allows for obtaining the corresponding voltage value. The accuracy of this model depends on exploitation of the approximation algorithm used for data not included in the tables.

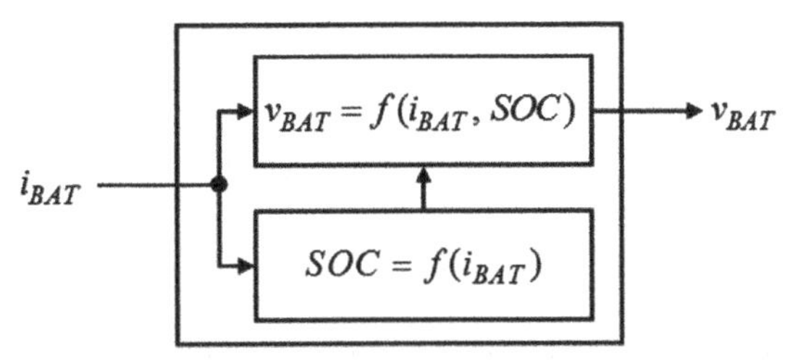

Figure 3.7 Representative schema of the purely experimental model of the 12-V/130-Ah Solar Sonnenschein battery.

2.5 Experimental Evaluation of Lead-Acid Battery Model

To compare the three models—static, dynamic, and purely experimental—the 12-V/130-Ah Solar Sonnenschein battery is operated by profiles of different currents (amplitude and duration). The estimated voltage response given by each model is then compared with that measured. As in Section 2.4.3, the tests are to be imposed on the battery, through a DC–DC converter controlled by a linear control, as the profiles of desired currents. For some *SOC* and some current profiles, it was found that the battery shows a response in nonsymmetrical voltage. This is due to the nonlinear dynamic behavior of the battery between the charging phase and the discharging phase. Resistance R_0 (static model) experimentally determined is not identical to the charging and discharging. For example, in the case of a current profile as shown in Fig. 3.8, two values can be determined: $R_{0_{CH}} = 17\ \text{m}\Omega$ and $R_{0_{DIS}} = 11\ \text{m}\Omega$.

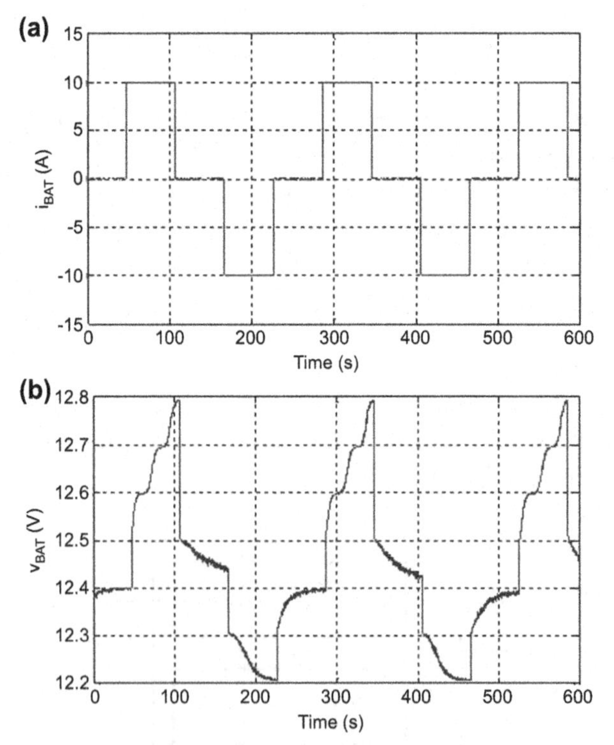

Figure 3.8 (a) Current profile and (b) its voltage response for a *SOC* = 40%.

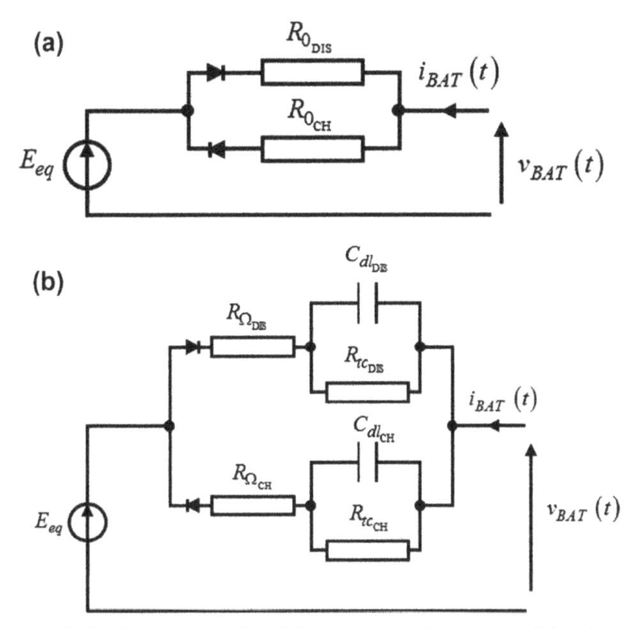

Figure 3.9 Nonlinear models of the battery: (a) static and (b) dynamic.

This observation leads us to bring the static model and the dynamic model to new equivalent circuit schemes shown in Fig. 3.9. In addition, the dynamic model is synthesized by considering one resistive-capacitive parallel circuit structure, which symbolizes all the networks illustrated in Fig. 3.5. This is because increasing the number of parallel circuits, with the used battery, does not improve the accuracy of the model.

Concerning the identification method, it is always based on the principle of optimization previously mentioned. The best parameter values are those that best fit possible with the measured voltage calculated. For the nonlinear dynamic model shown in Fig. 3.9b, the equation describing the voltage response to a negative current pulse (case of discharge) is defined by Eq. [3.14].

$$v_{BAT}(t) - E_{eq} = R_{\Omega_{DIS}} i_{BAT}(t) + R_{tc_{DIS}} i_{BAT}(t)\left(1 - \exp\left(\frac{t}{R_{tc_{DIS}} C_{dl_{DIS}}}\right)\right) \quad [3.14]$$

Thus the following parameters must be determined: $R_{\Omega_{DIS}}$, $R_{tc_{DIS}}$, and $C_{dl_{DIS}}$. It should also do the same for the charging phase. Fig. 3.10 shows the voltage response and estimated voltages by static and dynamic nonlinear models during a charging pulse and during a discharging pulse.

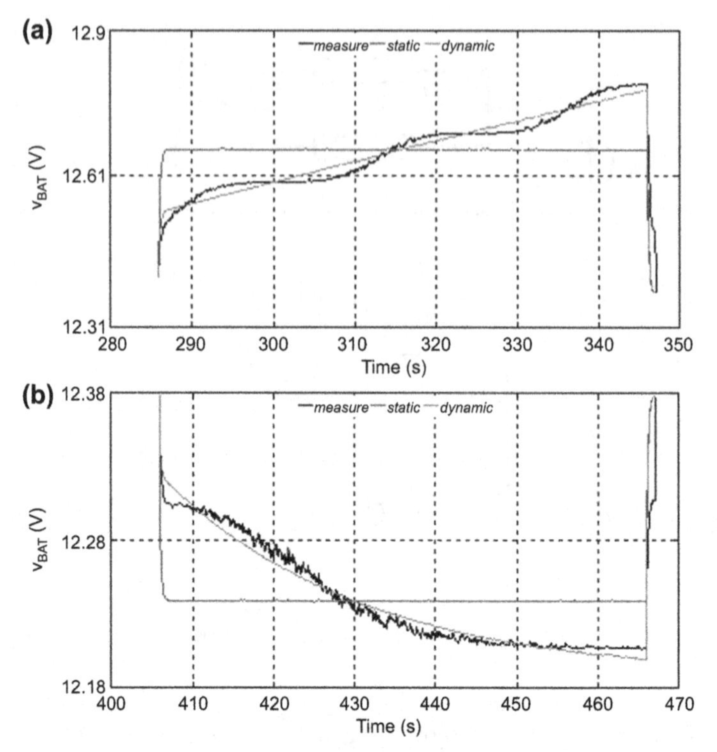

Figure 3.10 Parameter identification of nonlinear models (static and dynamic) for *SOC* = 40%: (a) charge and (b) discharge.

The obtained values are returned to recalculate the voltage response over the entire current pulse and then it is compared with the voltage derived from the purely experimental model (mapping steps), as shown in Fig. 3.11.

The dynamic model gives the best answer whereas the experimental model provides the answer with the largest error between the measured voltage and the calculated one; this difference is greater in charge regime than in discharge. The three models generate an error during the current zero crossing, where the power is zero. The evaluation of the accuracy of the models is performed by the mean absolute error (MAE). This MAE, calculated for several y measurement points, with the total number z of measurement points, in which v_{BATm} is the measured voltage and v_{BATc} the calculated voltage according to the model, is expressed by Eq. [3.15].

$$\mathrm{EMA} = \sum_{y=1}^{z} \frac{\left| v_{BATm}^{y} - v_{BATc}{}^{y} \right|}{z} \qquad [3.15]$$

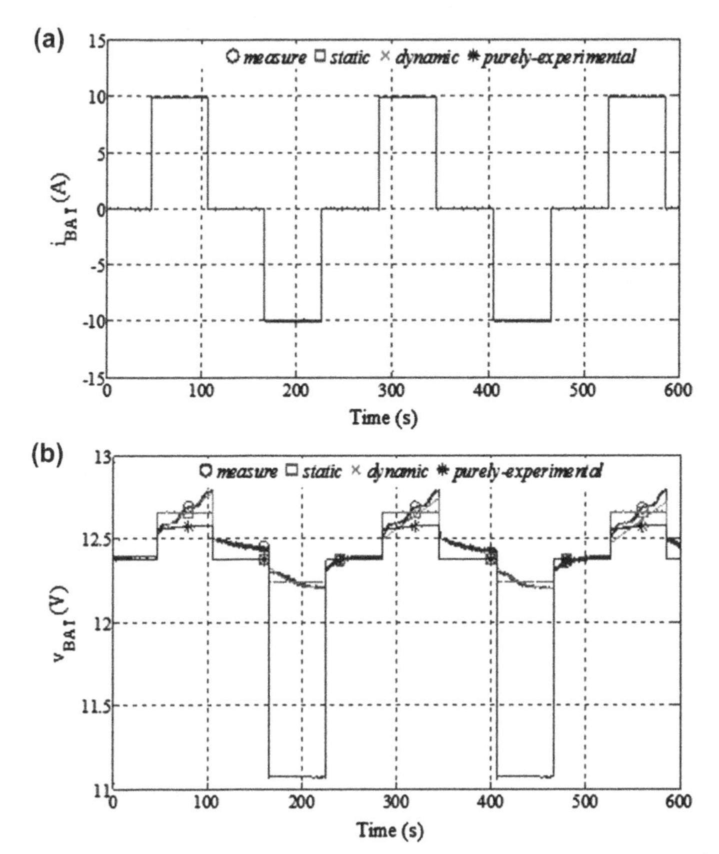

Figure 3.11 Comparison of the models for $SOC = 40\%$: (a) current profile and (b) voltage curves.

Table 3.2 Errors generated by models compared with the test shown in Fig. 3.11

Model	Error (mean absolute error)
Static	0.048
Dynamic	0.035
Purely experimental	0.34

This error, calculated over the entire simulation period in each case, is shown in Table 3.2.

According to the results presented in Table 3.2, the error of the static model is larger than that of the dynamic model. This is explained as follows. Because the resistors $R_{0_{CH}}$ and $R_{0_{DIS}}$ are identified from the charge cycle and discharge cycle, the static model gives the mean value of the voltage in both

Table 3.3 Values of the parameters of static and dynamic models

Static model				**Dynamic model**			
$R_{0_{CH}}$ (mΩ)	$R_{0_{DIS}}$ (mΩ)	$R_{\Omega_{CH}}$ (mΩ)	$R_{tc_{CH}}$ (mΩ)	$C_{dl_{CH}}$ (kF)	$R_{\Omega_{DIS}}$ (mΩ)	$R_{tc_{DIS}}$ (mΩ)	$C_{dl_{DIS}}$ (kF)
28.7	13.9	16.2	287	2.329	5.4	13.6	1.791

phases. Or, for the same current intensity, the value of the resistors of the static model depends strongly on the current direction, as shown in Table 3.3. Although similar findings can be drawn for the parameter values of the dynamic model, these parameters values are also presented in Table 3.3.

Likewise, and to take into account possible variations of *SOC*, four current profiles for different *SOC* were imposed and are illustrated in Fig. 3.12. The amplitude and period of the current are 5 A and 240 s, respectively, except for the final profile, which is a period of 120 s. Because the *SOC* is equal to 80%, the pulse current of 5 A could induce a voltage increase beyond the preset limit during all experimental tests. That is why the period of the last current profile is 120 s and not 240 s. For each profile, the output of each model is again calculated. Corresponding voltage responses are shown in Fig. 3.13.

The accuracy of the models, for the same current, depends on the nonlinear behavior of the battery, which is itself a function of the *SOC*. This behavior is taken into account by the static and dynamic models with nonlinear parameters, which take different values in charging and discharging. With this, the voltage curves calculated by these models are close to those measured (Fig. 3.13). Thus, although the error induced by the dynamic model is smaller than the error induced by the static model, both models allow close enough voltage response compared with the measured voltage. The voltage response of the static model is close to the average value of the measured voltage, especially during the charging regime, which shows a nonlinear effect that is difficult to model, whereas during the discharge regime the response is precisely as that given by the dynamic model (Fig. 3.13).

On the other side, in Fig. 3.13, during the switching charge-discharge, the static and dynamic models generate a significant error. This is due to a relaxation phenomenon, challenging behavior to be taken into account by these models. Knowing that the power is zero at the zero passage of the current, the error during relaxation is neglected.

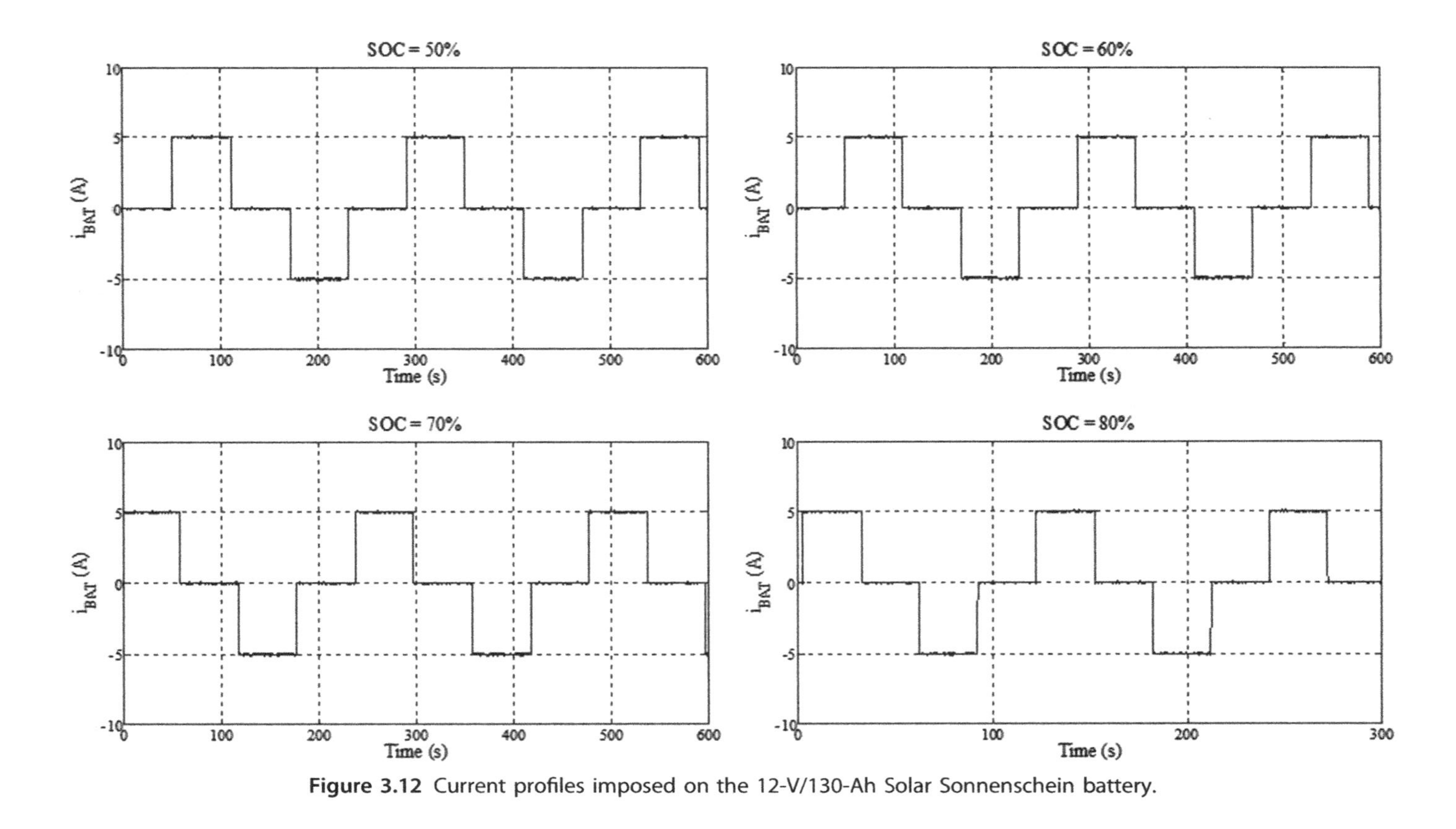

Figure 3.12 Current profiles imposed on the 12-V/130-Ah Solar Sonnenschein battery.

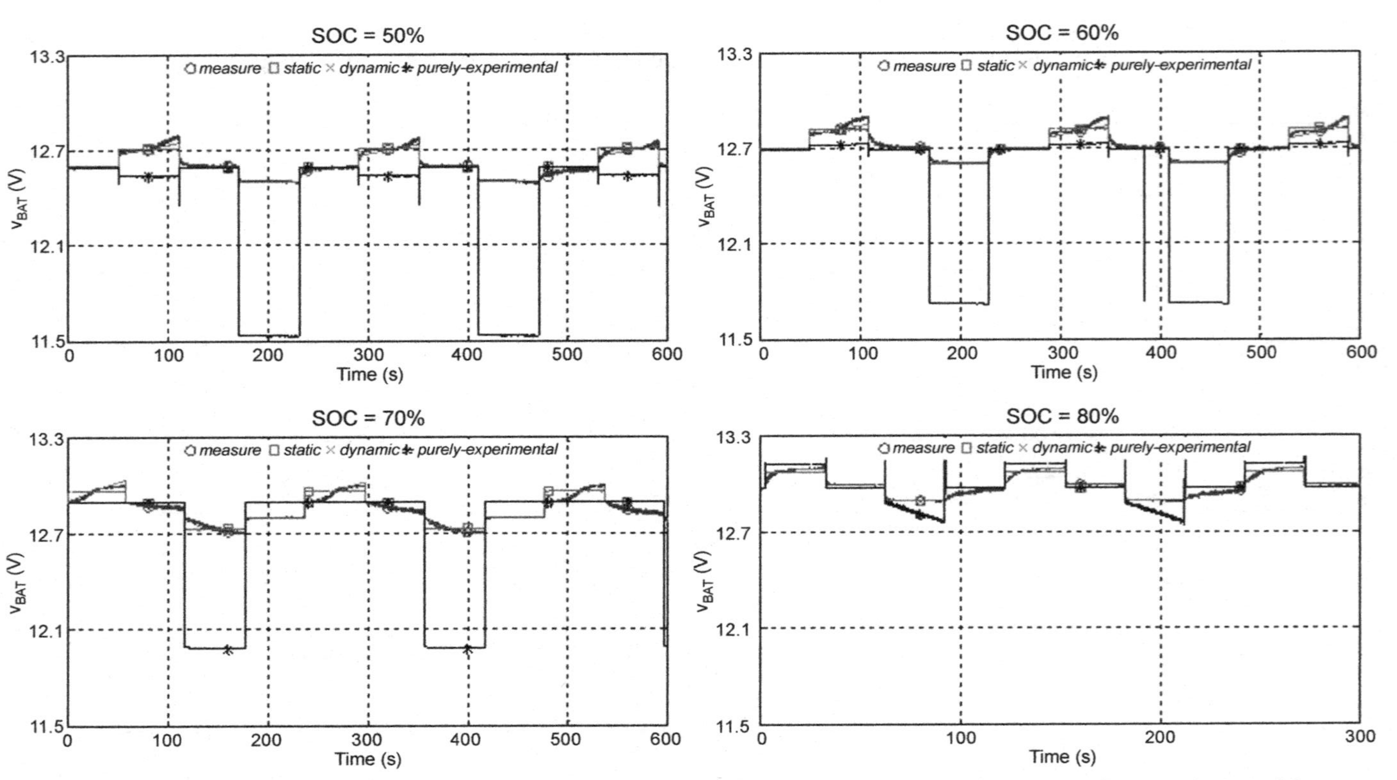

Figure 3.13 Measured voltage responses of the 12-V/130-Ah Solar Sonnenschein battery and those of the studied models.

The purely experimental model is a mapping of the voltage across the battery—voltage measured after the imposed currents and according to certain levels of *SOC*. The behavior of the purely experimental model, following the four profiles of current, shows the weakness and the level of error that this model can induce. The purely experimental model could be chosen for its simplicity, but because of the nonlinearity of the accumulator operating, it is clear that the state of health (*SOH*) must be taken into account. However, this requires further study for *SOH* modeling. The literature contains few studies on *SOH* modeling, although the battery aging is a constant concern in research. This mapping does not include the voltage E_{eq} measurements. Therefore the calculated voltage during recharging (Fig. 3.13) becomes higher than that calculated during charging, which is false.

This modeling study of an electrochemical battery allows firstly acquiring and deepening knowledge on the behavior, variable influences, transients, and other parameters that many articles discuss. The goal was originally to choose or build a reliable model that would, from a given current, calculate *SOC* and thus to obtain voltage. In the end, for the microgrid described at the beginning of this book, more reliable and more robust models should be studied.

The static model, already very simplified compared with the complexity of the dynamic phenomena, does not take into account the variation of the internal resistance according to *SOC* and temperature. The purely experimental model, as was proposed in this study, is not sufficiently developed to respond satisfactorily to the given objective.

The dynamic model may be a good compromise by taking into account the nonlinearity of its settings according to the *SOC* and the current direction. Moreover, despite the many parameters, the dynamic model needs no great calculation efforts. However, for a use as initially specified, a parameter mapping should be developed because the parameters vary depending on the *SOC* and the amplitude of the imposed current. This parameter mapping would result in numerous experimental tests, but *SOH* should also be taken into account in this mapping.

However, according to the experimental tests on three models, to study some applications such as microgrids, for which one must obtain information on the voltage value and the *SOC* for a given current, the static model can be an acceptable response because of its simplicity.

Finally, this model provides a model for knowledge; because of this, the experimental control of the storage system is more mastered.

3. DIESEL GENERATORS

Diesel generators convert fuel energy (diesel or biodiesel) into mechanical energy by means of an internal combustion engine and then into electric energy by means of an electric generator. Thus a diesel generator basically belongs to the internal combustion engines. Its name is linked to the used fuel. According to the sizing and level of its tank fuel, the use of the produced power may be assumed to be accessible anywhere at any time. A diesel generator integrated in a microgrid aims at making good compensations after any shortfalls in power production or a power failure due to any technical problems. Diesel generators are the most common electricity generator used in building-integrated microgrids because of their size (from 1 kVA to more than 1000 kVA), initial cost, simplicity, and the ordinary fuel being easy to buy.

3.1 Characteristics of Diesel Generators

A diesel generator is composed mainly of an internal combustion engine, an electric generator (usually a synchronous type), mechanical coupling, an automatic voltage regulator, a speed regulator, a support chassis, a battery for starting the motor that permits the diesel generator startup, a fuel tank, and a command panel. The diesel generator used in a microgrid may run as a continuous or prime generator and must be designed/sized to operate for significant durations at variable load.

To be part of a building-integrated microgrid, the diesel generator must respond to the requirements of reliability, rapid response time, fuel availability, and load supply capacity.

Compared with other generators based on an internal combustion engine that may take up to 2 min for startup, one notes that the diesel generator response time is satisfactory; it is able to start and supply the load within 10 s. However, for the stringent condition of real-time power balancing of a microgrid, time compensation is generally required.

Concerning reliability, a diesel generator is able to provide power quickly and continuously during an important time period. Nevertheless, it must be correctly sized and be associated with the necessary fuel tank, of which the supply frequency is well respected.

Regarding availability—it is easy for any person, business, or facility to select, finance, install, and service a generator in the United States because of a comprehensive system of local dealers and readily available supplies of clean diesel fuel.

Furthermore, for a high-quality diesel generator, the durability may reach tens of thousands of hours before their first overhaul. In addition, diesel generators are considered simple to implement and operate.

Diesel generators offer many advantages; however, except the needed compensation for the response time, another major disadvantage lies in pollutant emissions.

3.2 Operating Principle of a Diesel Generator

The diesel generator operating principle is almost the same as that of a microturbine except for the use of the air, so there is no compressor. Cogeneration is also possible; for this, a diesel generator requires two heat exchangers aiming at thermal transfer from exhaust gas and the cooling system. However, the main use of a diesel generator is to cover the needs of the power balancing in the public electricity grid or microgrid (feeding power either during peak periods or a shortage of main power generators) and to supply stand-alone applications as emergency power.

In normal operating mode, the diesel generator may be in "standby" mode and enters into activity to offset the power consumption peaks. This technique is mainly used in isolated microgrid operating mode when the main power becomes unavailable or insufficient to supply the load. Thus a diesel generator is most often used as a backup source to avoid the significant extra cost of the energy tariff during a limited time period or coupled to other generator devices aiming a high energy quality (ie, without fluctuations). One notes that for building electrical appliances, the diesel generators allow for reaching significant powers and operating time periods.

Regarding the urban DC microgrid, the difference between load consumption and renewable energy production causes fluctuations in the microgrid DC bus voltage. Thus, for off-grid operating mode or limited public grid availability, a power balance may be performed by adjusting the diesel generator power and/or storage for voltage stabilization. Although diesel generator large capacity backup power can provide long-term support for DC microgrid operation, its response speed is slow and the starting time lasts several seconds to reach stable output power.

Therefore, because of the slow dynamic behavior of the diesel generator, the DC microgrid power balancing needs compensation for a sudden load power increase or a sudden renewable power decrease. Otherwise the supercapacitors represent an electrochemical storage deemed to have a fast response and high instantaneous output power. Thus a supercapacitor may be a solution to compensate for dynamic power fluctuations before the steady operation of a diesel generator.

On the other hand, as for all generators based on an internal combustion engine, frequent start and stop cycles combined with low output power reduce diesel generator efficiency and introduce supplementary cost. To overcome these issues, the diesel generator may be assigned with two operating states—off and output rated power—with a working duty cycle. However, because the diesel generator works in a duty cycle, it may introduce the case that storage is charged by excess energy available from the microgrid. Because the high charging rate can largely affect the storage life, the storage power has to be limited.

3.3 Operating Cost Analysis of a Diesel Generator

To increase reliability in the electric power supply to end users, diesel generators are typically sized to meet the peak demand during the consumption peak hours, but they must run at very low power during off-peak consumption hours. This low-power operating mode leads to poor fuel efficiency and increased maintenance. To overcome this issue, one solution lies in implementation of two diesel generators—one for high power and one for low power. However, in this case the global cost will be very high and the diesel generator sizing becomes more complex and requires a fine knowledge about the load power demand. In addition, when a diesel generator's output is of low power, it runs too cold; therefore the combustion is inefficient.

Therefore, for all of these difficult issues, only one diesel generator is taken into account in the DC microgrid operation described in this study. It is controlled to continuously output the rated power during a minimum duty cycle, thus increasing the fuel efficiency and resulting in prolonged engine life. However, pollutant emissions are proportional to the diesel consumption generated by the chosen operating mode. To significantly reduce the emissions level of the diesel generator, it is possible to use biodiesel or biofuel blends in prime power applications. In addition, the use of a diesel generator in a microgrid is disadvantageous when facility cost and maintenance are taken into consideration. Therefore, in this study regarding an urban DC microgrid, the diesel generator is considered only for the off-grid operating mode serving as backup power for a limited period.

To summarize, the diesel generator is a highly nonlinear device. It presents downtime and delays, and its nonlinear behavior is difficult to control in an efficient manner. Indeed, to study the full dynamic of such a system, a high-order model is needed that takes into account several sometimes contradictory objectives for optimization. However, knowing that the urban DC microgrid study focuses on power balancing and smart grid communication, this model is not required to study the response of the system.

4. UTILITY GRID CONNECTION

In the case of insufficient power to supply the load, the system security is ensured by the public grid connection, electrochemical storage, and diesel generator. If the storage and diesel generator are not available or cannot provide enough power for different reasons, while the public grid supply mode is available the microgrid system operates in "absorption" related to the public grid. For excess PV power, if the *SOC* of electrochemical storage reaches its high limit, then the public grid connection gives the microgrid the possibility to inject energy and thus sell surplus production. In this case the microgrid system operates in "injection" related to the public grid. Therefore the connection to the public grid, through its control, is a very important element of a multisource system.

Fig. 3.14 illustrates the public grid connection. The chosen structure is classic, the filter is robust, and the command is called "simple."

The public grid connection will be good if at any time there is equality between the grid current i_G and its reference current i_G^*, as in Eq. [3.16]:

$$i_G(t) = i_G^*(t) = I_G^* \sqrt{2} \sin(\omega t + \varphi) \qquad [3.16]$$

where I_G^*, ω, and φ are the effective value of the current reference i_G^*, the public grid pulsation (ie, $\omega = 2\pi f$ with f the public grid frequency), and the phase shift between the grid voltage v_G and its current i_G, respectively.

To not generate reactive power and working at unity power factor, φ is 0. This value can be modified in real time to provide reactive power if it is requested by the public grid. Thus the studied multisource system can and should participate in system services that ensure at any time the balance of electricity flows on the grid and the safety, security, and efficiency of this grid.

The question to ask is this: How is i_G^* imposed using the leg converter B_3 and B_4?

Use a command by hysteresis correcting? This solution, very simple and quick to implement, still has a major drawback. Indeed, for the studied DC

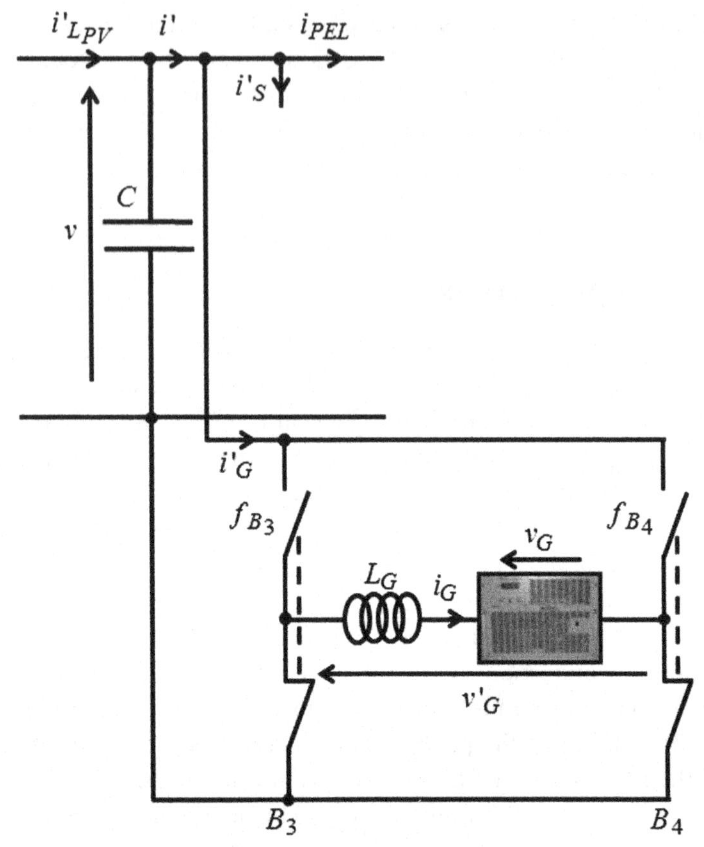

Figure 3.14 Electrical schema of the public grid connection.

microgrid, the DC bus voltage imposed is 400 V; therefore there would be a large voltage variation across L_G and v_G that would cause major variations of i_G. To limit this phenomenon would require a high inductance value (L_G) and/or imposition of a small calculation step. Given the existing equipment on the experimental platform, this solution cannot be upheld. A so-called linear drive by pulse width modulation (PWM) is proposed. It should be noted that although there are other types of control, the this problem does not require as much complexity.

The equation that models the "load" between the two-leg power converter B_3 and B_4 is given by Eq. [3.17]:

$$\frac{di_G}{dt} = \frac{1}{L_G}\left(v'_G - v_G - R_G i_G\right) \qquad [3.17]$$

with i_G the state variable, $v'_G$the control variable, v_G the disturbance variable, and R_G the internal resistance of the inductor L_G. Taking into account the principle that each inverted state variable requires a servo system and each timeless variable requires only direct inversion, the following reference control v'^*_G is obtained:

$$v'^*_G = C_G\left(i^*_G - i_G\right) + v_G \qquad [3.18]$$

with C_G a corrector to be defined.

The modulated current and voltage are defined by Eq. [3.19]:

$$\begin{bmatrix} i'_G \\ v'_G \end{bmatrix} = \alpha_G \begin{bmatrix} i_G \\ v_G \end{bmatrix} \qquad [3.19]$$

where α_G represents the average value over an operation period T from the difference of switching functions f_{B_3} and f_{B_4} (ie, $\alpha_G = \frac{1}{T}\int(f_{B_3} - f_{B_4})dt$, $f_{B_3} = \overline{f_{B_4}}$ and $\alpha_G \in [-1; 1]$).

From Eqs. [3.18] and [3.19] α^*_G can be determined:

$$\alpha^*_G = \frac{1}{v}\left(C_G\left(i^*_G - i_G\right) + v_G\right) \qquad [3.20]$$

Fig. 3.15 presents the used control synopsis. To obtain the switching functions $f^*_{B_3}$ and $f^*_{B_4}$, α^*_G must be compared to a reference voltage v_{REF} type triangle sine (eg, at 20 kHz).

To select the structure of the corrector and to calculate parameters, the same methodology as in chapter "Photovoltaic (PV) source modeling" is used (ie, the use of an integral proportional corrector). This is shown in Fig. 3.16 with K_{1_G} and K_{2_G} two constants determined by the desired dynamics.

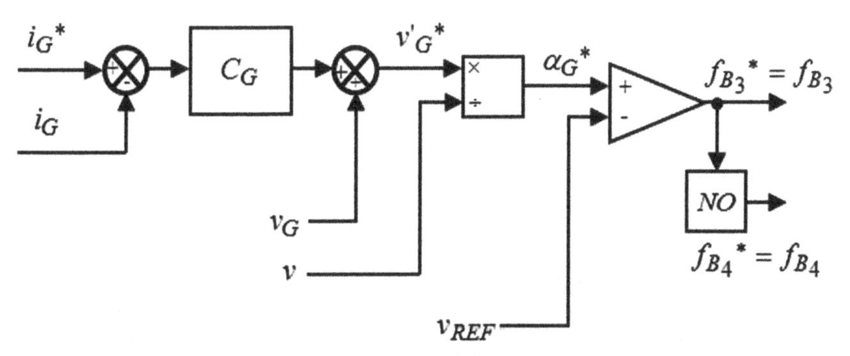

Figure 3.15 Structure of the used control.

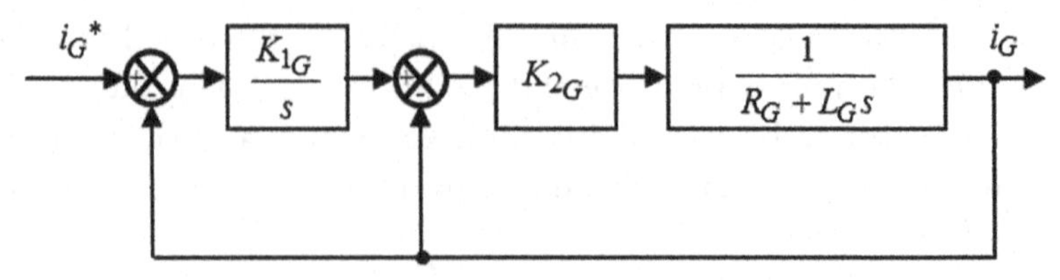

Figure 3.16 Integral proportional corrector structure in the current loop.

From Fig. 3.16 the closed-loop transfer function of the current loop is expressed by Eq. [3.21]:

$$\frac{i_G}{i_G^*} = \frac{1}{1 + \frac{K_{2_G} + R_G}{K_{1_G}K_{2_G}}s + \frac{L_G}{K_{1_G}K_{2_G}}s^2} = \frac{1}{1 + 2\zeta_G\tau_{BF_G}s + \tau_{BF_G}{}^2 s^2}$$

$$K_{1_G} = \frac{L_G}{(2\zeta_G L_G - \tau_{BF_G}R_G)\tau_{BF_G}} \; et \; K_{2_G} = \frac{2\zeta_G L_G}{\tau_{BF_G}} - R_G \qquad [3.21]$$

Depending on the damping coefficient ζ_G of the time constant in the closed loop τ_{BF_G} and the values of R_G and L_G, the coefficients K_{1_G} and K_{2_G} are easily computable.

From Figs. 3.15 and 3.16 it is possible to design the control synopsis to connect the public grid. Furthermore, to correctly synchronize the public grid voltage v_G and public grid current i_G, but also to know exactly the effective value V_G of v_G, the following are usually required:

- a simple measurement of v_G if it is always operated with $\varphi = 0$ when the voltage v_G does not have a harmonic and/or is not disturbed by the switching functions of B_3 and B_4;
- otherwise, the use of a phase-locked loop (PLL), as in this case study.

Fig. 3.17 presents the control synopsis for the public grid connection that implies the use of a PLL.

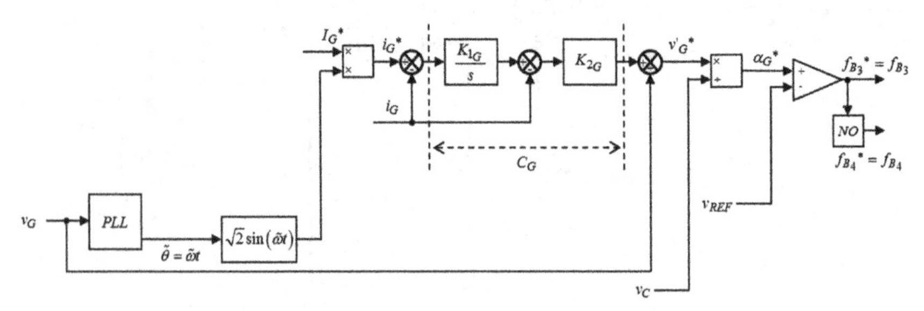

Figure 3.17 Control synoptic for the public grid connection.

4.1 Phase-Locked Loop System Control

The multisource system studied here has a public grid connection in single-phase mode. Note that the use of a PLL from a single-phase signal is more complex than for a three-phase signal. Indeed, in a balanced three-phase sine wave it uses just the transformation of Concordia to quickly obtain two quadrature signals, whereas in single phase, to create an orthogonal component, more sophisticated methods must be used. The four methods most currently used are [12]:

- delay on v_G,
- Hilbert transformation,
- inverse Park transformation, and
- the so-called "generalized second-order integrator" method.

Therefore in this study a conventional PLL structure is proposed, as shown in Fig. 3.18, and for reasons of robustness and experimental implementation simplicity, the fourth method is chosen, the so-called "generalized second-order integrator."

Fig. 3.19, drawn from the fourth method, gives the block structure "creating an orthogonal component" used in Fig. 3.18.

The transfer functions $\frac{\widetilde{v}_G(s)}{v_G(s)}$ and $\frac{q\widetilde{v}_G(s)}{v_G(s)}$ from Fig. 3.19 are equal to:

$$\begin{aligned} \frac{\widetilde{v}_G(s)}{v_G(s)} &= \frac{\widetilde{\omega}s}{s^2+\widetilde{\omega}s+\widetilde{\omega}^2} \rightarrow (a) \\ \frac{q\widetilde{v}_G(s)}{v_G(s)} &= \frac{\widetilde{\omega}^2}{s^2+\widetilde{\omega}s+\widetilde{\omega}^2} \rightarrow (b) \end{aligned} \quad [3.22]$$

Fig. 3.20 presents the Bode diagram in amplitude and phase for the transfer functions (a) and (b) Eq. [3.22]. This figure illustrates two points. First, the signal $\widetilde{v}_G$ represents the filtered voltage v_G, which may be

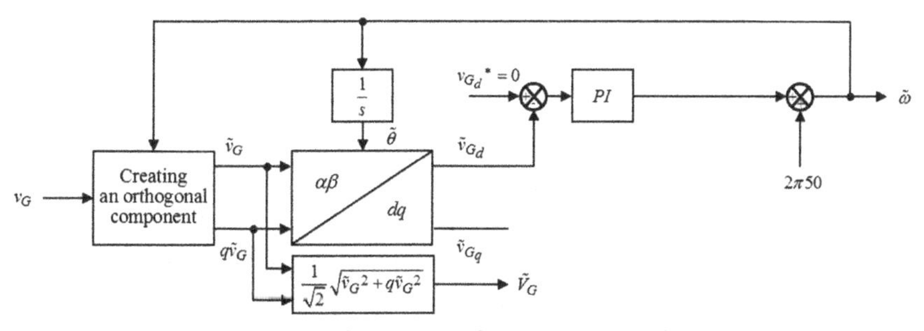

Figure 3.18 General structure of the used phase-locked loop.

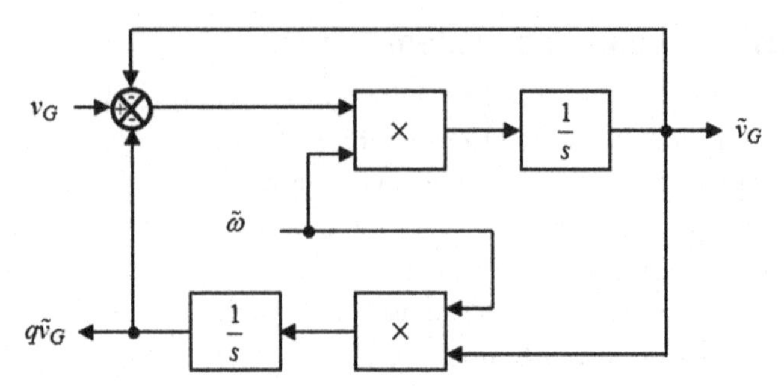

Figure 3.19 Block structure "creating an orthogonal component."

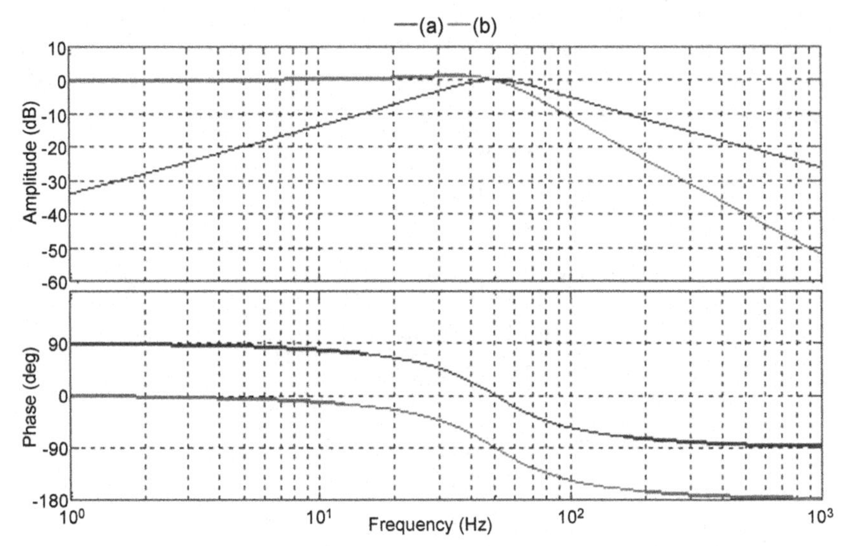

Figure 3.20 Bode diagram in amplitude and phase for the transfer functions (a) and (b).

advantageous when it is noisy. Second, the signal $q\widetilde{v}_G$ is in quadrature with the signals v_G and $\widetilde{v}_G$.

Fig. 3.21 illustrates temporal responses of $\widetilde{v}_G$ and v_G to a level of v_G (for 230 V effective value and $\omega = 314.16$ rad/s). Fig. 3.21 shows that it only takes one time period to get a signal in quadrature with v_G. The voltage $\widetilde{v}_G$ represents the filtered voltage v_G.

From Figs. 3.18 and 3.19, the structure of the used PLL is given in Fig. 3.22.

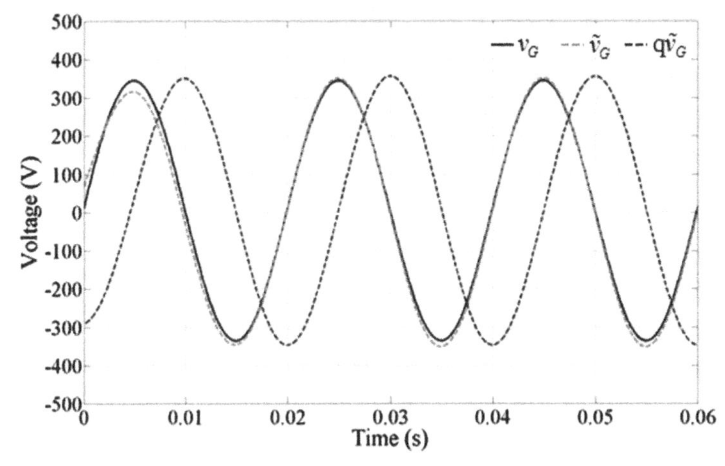

Figure 3.21 Experimental validation of the block structure "creating an orthogonal component."

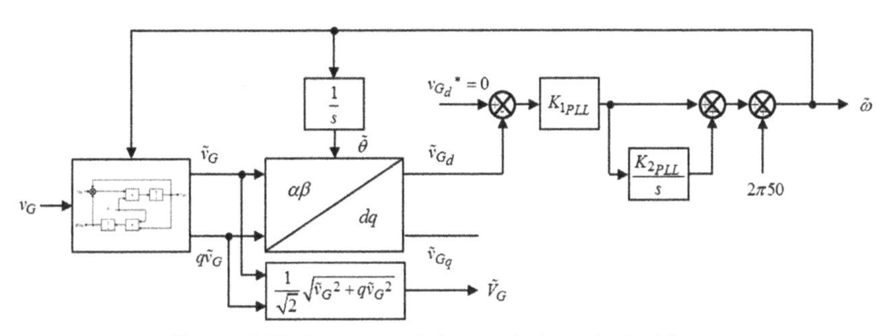

Figure 3.22 Structure of the used phase-locked loop.

The integral corrector gains $K_{1_{PLL}}$ and $K_{2_{PLL}}$ are not optimized in this study; this is not an essential point of the DC microgrid study. However, note that these gains were experimentally determined by tuning to obtain the best compromise between speed and accuracy ($K_{1_{PLL}} = 1$ and $K_{2_{PLL}} = 10$).

Fig. 3.23a and b, present temporal responses of $\widetilde{v}_{G_d}$, $\widetilde{v}_{G_q}$, and $\widetilde{\theta}$ to a level of v_G (for 230 V effective value). These figures show that it takes approximately three time periods to obtain a good estimate of θ ($\theta = \omega t$). Fig. 3.23a also shows that the proportional integral corrector (Fig. 3.22) cannot totally reject the disturbance, which is due to a slight difference in amplitude between $\widetilde{v}_G$ and $q\widetilde{v}_G$. This figure also shows that to properly use the $\widetilde{V}_G$ information, the signal must be filtered through a low-pass filter.

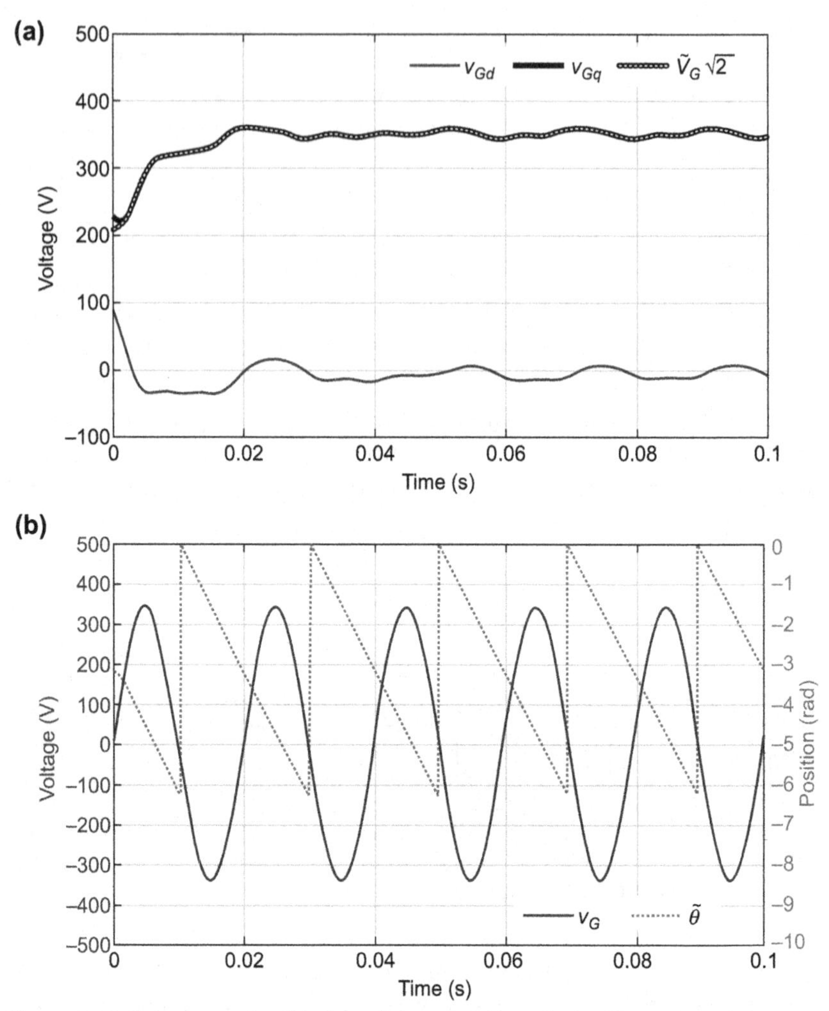

Figure 3.23 Experimental validation of the used phase-locked loop: (a) voltage temporal responses and (b) grid voltage and position.

4.2 Experimental Evaluation of the Phase-Locked Loop Implementation

It seems important to remember that because of safety of goods and people, the actual PV sources used cannot be directly connected to the public grid in the university laboratory. Therefore, for the experimental validation, a 3-kVA linear amplifier of four quadrants is used as a public grid emulator. This experimental validation is achieved with a damping coefficient ζ_G equal to $\sqrt{2}/2$ and a closed-loop cutoff frequency equal to 400 Hz (time constant in closed loop is $\tau_{BF_G} = 1/(2\pi 400)$ s). The damping coefficient was chosen for the best compromise between speed and accuracy. The closed-loop cutoff frequency has been experimentally obtained. Indeed, this value was increased until the gains K_{1_G} and K_{2_G} produce a noise at the edge of acceptability (ie, just before the limit of the instability). The theoretical approach does not take into account the noise related to the cutting of the power switches; it is a phenomenon that comes highly limit the earnings of all the system drives. Fig. 3.24 gives the evolution of the

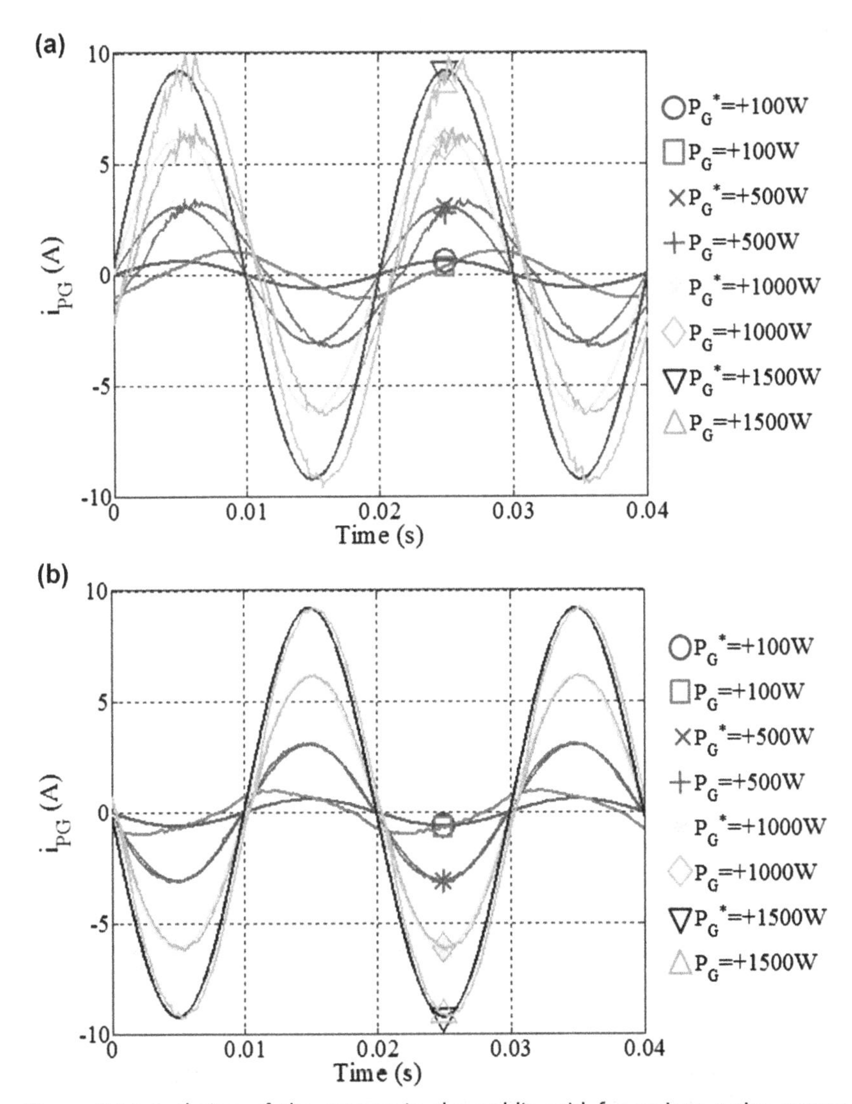

Figure 3.24 Evolution of the current in the public grid for various active power references: (a) absorption and (b) injection.

current i_G in the public grid for various active power references of absorption and injection.

Fig. 3.24a shows that for low active power reference, the current no longer follows the set point order. This has the following effects: $P_G \neq P_G^*$ and $Q_G \neq Q_G^* \neq 0$ (with $P_G^* = V_G I_G^* \cos \varphi^* = V_G I_G^*$). This phenomenon improves when P_G^* increases. Fig. 3.24b shows the same phenomenon but with a quicker and more significant improvement. Regarding the study specifications, it is possible to state that this connection works better in absorption than in injection. This problem is related to limiting the bandwidth of the 400-Hz corrector.

5. CONCLUSIONS

In this chapter the backup resources responding to the urban DC microgrid requirements were presented. Some general notions on electrochemical batteries and capacitors, fuel cells, and microturbines were first given, whereas important characteristics of lead-acid storage, diesel generators, and utility grid connection were more deeply analyzed and studied.

Regarding the electrochemical storage based on lead-acid batteries, three electric type models have been studied and analyzed: the static model, dynamic model, and purely experimental model. The main objective is to choose or build a reliable model that, from a given current, calculates the *SOC* and obtains the voltage. In the end, the following conclusions are made:

- The static model does not take into account the variation of the internal resistance according to *SOC* and temperature.
- The dynamic model takes into account the nonlinearity of its parameters based on *SOC* and the current direction; nevertheless, parameter mapping should be developed because the parameters vary according to *SOC, SOH,* and the amplitude of the current imposed.
- The purely experimental model proposed in this study is not sufficiently developed to respond satisfactorily to the given objective.

Therefore, to obtain information on the voltage value from a given current and taking account of the *SOC*, the static model is chosen to be used for the microgrid control. To summarize, this chapter provided a model for knowledge leading to a better experimental control of the storage system in the proposed microgrid.

The diesel generator is the most common electricity generator used in building-integrated microgrids. Although it is a highly nonlinear system, a brief analysis of operating modes and costs involved can draw conclusions about its use in a microgrid.

The diesel generator is a nonlinear device for which the downtime, delays, and nonlinear behavior are difficult to control; it requires a high-order modeling. However, knowing that the urban DC microgrid study focuses on power balancing and smart grid communication, it is considered that the system response can be studied without the development of this model. The other issues of the diesel generator are low dynamic response; poor fuel efficiency; and increased maintenance for low-power operating mode, pollutants emissions, facility cost and maintenance, and optimal sizing. Therefore, in this study on urban DC microgrids, the diesel generator is considered only for off-grid operating mode for backup power for a limited period.

The last backup resource is the utility grid connection presented as the main backup power for the grid-connected operating mode of the microgrid. The utility grid connection was studied through a PWM control and a conventional PLL structure used from a single-phase signal. To not generate reactive power, a unity power factor was imposed. However, the experimental validation of the proposed implementation highlighted that for low active power reference, the current does not follow the set point. This phenomenon is much more amplified in the case of energy absorption than energy injection. This is mainly because of the limitation of the bandwidth of the chosen controller. Therefore, so as to improve connection to the public grid during the absorption or injection at low power, the use of a resonant controller was proposed.

REFERENCES

[1] Linden D, Reddy TB. Handbook of batteries. McGraw-Hill; 2001.
[2] Larminie J, Lowry J. Electric vehicle technology explained. John Wiley & Sons; 2003.
[3] Berndt D. Valve-regulated lead-acid batteries. J Power Sources 2001;100(1−2):29−46.
[4] Barbarisi O, Vasca F, Glielmo L. State of charge Kalman filter estimator for automotive batteries. Control Eng Pract 2006;14(3):267−75.
[5] Gu WB, Wang CY, Li SM, Geng MM, Liaw BY. Modeling discharge and charge characteristics of nickel-metal hydride batteries. Electrochim Acta 1999;44(25):4525−41.
[6] Pudney P. Optimal energy management for solar-powered cars. 2000 [Ph.D. thesis of University of South Australia].
[7] Chan HL, Sutanto D. A new battery model for use with battery energy storage systems and electric vehicles power systems. In: IEEE power engineering society winter meeting conference; 2000.
[8] Dürr M, Cruden A, Gair S, McDonald JR. Dynamic model of a lead acid battery for use in a domestic fuel cell system. J Power Sources 2006;161(2):1400−11.

[9] Mauracher P, Karden E. Dynamic modelling of lead/acid batteries using impedance spectroscopy for parameter identification. J Power Sources 1997;67(1–2):69–84.
[10] Kuhn E, Forgez C, Lagonotte P, Friedrich G. Modelling Ni-mH battery using Cauer and Foster structures. J Power Sources 2006;158(2):1490–7.
[11] Piller S, Perrin M, Jossen A. Methods for state-of-charge determination and their applications. J Power Sources 2001;96(1):113–20.
[12] Ciobotaru M, Teodorescu R, Agelidis VG. Offset rejection for PLL based synchronization in grid-connected converters. In: Applied power electronics conference and exposition; 2008.

CHAPTER 4

Direct Current Microgrid Power Modeling and Control

1. INTRODUCTION

The direct current (DC) microgrid global system is composed of multisource power local generation coupled with a control system [1,2]. Fig. 4.1 shows this configuration. A DC microgrid operates in grid-connected mode and off-grid mode.

The multisource power system is composed of photovoltaic (PV) sources, storage, a diesel generator, and the public grid all connected to a common DC-link bus through their dedicated converters. The DC bus feeds a DC load, which could represent DC controllable building appliances. The PV sources' produced energy is intended for the priority of self-consumption, but excess power can be sold back to the public grid. To obtain a reliable power distribution, this system is safe because of the grid

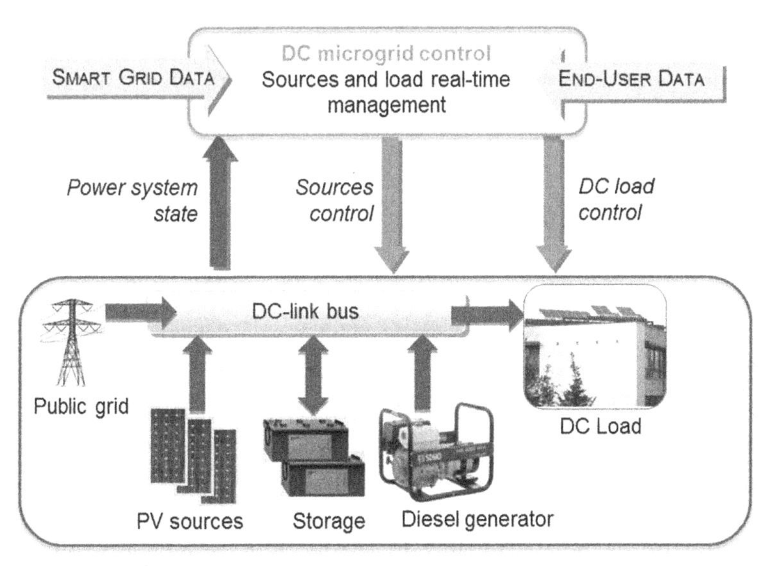

Figure 4.1 DC microgrid global overview. *DC*, direct current; *PV*, photovoltaic.

Urban DC Microgrid
ISBN 978-0-12-803736-2
http://dx.doi.org/10.1016/B978-0-12-803736-2.00004-9

connection for a further supply in case of need and by means of storage and a diesel generator that are involved in smoothing the power balance. The power system is designed to ensure continuous supply to the load, but load shedding actions may occur. Thus the energy management focuses on power balancing with respect to physical limits of the power sources and takes into account the grid availability and energy pricing transmitted by smart grid messages. Therefore, to easily meet objectives such as end-user demand, smart grid integration, dynamic pricing, energy cost minimization, continuous load supply, or load shedding/restoration, DC microgrid control must interact with metadata (smart grid data, weather data, building data, end-user data, etc.). However, this control configuration must be designed to represent a bottom-up approach that permits communication with the public grid and end users without increasing the basic algorithm implemented to ensure power balancing.

The DC microgrid global system can be seen as a hybrid dynamic system, defined as a system with continuous-time dynamics interacting with discrete-event dynamics. The hybrid behavior consists of continuous-time power system behavior controlled by discrete logic/event and control system outputs.

2. FUNCTIONS OF THE POWER SYSTEM CONTROL

DC microgrid control requires the following functions: power balancing, interacting with the smart grid and the end user, and energy management [3,4]. These aspects must be considered and integrated in the control strategy.

2.1 Power Balancing Principle

The multisource power system shown in Fig. 4.2 comprises a DC load; two reversible sources, grid and storage; and two unidirectional sources, a PV array (PVA) and a diesel generator. The power electronic interface is composed of a six-leg power converter (B_1 to B_6) and a set of inductors and capacitors to ensure compatibility between the different elements that are added. The system extracts maximum power from the PVA and manages the power transfer to directly feed the DC load (building) with respect to the available storage level and taking into account the public grid connection or the diesel generator. In case of insufficient energy toward the load, system security is ensured because of the grid connection and by means of storage and a diesel generator, which are primarily involved in

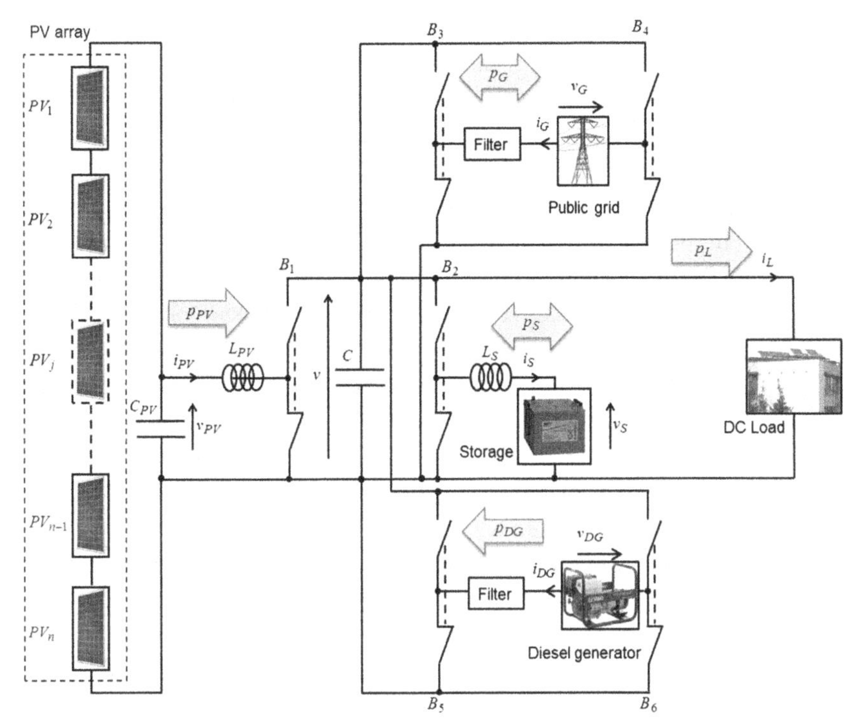

Figure 4.2 Multisource power system electrical schema. *DC*, direct current; *PV*, photovoltaic.

smoothing of the requested power. If there is any excess PV power, the grid connection provides the possibility to trade it back [5].

The PVA is composed of several PV panels electrically coupled in series. The power supplied by PVA depends on the solar irradiance, the PV cell temperature, the array voltage, and the current through the PVA. To maximize the produced energy from the PVA, a maximum power point tracking (MPPT) method is required to find and maintain the peak power. Chapter "Photovoltaic (PV) Source Modeling" presented an experimental comparison among four MPPT algorithms in strictly the same operating conditions and showed that the MPPT energy efficiencies are very similar. Because of the implementation simplicity and low cost, the perturb and observe (P&O) method is chosen to be applied. To extract the maximum power of PVA, an impedance adapter is used. It consists of a C_{PV} capacitor, an L_{PV} inductor, and a B_1 power converter leg; v_{PV} and i_{PV} are the PVA output voltage and current, respectively. The DC-link bus consists of a C capacitor, which is a common element for power sources and DC load.

The DC bus can directly supply the DC load. The DC load is modeled like a current source and i_L is the load current. The DC bus voltage v is considered to be allowed as the DC load voltage. The DC load is not controllable, but it can be limited by means of load shedding if necessary. The storage is considered as a voltage source v_S and is coupled with the DC bus through an inductor L_S and B_2 power converter leg; i_S is the storage current. The public grid is connected to the DC bus through the two-leg power converter (B_3 and B_4) and a filter; v_G and i_G are the public grid voltage and current, respectively. Regarding the power grid control, no reactive power control is involved: grid current i_G is always controlled in phase or in opposite phase with grid voltage v_G. The synchronization between i_G and v_G is achieved by phase-locked loop. The diesel generator is connected to the DC bus through the two-leg power converter (B_5 and B_6) and a filter; v_{DG} and i_{DG} are the diesel generator output voltage and current, respectively.

The following power expressions are used: $p_L = v \cdot i_L$ is the power absorbed by the DC load, $p_{PV} = v_{PV}\ i_{PV}$ is the power supplied by the PVA, $p_S = v_S \cdot i_S$ is the storage system transfer power, $p_G = v_G \cdot i_G$ is the grid connection transfer power, and $p_{DG} = v_{DG} \cdot i_{DG}$ is the diesel generator transfer power. For the power flow shown in Fig. 4.2, the following sign conventions are assumed: p_G and p_S are the positive value means to receive power and the negative value means to supply power, respectively; p_L, p_{PV}, and p_{DG} are always considered positive. DC microgrid system modeling is based on this sign convention.

In a steady-state operation, the output voltage v should be constant. According to the assumption that there are not losses through the inductors and neglecting the total losses of power converter legs (B_1 to B_6), the power balancing can be expressed by Eq. [4.1]:

$$p_{PV} = p_G + p_{DG} + p_S + p_L \qquad [4.1]$$

The PVA power p_{PV} changes with solar irradiance and temperature. The load power p_L changes according to the building's power demand. On the other side, the powers p_G, p_{DG}, and p_S can be controlled using the corresponding current reference given to the power system controller.

It is considered that for the grid-connected mode only the grid and storage should balance the power error between production and consumption. For off-grid mode the power error is balanced by means of the diesel generator and storage.

Power balancing is operated by stabilizing the DC bus voltage v. Thus the DC bus power adjustment p^* is defined by Eq. [4.2]:

$$p^* = p_{PV} - p_L - C_P(v^* - v) - C_I \int (v^* - v)dt$$
$$\text{with} \begin{cases} p^* = p_G^* + p_S^* \text{ for grid-connected mode} \\ p^* = p_{DG}^* + p_S^* \text{ for off-grid mode} \end{cases} \quad [4.2]$$

with v^* the DC bus voltage reference, C_P the proportional gain, and C_I the integral gain of the proportional integral controller. The PVA power p_{PV} and the load power p_L are considered as known disturbances for DC bus voltage control, and they are added in this equation to reject disturbances. In Eq. [4.2] the voltage v^* is imposed, the power p_{PV} is a function of solar irradiance and PV cell temperature, and the p_L is imposed by the DC load. The required power reference p^* is the output of the controller for stabilizing the DC bus voltage. Therefore, as mentioned before, the system should be able to manage powers; the power reference p^* must be distributed to the grid and storage for the grid-connected operating mode and to the diesel generator and storage for the off-grid operating mode. Thus, as a first approach, storage power reference p_S^* may be calculated in both cases according to a distribution coefficient K_D as in Eq. [4.3]:

$$p_S^* = K_D \cdot p^* \quad [4.3]$$

with $0 \leq K_D \leq 1$ and taking into account that p_S^*, p_G^*, and p_{DG}^* should not exceed the corresponding physical element limitation. The distribution coefficient K_D is constrained between 0 and 1, although other values still work. The reason is as follows: PV feed-in tariffs are still incentives nowadays, and grid charged storage and grid power injection by storage are forbidden by almost all power providers. In this study it is assumed that the energy storage does not inject power into the grid. Therefore constraining K_D between 0 and 1 ensures that power grid injection is only from PVA production.

In Eq. [4.3] the storage is considered always available within its physical limits—charge or discharge operating mode. Otherwise the storage power is $p_S^* = 0$. For given values of K_D an infinite number of solutions may exist, but for energy cost optimization K_D values must be calculated by optimization under constraints corresponding to techno-economic criteria, the weather forecast, end-user criteria, the load power forecast, etc. This optimization is proposed in Chapter "Direct Current Microgrid

Supervisory System Design." However, for this first approach the goal is to verify the feasibility of the proposed multisource energy management. Therefore the priority is given to the storage system with respect to the public grid [6] or diesel generator operating constraints. Therefore $p_S^* = K_D \cdot p^*$ is first calculated, and then p_G or p_{DG} are calculated with consideration to storage physical limitations, which means $p_S^* = 0$ if the storage reaches either its upper limit or its lower limit for state of charge (SOC) or its maximum power.

In this chapter, for the validation tests only two cases are retained: $K_D = 1$ and $K_D = 0.8$. If $K_D = 1$, then storage is a priority in charging and discharging mode and $p_S^* = p^*$. Once the storage reaches its upper limit or lower limit the public grid or the diesel generator take over for the power balancing. Thus, starting this instant, depending on the operating mode, $p_G^* = p^*$ or $p_{DG}^* = p^*$. For $K_D = 0.8$, the power reference p^* is shared between the storage and the public grid or the diesel generator. For example, during the grid-connected operating mode, $p_S^* = 0.8 \cdot p^*$ and $p_G^* = 0.2 \cdot p^*$. Once the storage reaches its upper limit or lower limit SOC, or its maximum power, the public grid takes over for the power balancing and $p_G^* = p^*$.

The distribution coefficient K_D describes the power distribution between grid and storage or diesel generator and storage for a general case. However, special situations must be taken into consideration. In those cases powers p_G^* or p_{DG}^* and p_S^* have to be modified according to their corresponding limits, either physical or imposed by the DC microgrid control, resulting in extra power balancing considerations. For example, as the strategy takes into account all operational constraints imposed by the public grid via the smart grid (eg, limitations on the power drawn from the grid to the system), if p_G^* is higher than the public grid limitations then p_G^* becomes equal to the imposed limitation and the load shedding should be operated to meet power balancing. These special situations are described in the subsequent DC microgrid power system modeling study.

2.2 Smart Grid Interaction

To better manage the energy produced by the PVA with respect to public grid availability, the multisource power control system must be able to communicate with the smart grid [7,8]. The communication is mainly based on a high-speed network integrated in the smart grid backbone. This main line connects many communication segments and applications, such as the territorial public service network control center (ancillary services, energy management and control), the data acquisition center, distributed energy resources, the substation of the public grid, etc. For low-speed communication devices, such as smart routers, a low-speed link can connect the urban microgrids and many smart meters to the backbone. Thus microgrid power system control is seen as a node of one of these segments and does not require the bandwidth and performance needed in the backbone [9].

It is considered that the main messages to send or to receive by means of the future smart grid concern the following subjects: the peak consumption period, demand-side management, demand response constraints, dynamic energy pricing, dynamic time-of-use tariff, public grid availability regarding supply and injection, and load shedding [10–12].

The demand-side management focuses on energy efficiency, such as more efficient appliances, and refers to modifying end users' energy demand through various ways, either explicit orders or financial incentives, such as demand response and time-of-use tariff. In general, the grid power capability must satisfy the peak consumption and provide a supplementary security margin for unforeseen events. By reducing peak consumption, demand-side management can significantly reduce the need for expensive new power plants and other spinning reserves as well as decrease the peak energy price and volatility of price in the electric market. This phenomenon is because the generation cost increases exponentially near maximum generation capacity. A small reduction in demand may result in a big reduction in generation cost and in turn a reduction in the price of electricity. To summarize, simply reducing or shifting peak consumption is much less expensive than building new power plants and introducing energy storage.

The demand response is defined as a mechanism that may affect the load using an explicit signal (message) broadcasted by transmission or distribution system operators. Most of the time the signal is not necessarily instantaneous, which may refer to a situation for the next hour(s) or day. It is mainly performed for emergency situations before load shedding and for ancillary services. Demand response actually encourages consumers to reduce their power demand. Nevertheless, it can be used sometimes to increase power demand in the case of high production and low demand; therefore it can be also an encouragement to store energy.

Other useful worldwide mechanisms to perform peak shaving are dynamic pricing or time-of-use tariffs. This refers to the energy tariff variations versus time: energy price is more expensive in peak hours than in off-peak hours. Therefore the end users are encouraged to shift uncritical load in the off-peak hours. Energy consumption for peak hours and off-peak hours is recorded separately to calculate the energy invoice. This requires a meter able to respond to grid switching signals between peak and off-peak hours. In the case of energy dynamic pricing, expected to be largely spread around the world in the coming years (starting in January 2016 for the French power grid), a smart meter should be designed and implemented; its main ability is linked to the smart grid communication that is supposed to transmit messages concerning energy tariffs.

Another issue of the current public grid is linked to the power injection from distributed renewable sources. Thus, because of the intermittency and fluctuation characteristics of renewable production, considerations must be taken to handle injection problems facing future high-level penetration of renewable sources. Volatile renewable productions can result in new challenges in the grid for power balancing and stability problems. Hence it is also essential to regulate the injected power. In addition, the distributed energy production should be able to inform the public grid of power demand and power injection forecasts.

Therefore, to provide capability for grid regulation in supply and injection cases, smart grid communication is considered and involved in the proposed DC microgrid power control. It is assumed that the smart grid informs the power system control at different times on the operational conditions and constraints of the public grid, including dynamic energy pricing, dynamic time-of-use tariffs, the peak consumption period, demand response constraints, demand-side management, maximum power supply limits, and/or power injection for the end user, etc. The dynamic mechanism is proposed in the form of bidirectional smart grid messages.

3. DIRECT CURRENT MICROGRID POWER SYSTEM MODELING CONSIDERING CONSTRAINTS

To analyze the system with a goal of establishing the overall control strategy, modeling and simulation are necessary. The DC microgrid power system is considered a hybrid dynamic system that combines continuous-time dynamics with discrete-event dynamics. Therefore the behavioral modeling of a global system is proposed by means of interpreted Petri Net (PN) formalism, which is a powerful graphical and mathematical tool for describing and analyzing different hybrid dynamic systems.

Thus interpreted PN modeling is used to split continuous element behavior into discrete states, permitting a bottom-up modeling to facilitate the design of overall control strategy by studying each detailed, discrete state [13]. Control functions are then studied and integrated into each discrete state.

The simulation of power system control has a hybrid aspect. The continuous and discrete aspects coexist and interact with each other. The power system components behave with continuous dynamics whereas the control strategy by PN modeling behaves in a discrete way [14]. To perform numerical hybrid simulation tests, Simulink and Stateflow developments in the MATLAB environment are proposed. Simulink helps to model the continuous dynamics of the system and Stateflow is used to specify the discrete control logic and the modal behavior of the system. Stateflow design language is based on the concept of hierarchical automata from Statecharts and allows developing the PN model through the approximation made between states/places and switching/firing transition conditions. Thus, in this section, at the same time of PN modeling, Stateflow modeling is given, which permits simulating the PN model in the MATLAB-Simulink environment.

3.1 Introduction to Petri Net Modeling

Defined for the first time in 1962 by Carl Adam Petri, PN modeling is a powerful graphical and mathematical tool for describing and analyzing different hybrid dynamic systems. Wide PN application areas are found in performance evaluation, communication protocols, discrete-event systems, flexible manufacturing/industrial control systems, digital filters, and asynchronous circuits and structures [15]. PN modeling allows behavioral analysis, discrete-event simulator and controller design, and performance evaluation. PN formalism offers an exact mathematical definition of their execution semantics with a well-developed mathematical theory for process analysis. It is possible to set up state equations, differential algebraic equations, and other mathematical models governing the behavior of systems.

To quickly summarize, a PN is defined as a five-tuple, $PN = \langle P, T, F, W, M_0 \rangle$, where $P = \{P_1, P_2, \cdots, P_m\}$ is a finite set of places, $T = \{T_1, T_2, \cdots, T_n\}$ is a finite set of transitions, $F \in (P \times T) \cup (T \times P)$ is a set of arcs (flow relation), $W : F \rightarrow \{1, 2, 3, \cdots\}$ is a weight function that assigns a weight to each arc (numbers superior to one over flashes), and $M_0 : F \rightarrow \{1, 2, 3, \cdots\}$ is the initial marking (graphically token within places) $P \cap T = \emptyset$ and $P \cup T \neq \emptyset$.

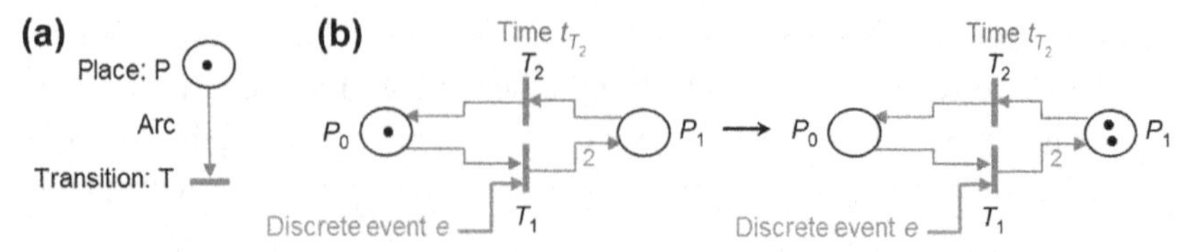

Figure 4.3 (a) Autonomous PN and (b) timed-interpreted Petri Net.

Thus a PN graphic contains four kinds of objects: place (state variable), transition (state transformer), token (indicator for active state), and arc. The directed arcs connect places and transitions and indicate the token flow when transition is fired. A transition can be enabled if its upstream or input places have enough tokens according to corresponding arc weights. An enabled transition may fire at any time. Fig. 4.3a presents a simple autonomous PN.

As a timed-interpreted PN, each transition could have conditions related to it, such as a discrete event or an associated time, and the transition is fired only if the transition is enabled and its conditions are satisfied. When fired, the token(s) in the input place(s) are moved to output place(s), according to arc weight and place capacities, as shown in Fig. 4.3b. In this case places occupied by tokens are interpreted as indicating the state of the system or available resources in the system. As aforementioned, an enabled transition may or may not be fired depending on the additional firing condition related to the transition. For example, the firing condition relating to T_1 may be a discrete event *e*. The transition T_1 is enabled with enough tokens presented in input place P_1 in Fig. 4.3b but cannot be fired while the transition condition is not satisfied. Firing transition T_1 means removing the tokens in the input place, the number of which depends on the arc weight, and adding tokens to the output place, for which the number also depends on the corresponding arc weight, as shown in Fig. 4.3b. After firing the system status changes. For the same example, in the case of timed transition as for T_2, after t_{T_2} time duration, the transition T_2 is fired but only one token leaves the place P_2 arriving in place P_1 and so on.

Starting with an initial marking, given by initial tokens of places, the graph-based structure simulates the dynamic behavior of a system by continuously firing enabled transitions. Many static and dynamic properties of a PN may be mathematically proven. Concerning the microgrid modeling study, the directed arc's weight is considered to be 1. Furthermore, it is not necessary to detail and present the basic analysis of PN models in this modeling study. However, regarding the PN models presented hereafter, based on place invariants and transition invariants, we note that the structural verification proves that each model is live (there exists an initial marking for which the PN is live), bounded (token limitation for all initial marking), and without deadlocks or conflicts.

3.2 Smart Grid Interaction Modeling

To use the full production of the PV energy, with respect to the grid availability, the DC microgrid control system must be able to communicate with the public grid. Hence, by means of a smart grid communication device, the control system is informed on the energy price and the limitations of power supply and the power injection at different times; some grid-operating modes could be defined. In addition, this device is able to communicate information back to the public grid concerning monitoring and forecasting injection or consumption.

The smart grid interaction is modeled with a timed-interpreted PN represented by two places (P_0, P_1) and two transitions (T_1, T_2), as illustrated in Fig. 4.4 [16]. The P_0 place represents the smart grid in a standby mode and the P_1 place is reached when the operator gives the grid-operating mode. This event is the firing condition of the transition T_1. After the t_G time, measured in minutes or hours from the arrival of the token in place P_1, the timed transition T_2 can be fired and the power control system is informed of the switching operating mode instant.

Once the control system is informed about the grid-operating mode, it controls the power system according to the specific grid-operating mode and each element constraint.

3.3 Grid-Operating Mode Modeling

The public grid companies limit the usage of energy in some areas during the peak-hour consumption period. Another fact is that sometimes the

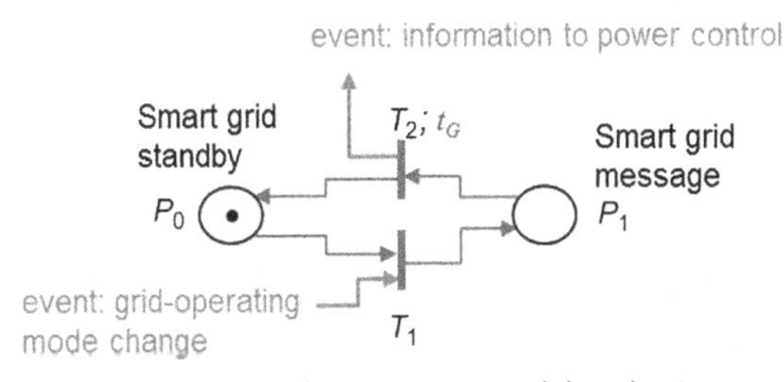

Figure 4.4 Smart grid interaction modeling by Petri Net.

price of electricity could be higher than normal; therefore end users are supposed to use less electricity. These two grid-operation modes are taken into consideration.

To illustrate the differences between grid-operating modes, it is assumed that the load has P_{L_max} as contractual subscribed power and the PVA has P_{PV_STC} as peak power provided by the PV panel manufacturer under industrial standard test conditions (STC). The values of P_{L_max} and P_{PV_STC} are constant values. Hence, the grid power evolution for the supply operating mode, p_{G_S}, is defined as $0 \leq p_{G\_S} \leq P_{G\_S\_lim}$, with $P_{G\_S\_lim} \leq P_{L\_max}$. The $P_{G\_S\_lim}$ is imposed by the grid-operating mode. Regarding the injection-operating mode, the injected grid power p_{G_I} evolution is defined as $0 \leq p_{G\_I} \leq P_{G\_I\_lim}$ with $P_{G\_I\_lim} \leq P_{PV\_STC}$. It is assumed that the storage energy is not used for injection into the grid. The injected power may have constrained limits, but to simplify the model at this primary step it is assumed that there is no imposed injection limit. Thus the two grid-operating modes are:

- *Normal grid-operating mode*: $P_{G\_S\_lim} \geq P_{L\_max}$ and $P_{G\_I\_lim} \geq P_{PV\_STC}$. The load may use maximum power P_{L_max} from the grid, and the system may inject the maximum power into the grid. During this mode grid supply and injection powers are not limited. The grid is always able to supply the load, and it is controlled to cooperate with storage supply. On the other hand, the grid power injection can absorb the total PVA power. In addition, energy price is considered normal.
- *Constrained grid-operating mode*: $P_{G\_S\_lim} < P_{L\_max}$ and $P_{G\_I\_lim} < P_{PV\_STC}$; therefore the smart grid message constrains the grid usage. During grid-constrained mode the grid power should respect the supply and injection limits. Once the calculated reference power p_G^* exceeds the supply or injection limits, p_G^* is constrained to the corresponding limit. If grid power is constrained, then the load power demand may suffer a shortage of power supply and should be constrained to ensure power balance in the system. Otherwise the load can use maximum power P_{L_max} from the grid, but the energy price is normal only for the part less or equal to $P_{G\_S\_lim}$. For the part that exceeds $P_{G\_S\_lim}$ the energy price becomes higher. The end user is encouraged to use less than $P_{G\_S\_lim}$. The system may inject maximum power into the grid.

The grid PN model is shown in Fig. 4.5, and definitions of places that are associated with the grid-operating modes are given in Table 4.1.

Because the smart grid gives information about the grid-operating mode, this message is regarded as an event associated with the T_2

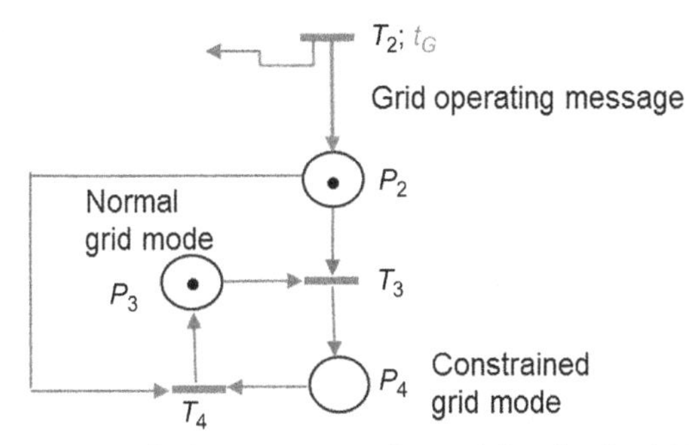

Figure 4.5 Grid-operating mode modeling by Petri Net.

Table 4.1 Definitions for grid-operating mode in the Petri Net model

Name	Description
P_2	Grid-operating message after t_G
P_3	$P_{G\_I\_\lim} \geq P_{PV\_STC}$ and $P_{G\_S\_\lim} \geq P_{L\_\max}$
P_4	$P_{G\_I\_\lim} < P_{PV\_STC}$ and $P_{G\_S\_\lim} < P_{L\_\max}$
T_3	T_3 is enabled if P_2 and P_3 are both occupied with token
T_4	T_4 is enabled if P_2 and P_4 are both occupied with token

transition. Moreover, the switching operating mode event will occur after the time t_G. In this model the transition T_3 and T_4 firing conditions are given only by the token presence in upstream places. The Stateflow model of the grid-operation modes and the transition conditions are illustrated in Fig. 4.6.

3.4 Storage Operating Mode Modeling

The storage is able to charge and discharge when its SOC is in the normal range $SOC_{\min} \leq SOC \leq SOC_{\max}$, with $SOC_{\min}$ and $SOC_{\max}$ representing, respectively, the minimum and maximum SOC limits for storage operation. Because storage capacity is limited, the SOC, indicating the storage charging level ranging from 0% to 100%, is an important parameter to avoid storage damage by overcharging and overdischarging. Many complex SOC calculation algorithms have been proposed using the Extended Kalman Filter, an H_∞ observer, and neural networks; however, these methods depend on several parameters' identification that is difficult to tune. For the power control of a microgrid, a simplified SOC calculation

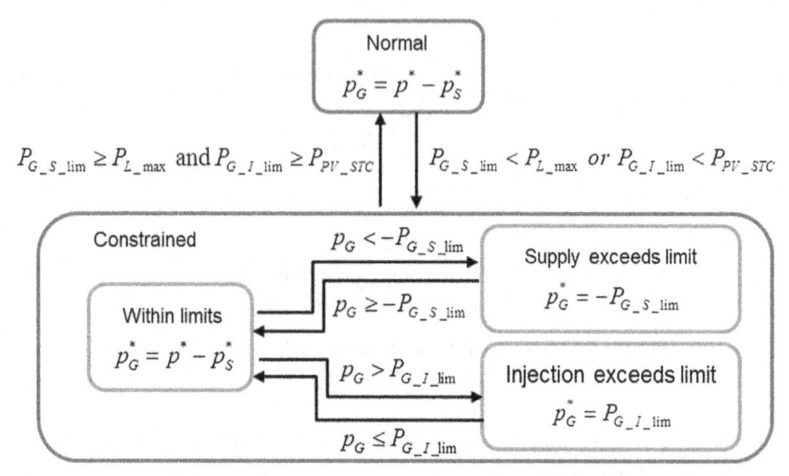

Figure 4.6 Grid behavior Stateflow model.

is expressed by Eq. [4.4], which is robust yet satisfies the requirement of the proposed control:

$$SOC = SOC_0 + \frac{1}{3600 \cdot v_S \cdot C_{REF}} \int p_S \cdot dt \qquad [4.4]$$

with C_{REF} the storage nominal capacity (Ah), v_S the storage voltage, and SOC_0 as the initial SOC.

In addition, because the high charging rate can largely affect the storage life, the storage power p_S is limited as in Eq. [4.5]:

$$-P_{S\_max} \leq p_S \leq P_{S\_max} \qquad [4.5]$$

with P_{S_max} the storage power limitation.

In grid-connected mode the storage charging is operated only by PVA production, whereas in off-grid mode the storage can also be charged by a diesel generator. There are three storage operating modes: charge, discharge, and turnoff.

- *Storage in charge mode*: $SOC < SOC_{max}$. During the charge, before the SOC_{max} is reached, the storage power should be controlled following K_D, which is supposed to be determined and transmitted by the power control, to balance power in the system as $p_S^* = K_D \cdot p^*$.
- *Storage in discharge mode*: $SOC > SOC_{min}$. During discharge, before the SOC_{min} is reached, the storage power should be controlled following K_D, given by the power control, to balance power in the system as $p_S^* = K_D \cdot p^*$.

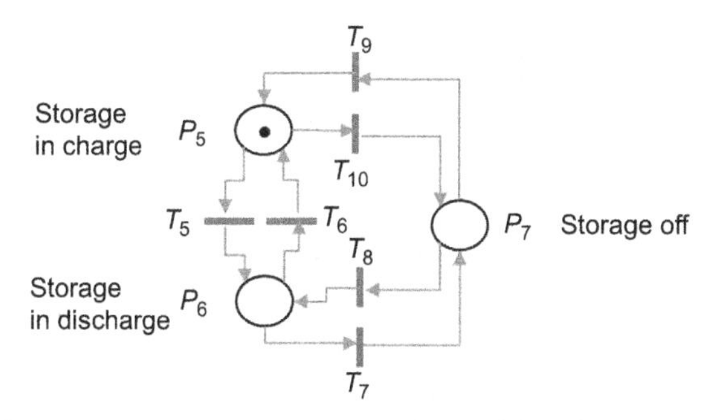

Figure 4.7 Storage operating mode modeling by Petri Net.

- *Storage turn-off mode.* This mode is activated when reaching $SOC_{\min}$ in the discharge case or $SOC_{\max}$ in the charge case but also when reaching P_{S_max}. If necessary, it is also possible for the power control system to send a "turn-off" order. When $SOC_{\max}$ is reached, the storage should be protected from overcharge. Thus, in the case of positive power reference p^*, the storage power reference p_S^* should not be calculated according to K_D but should be imposed on $p_S^* = 0$. In the case of discharge, when $SOC_{\min}$ is reached, $p_S^* = 0$ as well. In the case of $p_S^* = 0$, the grid is supposed to supply or absorb the adjustment power without respect to the K_D value. This is why in grid-connected operation grid power is calculated by $p_G^* = p^* - p_S^*$ instead of $p_G^* = (1 - K_D) \cdot p^*$. Furthermore, for off-grid operation, the three operating modes of the storage are modeled by PN and are shown in Fig. 4.7.

Definitions of places and transition firing conditions are given in Table 4.2.

Table 4.2 Definitions for storage for the Petri Net model

Name	Description
P_5	Charging operating mode
P_6	Discharging operating mode
P_7	Storage turn-off mode
T_5	$SOC > SOC_{\min}$ and $p_{PV} < p_L$
T_6	$SOC < SOC_{\max}$ and $p_{PV} > p_L$
T_7	$SOC \leq SOC_{\min}$ or $p_S = P_{S\_max}$ or "turn-off" order
T_8	$SOC > SOC_{\min}$ and $p_{PV} < p_L$
T_9	$SOC < SOC_{\max}$ and $p_{PV} > p_L$
T_{10}	$SOC \geq SOC_{\max}$ or $p_S = P_{S\_max}$ or "turn-off" order

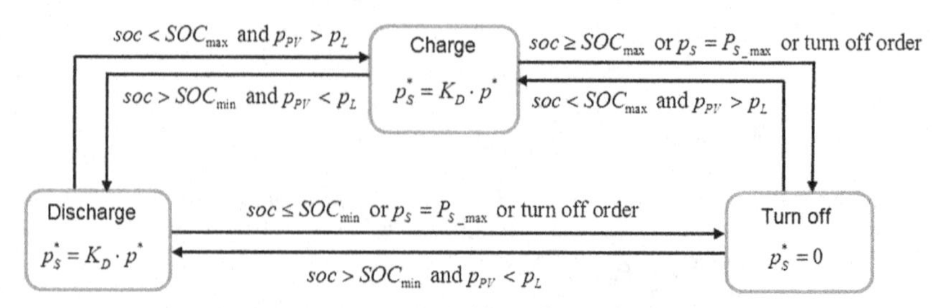

Figure 4.8 Storage behavior of Stateflow model.

The Stateflow model of the storage operation modes and the transition conditions are illustrated in Fig. 4.8.

3.5 Photovoltaic Operating Mode Modeling

A PVA operates mainly at its maximum power point, whatever the solar irradiance g and cell temperature variations, following an MPPT method. In this study the P&O algorithm is applied to impose the current reference. However, when there is not enough solar irradiation or no power demanded by the other elements, the PVA is turned off and $p_{PV} = 0$; otherwise it is supposed to work in MPPT mode with $p_{PV} = p_{PV\_MPPT}$, which represents the MPPT PVA power, or in constrained mode with $p_{PV} = p_{PV\_CONS}$, which represents the PVA constrained power. Hence, three PVA operating modes are considered:

- *Off mode*: $p_{PV} = 0$. When there is not enough solar irradiance, $g < G_{min}$, or no power demanded by the other elements, PVA is turned off; therefore $p_{PV} = 0$. The minimum of solar irradiance, G_{min} (W/m^2), represents the least level for which the $p_{PV} \neq 0$.
- *MPPT mode*: $p_{PV} = p_{PV\_MPPT}$. For $g > G_{min}$ and power demanded by at least one element, the PVA is supposed to operate with the MPPT algorithm.
- *Constrained mode*: $p_{PV} = p_{PV\_CONS}$. In the case of grid injection limitation $P_{G\_I\_lim} < P_{PV\_STC}$ the PVA constrained mode may occur and must be considered. It happens when storage is full and the PVA produces more power than the load power demand plus the grid injection limit. Thus PVA must be able to limit output power to the just needed power represented by the PVA constrained power p_{PV_CONS} instead of outputting the MPPT power p_{PV_MPPT}. PVA constrained power is calculated by the power control $p_{PV\_CONS} = p_L + P_{G\_I\_lim}$ following the PVA power constrained algorithm.

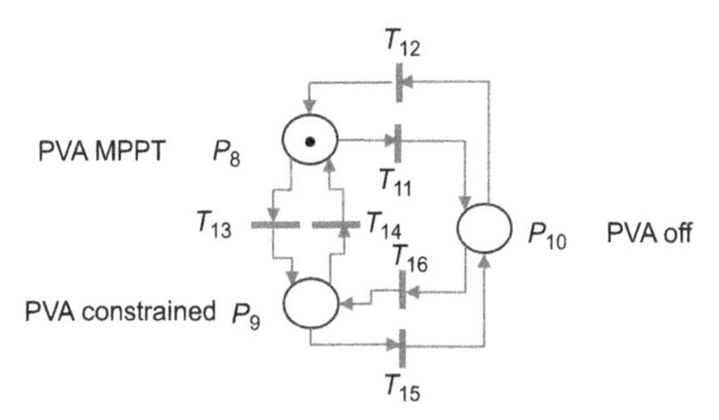

Figure 4.9 PVA operating mode modeling by Petri Net. *MPPT*, maximum power point tracking; *PVA*, photovoltaic array.

The PN model is shown in Fig. 4.9 with the definitions given in Table 4.3.

The Stateflow model of the PVA operation modes and the transition conditions are illustrated in Fig. 4.10.

3.6 Diesel Generator Operating Mode Modeling

The diesel generator is used as a backup source only in the off-grid mode of the DC microgrid. Note that the cogeneration aspect of the diesel generator is not considered in this study. As for all fuel generators based on a thermal machine, because of the low dynamic response, frequent start-stop cycles, and low output power, the diesel generator efficiency is reduced and

Table 4.3 Definitions for the photovoltaic array Petri Net model

Name	Description
P_8	$p_{PV} = p_{PV\_MPPT}$
P_9	$p_{PV} = p_{PV\_CONS}$
P_{10}	$p_{PV} = 0$
T_{11}	$g < G_{\min}$ or ($p_L = 0$ and $SOC \geq SOC_{\max}$ and $P_{G\_I\_\lim} = 0$)
T_{12}	$g \geq G_{\min}$ and $p_{PV} < p_L + \lvert p_S \rvert + P_{G\_I\_\lim}$
T_{13}	$g \geq G_{\min}$ and $p_{PV} > p_L + P_{G\_I\_\lim}$ and $SOC \geq SOC_{\max}$
T_{14}	$g \geq G_{\min}$ and $p_{PV} < p_L + \lvert p_S \rvert + P_{G\_I\_\lim}$
T_{15}	$g < G_{\min}$ or ($p_L = 0$ and $SOC \geq SOC_{\max}$ and $P_{G\_I\_\lim} = 0$)
T_{16}	$g \geq G_{\min}$ and $p_{PV} > p_L + P_{G\_I\_\lim}$ and $SOC \geq SOC_{\max}$

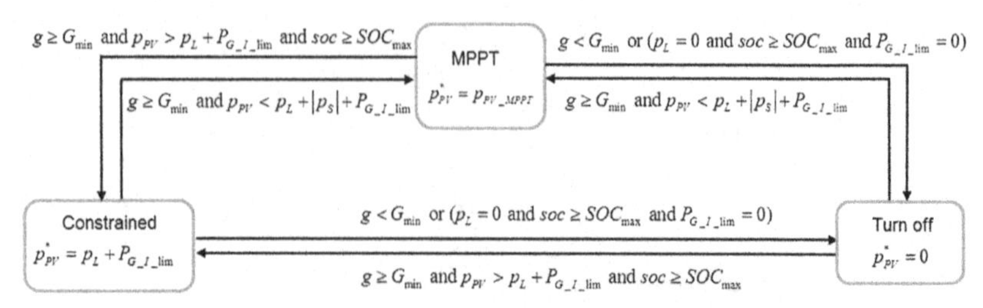

Figure 4.10 Photo Voltaic array behavior Stateflow model.

supplementary cost is introduced. To overcome these issues, in this study the diesel generator is assigned with two operating states: off and output rated power given by Eq. [4.6]:

$$p_{DG} = \lambda \cdot P_{DG\_P} \text{ with } \lambda \in \{0, 1\} \quad [4.6]$$

where P_{DG_P} is the diesel generator rated power. Considering a discrete time instant t_i, from initial t_0 to final t_F, with time interval Δt, a working duty cycle is assigned to diesel generator control as in Eq. [4.7]:

$$p_{DG}(t_i) = p_{DG}(t_{i-1}) \text{ if } rem(t_i/dt_{DG}) \neq 0, \; t_i = \{t_0 + \Delta t, t_0 + 2\Delta t, \ldots, t_F\} \quad [4.7]$$

with dt_{DG} the time duration of the diesel generator working duty cycle and *rem* the function that returns the remainder of the division. Because the diesel generator works in duty cycle, it introduces the case that storage is charged by rated PVA power plus rated diesel generator power. Thus the storage power p_S is limited as given in Eq. [4.5].

Hence, two operating modes are assigned to diesel generator:

- *Off mode*: $p_{DG} = 0$. If there is enough available power, either produced by the PVA or provided by the storage, or there is no power demanded by the load, then the diesel generator should not be started; therefore $p_{DG} = 0$. Otherwise, when the diesel generator is being run, it should be turned off after the working cycle time duration dt_{DG} or for $p_S \geq P_{S\_max}$; therefore in both cases $p_{DG} = 0$.
- *Rated power mode*: $p_{DG} = P_{DG\_P}$. If there is not enough available power, either produced by the PVA or provided by the storage, to supply the load, then the diesel generator is turned on. Therefore for $p_{PV} + p_S < p_L$ the diesel generator is supposed to output its rated power during working at least one duty cycle dt_{DG}.

The PN model of the diesel generator operating mode is shown in Fig. 4.11 with the definitions given in Table 4.4.

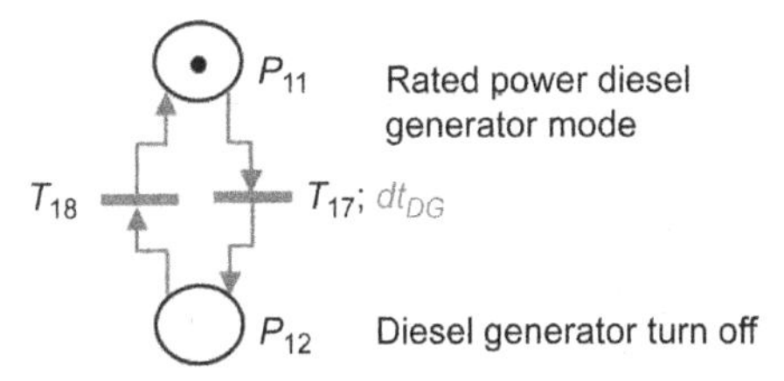

Figure 4.11 Diesel generator operating mode modeling by Petri Net.

Table 4.4 Definitions for diesel generator Petri Net model

Name	Description
P_{11}	$p_{DG} = P_{DG\_P}$
P_{12}	$p_{DG} = 0$
T_{17}	$(p_{PV} + p_S > p_L$ and $dt_{DG})$ or $(p_S \geq P_{S\_max})$
T_{18}	$p_{PV} + p_S < p_L$

The Stateflow model of the diesel generator operation modes and the transition conditions are illustrated in Fig. 4.12.

3.7 Load Operating Mode Modeling

Concerning the load, a limited power p_{L_lim} is introduced to restrict the load power demand for special situations. The DC load consists of critical appliances and interruptible appliances. The critical appliances require a continuous power supply whereas the interruptible appliances can be temporarily shed. Because the load is mainly demanded by the end user, the

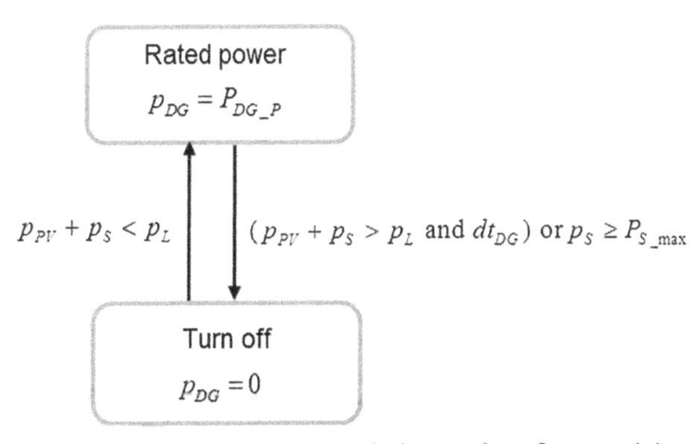

Figure 4.12 Diesel generator behavior Stateflow model.

power system can only control the load by load shedding. To maintain safe supply for the critical appliances, the power system control can send a load shedding signal to disconnect some appliances; therefore load power demand can only reach a limited power level p_{L_lim}. To describe the limit imposed to the load, despite system scale for different building applications, the load shedding coefficient $K_L = p_{L\_lim}/P_{L\_max}$ is defined. Many values of K_L could be physically possible, with $K_L \in [0,1]$. Thus two load operating modes are considered:

- *Normal load mode*: $0 \le p_L \le P_{L\_max}$. The load can demand the maximum power P_{L_max}; therefore load shedding is not needed and $K_L = 1$.
- *Constrained load mode*: $0 \le p_L \le p_{L\_lim}$. According to smart grid messages, in the case when storage is empty and grid supply is constrained, the demanded load power cannot reach P_{L_max}; its evolution is only within a limited power as $0 \le p_L \le p_{L\_lim}$. Therefore if $p_L > p_{L\_lim}$, then the microgrid control must turn off part of the DC load, resulting in load shedding. The load power limit refers to the case when the storage is empty and the capacity of supplying the load by PVA production and grid maximum supply is inferior to load power demand. Therefore $p_{L\_lim} = p_{PV} + P_{G\_S\_lim}$ in grid-connected mode or $p_{L\_lim} = p_{PV} + p_{DG}$ in off-grid mode. Then the power system compares the current load demand with p_{L_lim}; if the load demand is greater than the limit, then the load should be shed. Therefore K_L changes according to the available energy for supplying the load as $K_L = (p_{PV} + P_{G\_S\_lim} + p_{DG})/P_{L\_max}$.

The interpreted PN model of discrete states of load operating mode is shown in Fig. 4.13.

The definitions of places and transition firing conditions for the load PN model are presented in Table 4.5.

The Stateflow model of the load operation modes and the transition conditions are illustrated in Fig. 4.14.

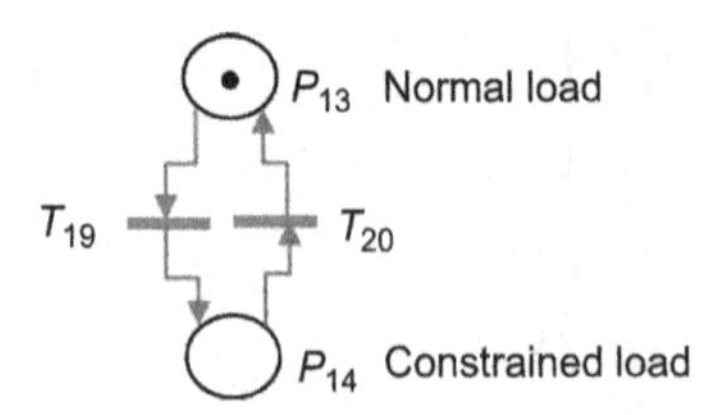

Figure 4.13 Load operating mode modeling by Petri Net.

Table 4.5 Definitions for load Petri Net model

Name	Description
P_{13}	$0 \leq p_L \leq P_{L\_max}$
P_{14}	$0 \leq p_L \leq K_L \cdot P_{L\_max}$
T_{19}	$SOC \leq SOC_{min}$ and $(P_{G\_S\_lim} + p_{DG} + p_{PV} < p_L)$
T_{20}	$SOC > SOC_{min}$ or $(P_{G\_S\_lim} + p_{DG} + p_{PV} \geq p_L)$

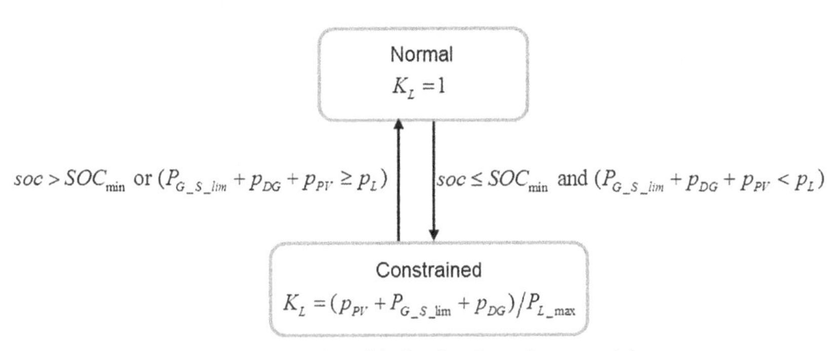

Figure 4.14 Load behavior Stateflow model.

3.8 Direct Current Microgrid Power System Global Behavior by Interpreted Petri Net Modeling

The power control DC microgrid structure has to be designed following PN modeling. This interface includes the operation criteria such as on-grid or off-grid as well as smart grid message communication. The implemented algorithms are supposed to calculate power references for sources with respect to their limitations and the load shedding coefficient.

Fig. 4.15 presents the power system global behavior modeling by PN for grid-connected mode indicated by the place P_i occupied by its token. Let us suppose that the grid is in normal operating mode, the PVA is in MPPT operating mode, the storage is in charging mode, and the smart grid is sending the message to announce the grid-operating mode switching (token in all respective places as initial marking). Thus the T_1 transition is fired and a token arrives in place P_1. After t_G time, the T_2 is fired and P_2 wins a token. Place P_4 is empty; thus the transition T_4 is not enabled. Because place P_3 is already occupied by a token, places P_2 and P_3 enable T_3, the firing of which moves these tokens from the input places P_2 and P_3 in the output place P_4. That is the moment when the grid-operating mode is

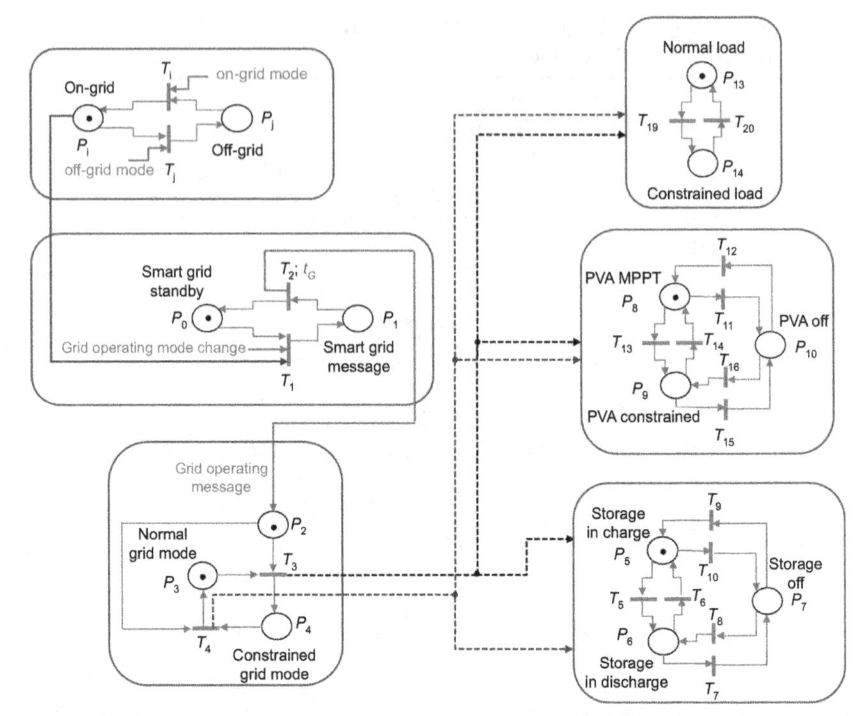

Figure 4.15 Power system global behavior modeling by Petri Net for grid-connected mode. *MPPT*, maximum power point tracking; *PVA*, photovoltaic array.

switched; the place P_4 is occupied by a token; and the power control interface is informed of grid power limits $P_{G\_S\_\lim}$ and $P_{G\_I\_\lim}$, energy time of use, actual pricing, etc. This information is taken into account for the power balancing control.

Concerning the load, that could be possible but not immediately required to switch in constrained mode. This happens only if the T_{19} firing conditions are true. In this case the power control interface transmits the coefficient K_L to the real load shedding device.

Fig. 4.16 shows the power system global behavior modeling by PN for off-grid mode indicated by the place P_j occupied by its token. Let us suppose that the diesel generator is in rated power operating mode, the PVA is in MPPT operating mode, the storage is in charging mode, and the load in normal operating mode. Concerning the load, that could be possible but not immediately required to switch in constrained mode. If $p_{PV} + p_S > p_L$, then only after the dt_{DG} time duration, the T17 transition is fired and a token arrives in place P_{12}. In addition, for $p_S \geq P_{S\_\max}$ this could

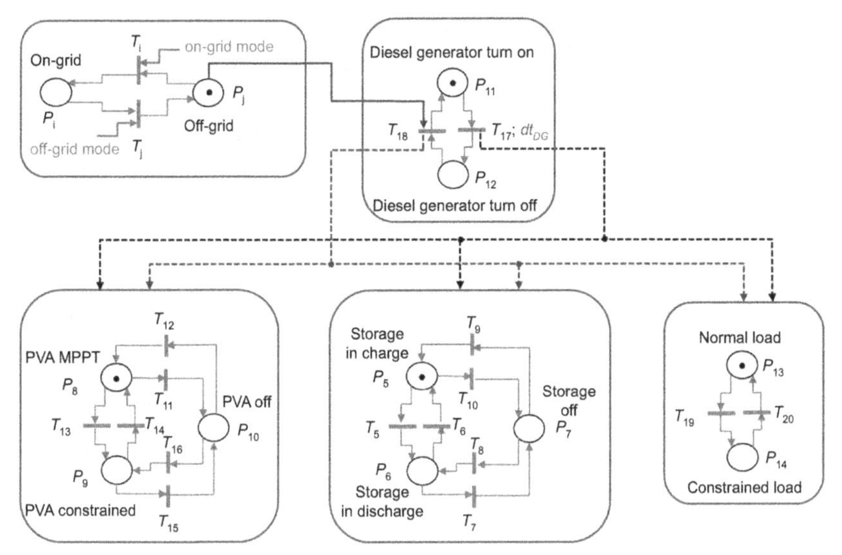

Figure 4.16 Power system global behavior modeling by Petri Net for off-grid mode. *MPPT*, maximum power point tracking; *PVA*, photovoltaic array.

also happen regardless of the time duration dt_{DG}. That is the moment when the diesel generator operating mode is switched and the power control interface involves only the PVA and the storage to balance the load power. This information is taken into account by the power balancing control.

On the other side, if the signal for on-grid mode appears, because the place P_j is occupied by a token, the transition T_i is fired, resulting in place P_i occupied by a token; therefore the on-grid mode is being operated following Fig. 4.15.

The described system has a hybrid aspect from the higher level, which is the operation criteria interface, to the lower one, which is power control. The continuous and discrete aspects coexist and interact with each other. As a method for validating such models the hybrid simulation is proposed. It is based on the connection and interaction of two submodels for which continuous and discrete simulation progress in alternation. Continuous simulation takes care of the continuous dynamic and is executed while no event has been detected. The main problem of the hybrid simulation is the synchronization between the models. To perform numerical hybrid simulation tests, Simulink and Stateflow development of the MATLAB environment are proposed. Simulink helps to model the continuous dynamics of the system and Stateflow is used to specify the discrete control logic and the modal behavior of the system. Stateflow design language is based on the concept of hierarchical automata from Statecharts and allows for developing the PN model through the approximation made between states/places and switching/firing transition conditions. Thus the proposed power system global behavior modeled by PN is translated into the Stateflow charts model. To exemplify this hybrid dynamic system modeling and simulation by Stateflow in the MATLAB environment during its interaction with Simulink, Fig. 4.17 (screen-printing where the nomenclature notation is adapted as required by Stateflow) shows the on-grid mode operation at a given instant [16]. In this figure the overall behavior of the power system is shown in the way that activated states or transitions are highlighted. This figure shows the instant when the load is in normal mode, the storage is fully charged, grid power injection reaches the injection limit, and PVA is thus constrained. Hence, with the help of Stateflow it is possible to verify the proposed control strategy with visualization of detailed element behavior at each time instant.

The interpreted PN modeling helps better analyze and understand the phenomena and modes of operation that have to be considered in the design of the power control interface and to validate a comprehensive approach that permits analyzing relations between different discrete states by numerical simulation.

Therefore a set of minimum requirements can define the power control interface design:

- limits imposed by the utility grid, $P_{G\_S\_\mathrm{lim}}$ and $P_{G\_I\_\mathrm{lim}}$, have to be taken into account.
- distribution coefficient K_D has to be an optimization result for which the objective function is the global energy cost minimization for a given time period. This optimization is under constraints that are related to the energy tariff, energy time of use, grid power limits, storage physical limits, diesel generator rated power and working duty cycle, power required by the load in real time, and the load shedding level accepted by the end user.
- Grid power and storage power references, p_G^* and p_S^*, have to be calculated.
- PVA has to be controlled following an MPPT method but also with a constrained algorithm with respect to a limited output power; therefore the PVA power limiting reference $p_{PV\_CONS}^*$ has to be calculated.
- Load shedding coefficient K_L has to be calculated and transmitted to the shedding device.

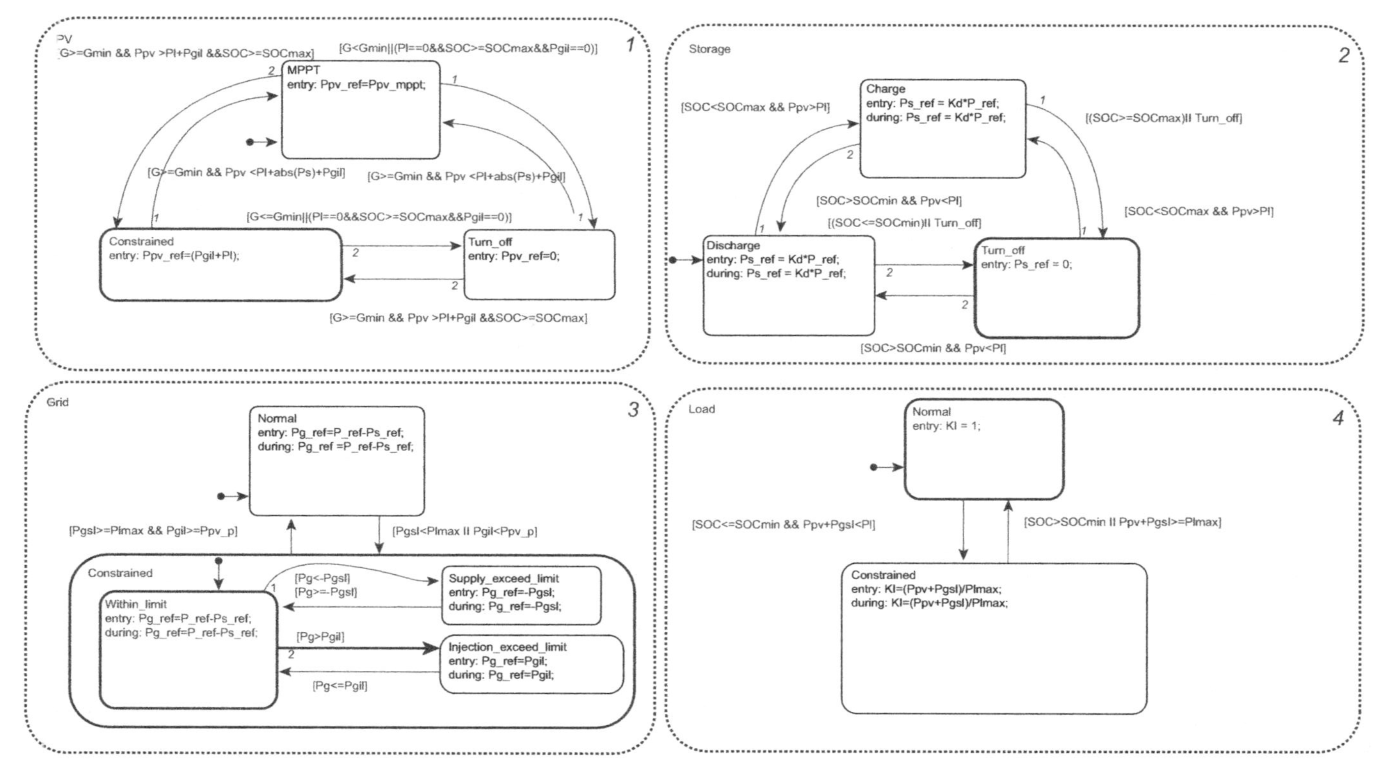

Figure 4.17 Discrete state system simulation by Stateflow for the on-grid mode operation. *MPPT*, maximum power point tracking; *PV*, photovoltaic.

To summarize, because of the interpreted PN modeling transferred to the Stateflow tool of MATLAB-Simulink, power system behavior simulation allows for validation before experimental tests.

4. DIRECT CURRENT MICROGRID POWER SYSTEM CONTROL

The power system behavior modeling leads to the whole control strategy design, which concerns the power balancing, load shedding, PVA power limiting, diesel generator constraints, and imposed limits by the utility grid. To obtain real-time control, power system elements must be ranged in specified sequence and an algorithm could be obtained. Hence, the continuous dynamics of the system are operated through an implemented algorithm that calculates power system references with respect to imposed limitations and gives the load shedding level. This algorithm focuses on the power system control strategy.

4.1 Power System Control for Grid-Connected Mode

As mentioned at the beginning of this chapter, concerning the distribution coefficient for the validation tests only two cases are retained: $K_D = 1$ (until its possible limits, storage is a priority in charging and discharging mode and $p_S^* = p^*$) and $K_D = 0.8$ (p^* is shared between the storage and the public grid).

4.1.1 Simple Strategy for Grid-Connected Control Algorithm

The continuous dynamics of the system are operated by the power control through an implemented algorithm that calculates reference power values p_S^* and p_G^* following Eqs. [4.2] and [4.3], with respect to their limitations, and gives the value of K_L. The algorithm flowchart for grid-connected operation is presented in Fig. 4.18 [17].

This algorithm focuses on the power system control. For the first approach presentation it is assumed that grid-operating mode does not impose limits for injection; therefore $P_{G\_I\_\lim} = P_{PV\_STC}$. On the other hand, to respect the public grid supply availability, if the calculated p_G^* is higher than the limit imposed by the smart grid, then $p_G^* = P_{G\_S\_\lim}$ and a load shedding could be imposed, depending on the p_{PV} value and the storage availability. Note that p_{LD} is the load power demand and p_L is the real load power.

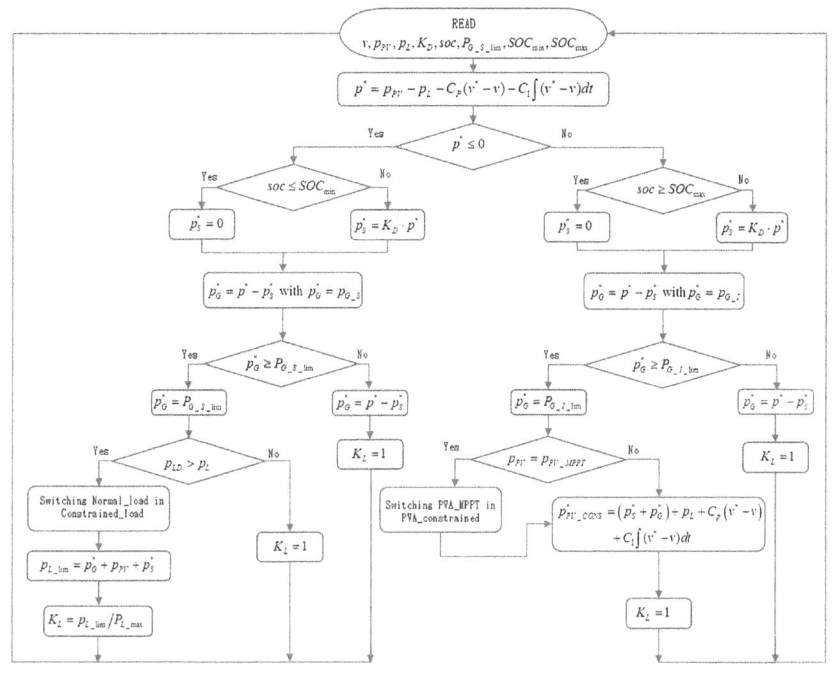

Figure 4.18 Strategy control flowchart for grid-connected operation.

4.1.2 Experimental Tests to Evaluate the Grid-Connected Mode

The experimental tests are based on the electrical schema presented in Fig. 4.2 and the experimental platform, the main parameters of which are given in Table 4.6 [17].

The first experimental test illustrated in Fig. 4.19 was recorded on April 24, 2011, in Compiegne, France. For the power flow shown in Fig. 4.2 and for the PN modeling it was assumed that p_G and p_S positive values mean to receive power and negative values mean to supply power; p_L, p_{PV}, and p_{DG} are always considered positive. To illustrate a real study case, recorded curves that do not respect this rule are given in Fig. 4.19. Thus, in this case only, the results representing the power evolution are given with respect to the following rules: power grid supply and power storage discharging mode are always positive; power grid injection and power storage charging mode are always negative.

The numerical used limits and values are given as follows: $K_D = 1$, $P_{G\_S\_lim} = 800$ W, $P_{G\_I\_lim} = P_{PV\_STC} = 2000$ W, $P_{L\_max} = 1500$ W, $v^* = 400$ V, $P_{S\_max} = 1200$ W, $C_{REF} = 130$ Ah, and arbitrary values for

Table 4.6 Principal element description for experimental test

Element	Parameter	Device
Storage (serial 8 battery units)	96 V/130 Ah	Sonnenschein solar S12/130 A
PVA (16 PV panel in series, with given *STC* parameters)	$I_{MPP} = 7.14$ A, $V_{MPP} = 280$ V	Solar-Fabrik SF-130/2-125
Grid emulator	3 kVA	Puissance+; bidirectional linear amplifier
Programmable DC electronic load	2.6 kW	Chroma 63202
Controller board		dSPACE 1103
Power electronic converter	600 V to 100 A	SEMIKRON SKM100GB063D

DC, direct current; *PV*, photovoltaic; *PVA*, PV array; *STC*, standard test condition.

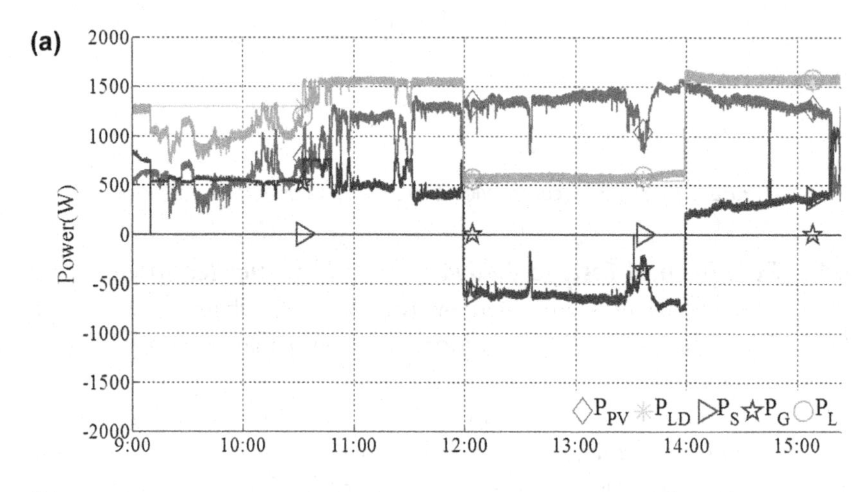

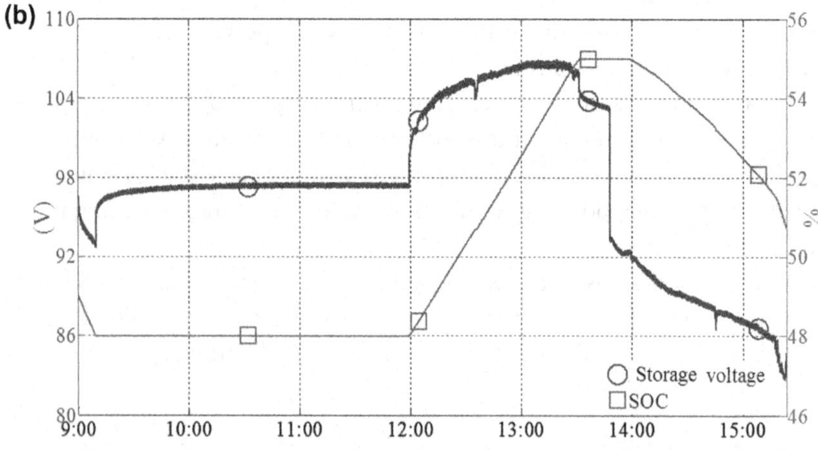

Figure 4.19 Evolution of (a) powers, (b) storage voltage and its *SOC*, (c) control coefficients, and (d) DC bus voltage recorded on April 24, 2011. *SOC*, state of charge.

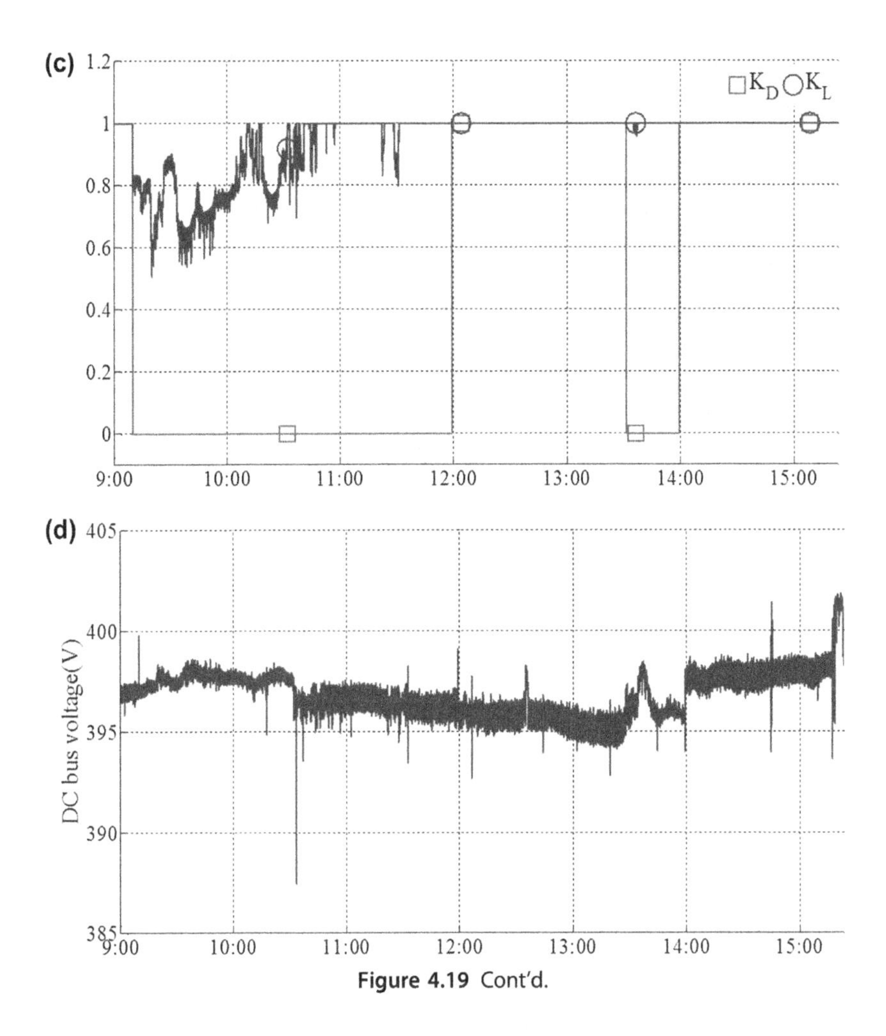

Figure 4.19 Cont'd.

$SOC_{min} = 48\%$ and $SOC_{max} = 55\%$. Note that considering the storage capacity installed on the platform, the *SOC* limits are selected to show the system behavior with relevant storage events, as full and empty, in a day run, although the values indicated by the manufacturer are far larger. As indicated by $K_D = 1$, the storage is supposed to supply or inject 100% of the demanded power p^* unless the *SOC* limits are reached.

Fig. 4.19a–c, show that the storage supplies the load, then as soon as one of the *SOC* limits is reached, the grid takes over. Following the evolution of K_L, the load is limited to p_L (Fig. 4.19a) while $P_{G\_S\_lim} + p_{PV} < p_{LD} + \Delta p$; the Δp is used as a security margin in power fluctuation and is given the value of 200 W. While $p_{PV} > p_{LD}$ the storage is available for charging until the SOC_{max} is reached; then the grid connection is used for selling excess energy.

The steady DC bus voltage illustrated in Fig. 4.19d proves that the strategy works well to balance the power production and consumption. The DC bus voltage remains almost constant with steady-state error that is inherent with a proportional controller. However, the maximum DC bus voltage fluctuation is approximately 3.25% of the rated DC bus voltage.

During the dynamic process around 9:09 shown in Fig. 4.20a and b, the storage reaches the limit SOC_{min} and the grid takes over. However, the voltage controller is able to make the DC bus voltage steady with an acceptable fluctuation.

The second experimental test, illustrated in Fig. 4.21, was recorded on April 26, 2011. The numerical used limits and values are given as follows: $K_D = 0.8$, $P_{G\_S\_lim} = 800$ W, $P_{G\_I\_lim} = 2000$ W, $P_{L\_max} = 1500$ W,

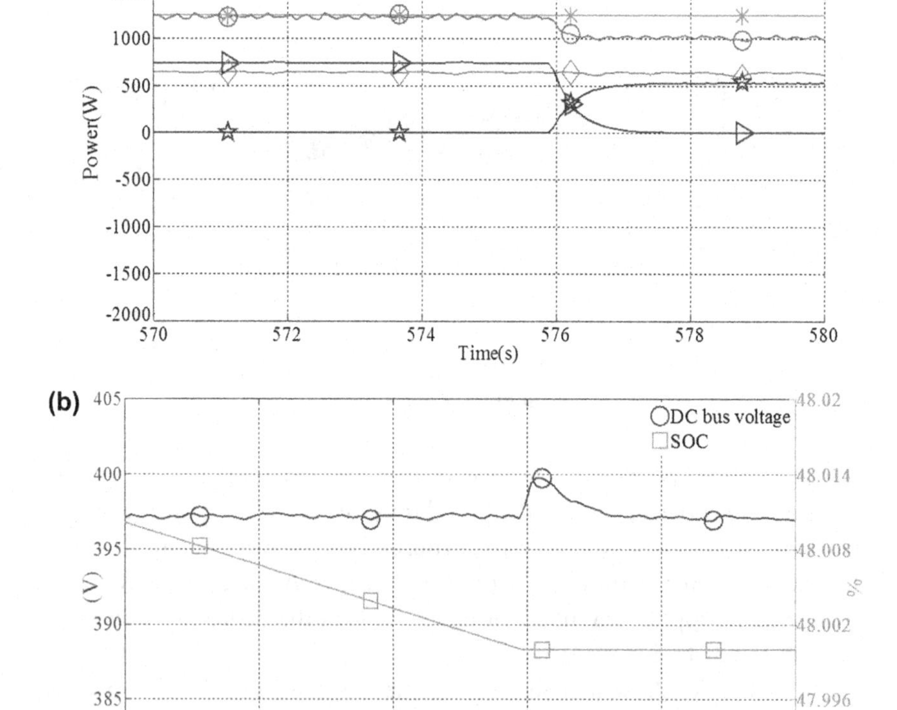

Figure 4.20 Detailed evolution of (a) powers and (b) DC bus around 9:09 on April 24, 2011. *DC*, direct current; *SOC*, state of charge.

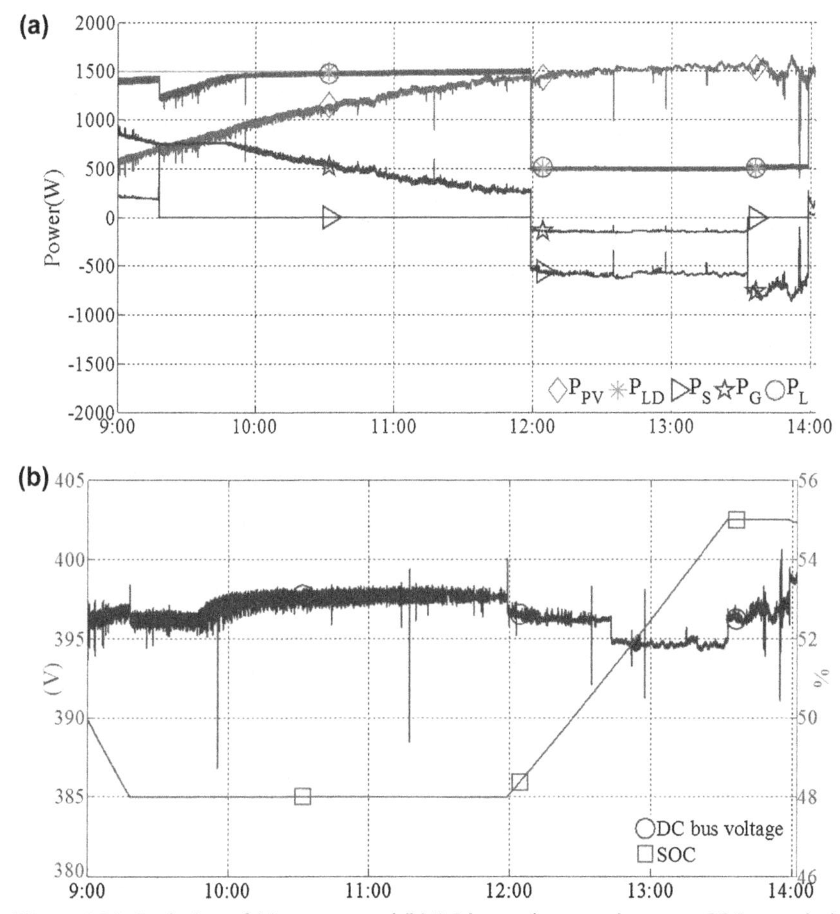

Figure 4.21 Evolution of (a) powers and (b) DC bus voltage and storage *SOC* recorded on April 26, 2011. *DC*, direct current; *SOC*, state of charge.

$v^* = 400$ V, $P_{S\_max} = 1200$ W, $C_{REF} = 130$ Ah, and arbitrary values for $SOC_{min} = 48\%$ and $SOC_{max} = 55\%$. Fig. 4.21a shows that the storage is used to supply or inject 80% of p^* when available and the load power is limited following K_L as aforementioned. Fig. 4.21b illustrates the storage availability and the DC bus voltage that remains almost constant with some negligible fluctuations.

Finally, taking into account that the goal was to verify the feasibility of the proposed system control under public grid constraints, the DC microgrid in grid-connected operation, as has been modeled, responds satisfactorily with the outlined strategy. However, to use fully and correctly the available storage, or to shed fewer appliances, the control must be improved by introducing distribution coefficient optimization. In Chapter "Direct Current Microgrid Supervisory System Design" limits for injection will be imposed and algorithms will calculate solutions corresponding to techno-economic criteria, weather forecasts, conditions of use, power demand forecast, etc.

4.2 Power System Control for Off-Grid Mode

DC microgrid power control, for the off-grid case, achieves power balancing because of the diesel generator and storage powers [18,19]. The adjustment power p^* is shared between the storage and the diesel generator with priority for the storage. If the diesel generator is already started, for this situation, if there is enough PVA power to supply the load, then storage can be charged within its physical limits.

4.2.1 Simple Strategy for Off-Grid Control Algorithm

A diesel generator works in bang-bang mode and only the storage power can be continuously controlled, which induces some difficulty in power balancing. As aforementioned, it could happen that the PVA outputs rated power while the diesel generator is turned on and load power demand is low. Charging storage with power superior to its limit will shorten storage life; therefore at least one source needs to be limited or cut off. The priority is defined as shedding PVA first, and then cut off the diesel generator if SOC_{max} is reached. The power balancing control algorithm is illustrated in Fig. 4.22.

The proposed algorithm considers some a priori rules and correlations to choose the best course of action and estimate the parameter values. First a comparison between the demanded load power and PVA power is done, resulting in positive or negative reference power p^*. Then the algorithm tests the SOC value; if $SOC \leq SOC_{min}$, then power from the diesel generator is required to supply the load. Once the diesel generator has been initialized, it will be kept on for at least one duty cycle. On the other side, for PVA power superior to the load power demand, the load is supplied by PVA power and the excess energy is stored, if possible. The part of PVA power that must be shed and the part of load power that must be shed are calculated so as to ensure the power balance of the system. In addition, the algorithm is supposed to take into account the value imposed for a final SOC equal to the initial SOC (for more clarity this calculus is not shown in the flowchart); this is done only during the iterations of the last hour of the operation duration. If SOC is inferior to

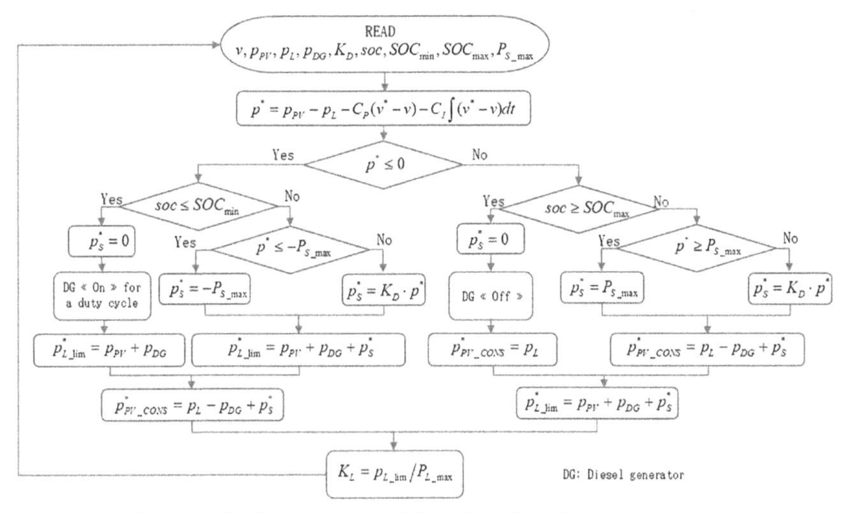

Figure 4.22 Strategy control flowchart for off-grid operation.

its imposed final value, then the storage is charged by the diesel generator if needs it.

4.2.2 Experimental Test to Evaluate the Off-Grid Mode

The experimental test is based on the electrical schema presented in Fig. 4.2 and the experimental platform for which the main parameters are given in Table 4.6, with the difference that, in this case, the linear amplifier (3 kVA) is used as a diesel generator emulator. The off-grid test was performed for operation on September 4, 2013, in Compiegne, France. The numerical used limits and values are given as follows: $K_D = 1$, $P_{DG\_P} = 300$ W, $dt_{DG} = 1200$ s, $P_{PV\_STC} = 1200$ W, $P_{L\_max} = 1200$ W, $v^* = 400$ V, $P_{S\_max} = 400$ W, $C_{REF} = 130$ Ah, and arbitrary values for $SOC_{min} = 49\%$ and $SOC_{max} = 51\%$. Once again the SOC limits are selected to show the system behavior with relevant storage events, as full and empty, in a day run, although the values indicated by the manufacturer are far larger. The same observation has to be noted for the chosen limits concerning P_{DG_P}, P_{PV_STC}, P_{L_max}, and P_{S_max}. As indicated by $K_D = 1$, the storage is supposed to supply or inject 100% of the demanded power p^* unless the SOC or P_{S_max} limits are reached.

The operation is performed based on the power balancing control algorithm. The experimental real-time power flow is shown in Fig. 4.23a. The experimental SOC and the control coefficients K_D and K_L are presented in Fig. 4.23b and the DC bus voltage evolution is given in Fig. 4.23c.

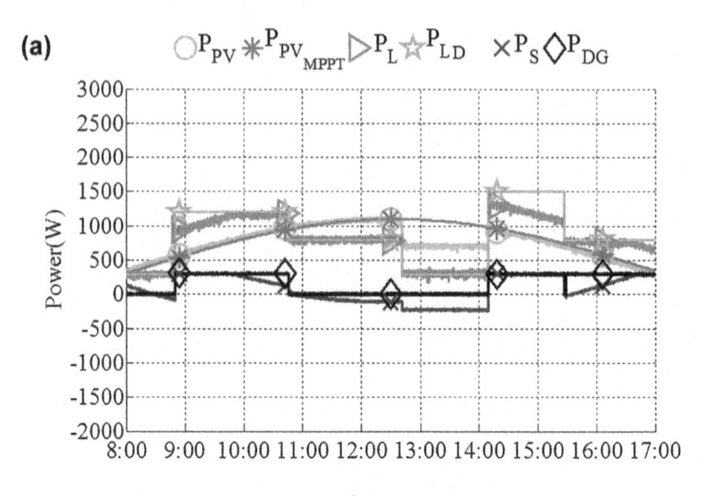

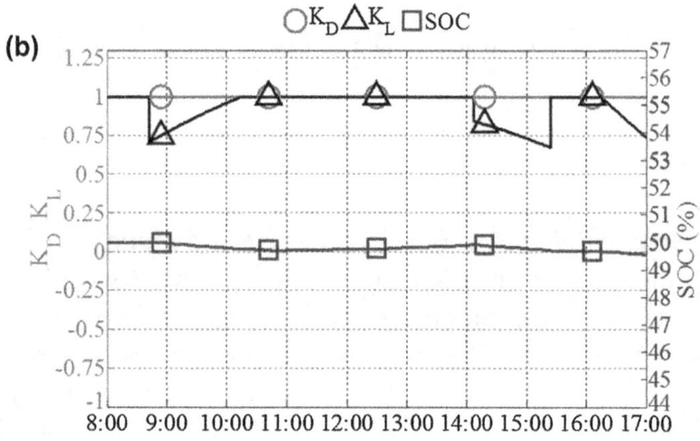

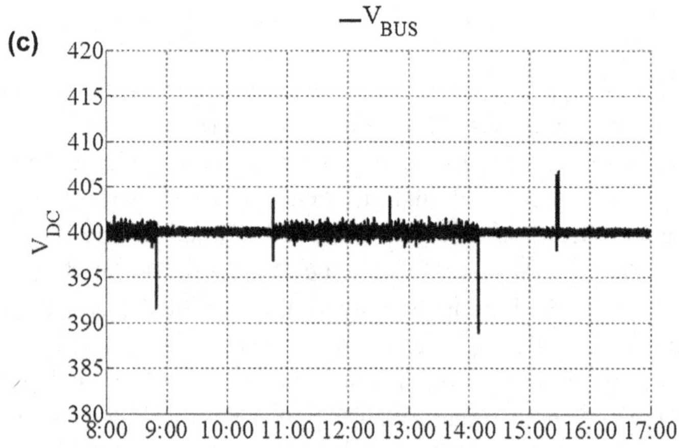

Figure 4.23 Evolution of (a) powers, (b) control coefficients and *SOC*, and (c) DC bus voltage recorded on November 12, 2013. *SOC*, state of change.

As mentioned earlier, for the PN modeling power flow it was assumed that p_G and p_S positive values mean to receive power and negative values mean to supply power; p_L, p_{PV}, and p_{DG} are always considered positive. Concerning the power flow curves shown in Fig. 4.23, for more graphical clarity the charging storage power p_S is represented by negative values whereas the discharging storage power p_S is represented by positive values.

In this test, knowing that the PVA production is low early in the morning, the storage also supplies the load so that the complete load power demand p_{LD} is satisfied. Before 8:30 the $p_{PV} > p_{LD}$ and the storage is in charging operating mode for a few minutes until the load power demand rises to its maximum value. Starting this moment the power is balanced because of the diesel generator, the storage power, and load shedding. Even if the $SOC > SOC_{\min}$, because of the storage power limitation the diesel generator is started. Because the sum of these two powers is not high enough to meet the load power demand, the load power is limited. The evolution of the control coefficient K_L given in Fig. 4.23b highlights this event. During 10:10−14:05 there is no more load shedding performed, resulting in $p_{LD} = p_L$.

Because of the storage power limit value, during 12:40−14:05 there is no other possibility that can absorb PVA production and the PVA production is constrained; it can be observed that $p_{PV\_MPPT} > p_{PV}$, inducing storage charging with low power. The PVA is recovered to produce MPPT power again when the load power demand increases again. Nevertheless, as PVA power decreases the power is balanced because of the diesel generator, which is started again; the storage power; and load shedding, which is performed until 15:20.

The power is well balanced during the operation, as shown in Fig. 4.23c by the steady DC bus voltage. The DC bus voltage fluctuates approximately 3% at the instants of limiting PVA power and starting the diesel generator control, which is related to the corresponding control dynamics and is acceptable. The DC bus voltage pulse is generally coming from two sides: on the one hand, the voltages and currents are filtered and the powers are calculated by the filtered signals; on the other hand, power references to stabilize the DC bus voltage are compensated by the filtered power, which is also filtered, therefore it has a delay in time. This problem, the development of which is not considered here, could be solved by using less filtered signal or improving the performance of correctors to cancel the compensation.

5. CONCLUSIONS

Integrated renewable electricity and smart grid interaction concern the utility companies and the end users. In the next 10 years the smart grid could concern tertiary and residential buildings with power "routers," the goal of which is to manage the real-time power demand to adjust the electricity generation. For the end user this device could control energy use, operating hours, and pricing. In this chapter, based on power system behavior modeling by interpreted PN and Stateflow, the power system control strategy is designed with consideration to each element's constraints and their behavior. A simple and quick-to-implement power control strategy is then applied. This strategy was not necessarily developed to improve global efficiency or the life cycle of the storage system; rather, it was intended to design energy management for a multisource power system including a PVA, storage, diesel generator, and grid connection. The major contribution of this system modeling is linked to the proposed control design allowing better PV production grid integration with load shedding and respecting grid supply limitations to reduce grid peak consumption. In addition, the power control should provide an interface for optimizing local power flow to reduce negative impact on the utility grid and to reduce energy cost. The modeling of the system provides discrete event analysis and facilitates the system design.

Experimental results show that the system is able to maintain stable operation and responds to the grid limits, calculates the powers reference of storage and the grid, and constrains the load. With one key interface parameter, different interface values show how it can affect the whole energy operation. Following the message received from the smart grid, the energy management system takes into account the power grid limitation, calculates the power reference of storage and the grid, and constrains the load if necessary. According to the on-grid strategy, power flow is mainly affected by grid power limits and control parameter K_D. The grid power limits are given by the smart grid message; therefore the control strategy can offer the possibility for the smart grid to manage grid usage, acting for demand response to help make better use of grid assets and improve the overall grid performance. The off-grid control strategy highlights the difficulty to take into account the dynamic response of the diesel generator, which shares with the storage the power adjustment reference. However, the control strategy can maintain power balancing with any K_D value. By giving different distribution coefficient K_D values, the power flow in the power system can be different. A large K_D value during peak hours can reduce the power demand from the grid as load shedding. The K_D value should be calculated by the power control interface, which leads to the development of local power flow optimization.

In the next chapter the design of the DC microgrid power control interface is presented as advanced energy management. Predictions and uncertainties on power sources and load demand will be considered and cost energy minimization will be studied. Optimization under constraints related to the risk of discrepancy among the PVA production, grid availability, and load management prediction should permit the calculation of K_D.

REFERENCES

[1] Hatziargyriou N. Microgrids: architectures and control. Willey-IEEE Press; 2014.
[2] Hassan MA, Abido MA. Optimal design of microgrids in autonomous and grid-connected modes using particle swarm optimization. IEEE Trans Power Electron 2011;26(3):755–69.
[3] Dragicevic T, Vasquez JC, Guerrero JM, Skrlec D. Advanced LVDC electrical power architectures and microgrids: a step toward a new generation of power distribution networks. IEEE Electrification Mag 2014;2(1):54–65.
[4] Patterson BT. DC, come home: DC microgrids and the birth of the "Enernet". IEEE Power Energy Mag 2012;10(6):60–9.
[5] Locment F, Sechilariu M, Houssamo I. Multi-source power generation system in semi-isolated and safety grid configuration for buildings. In: The 15th IEEE Mediterranean electrotechnical conference MELECON; 2010.
[6] Locment F, Sechilariu M, Houssamo I. DC load and batteries control limitations for photovoltaic systems. Experimental validation. IEEE Trans Power Electron 2012; 27(9):4030–8.
[7] Sechilariu M, Wang BC, Locment F. Building-integrated microgrid: advanced local energy management for forthcoming smart power grid communication. Energy Build 2013;59:236–43.
[8] Kanchev H, Lu D, Colas F, Lazarov V, François B. Energy management and operational planning of a microgrid with a PV-based active generator for smart grid applications. IEEE Trans Industrial Electron 2011;58(10):4583–92.
[9] Wang BC, Sechilariu M, Locment F. Intelligent DC microgrid with smart grid communications: control strategy consideration and design. IEEE Trans Smart Grid 2012;3(4):2148–56. Special issue on Intelligent buildings and home energy management in a smart grid environment.
[10] Guerrero JM, Chandorkar M, Lee T-L, Loh PC. Advanced control architectures for intelligent microgrids—part I: decentralized and hierarchical control. IEEE Trans Industrial Electron 2013;60(4):1254–62.
[11] Guerrero JM, Loh PC, Lee T-L, Chandorkar M. Advanced control architectures for intelligent microgrids—part II: power quality, energy storage, and AC/DC microgrids. IEEE Trans Industrial Electron 2013;60(4):1263–70.
[12] Savaghebi M, Jalilian A, Vasquez JC, Guerrero JM. Secondary control scheme for voltage unbalance compensation in an islanded droop-controlled microgrid. IEEE Trans Smart Grid 2012;3(2):797–807.

[13] Wang BC, Sechilariu M, Locment F. Energy management modelling for building integrated multi-source power system with smart grid interaction. In: International conference on theory and application of modeling and simulation in electrical power engineering ELECTRIMACS; 2011.
[14] Lu D, Fakham H, Zhou T, François B. Application of Petri Nets for the energy management of a photovoltaic based power station including storage units. Renew Energy 2010;35(6):1117–24.
[15] David R, Alla H. Discrete, continous, and hybrid Petri Nets. 2nd ed. Berlin: Springler-Verlag; 2010.
[16] Wang BC, Sechilariu M, Locment F. Power flow Petri Net modelling for building integrated multi-source power system with smart grid interaction. Math Comput Simul 2013;91:119–33.
[17] Sechilariu M, Wang BC, Locment F. Building integrated photovoltaic system with energy storage and smart grid communication. IEEE Trans Industrial Electron 2013;60(4):1607–18. Special issue on Distributed generation and micro-grids.
[18] Sechilariu M, Locment F, Wang BC. Photovoltaic electricity for sustainable building. Efficiency and energy cost reduction for isolated DC microgrid. Energies 2015;8(8): 7945–67. Special issue on Solar photovoltaics trilemma: efficiency, stability and cost.
[19] Sechilariu M, Wang BC, Locment F. Power management and optimization for isolated DC microgrid. In: The 22nd IEEE International symposium on power electronics, electrical drives, automation and motion SPEEDAM; 2014.

CHAPTER 5

Direct Current Microgrid Supervisory System Design

1. MULTILAYER SUPERVISORY DESIGN OVERVIEW

A microgrid that interacts strongly with its environment (communication, intelligent network, metadata exchange, etc.) is a complex physical system that can be seen, defined, and represented as a dynamic hybrid system [1] following the Petri Net (PN) and Stateflow modeling [2] that is detailed in chapter "Direct Current Microgrid Power Modeling and Control." Eq. [5.1] gives the mathematical definition of dynamic hybrid systems [3]:

$$\begin{aligned} \dot{x}(t) &= F(x(t), q(t), u(t)) = A(q)\cdot x(t) + B\cdot u(t) \\ y(t) &= C\cdot x(t) \\ [x(t^+), q(t^+)] &= G(x(t), q(t), v(t)) \text{ if } v(t) \in \Omega_c \text{ occurs} \\ x(t_0) &= x_0, \quad q(t_0) = q_0 \end{aligned} \qquad [5.1]$$

with $x(t) \in X \subset \mathbb{R}^n$, $q(t) \in Q \subset \mathbb{N}^m$, $u(t) \in U \subset \mathbb{R}^c$, $v(t) \in \Omega_c \subset \mathbb{N}^d$, n, m, c, and d known; $t \in \mathbb{R}^+$ represents the time; X is the set of continuous states; Q is the set of discrete states; U is the set of continuous commands; Ω_c is the set of discrete commands; F represents the continuous dynamic of the system; G represents the hybrid dynamic of the system; and A, B, and C are the system matrix.

The representation of the system in the form of PN or Stateflow allows having a functional view of the dynamic hybrid system that the system of equations given in Eq. [5.1] does not provide. Indeed, PN and Stateflow have the advantage of highlighting the system operating modes and the transition from one mode to another. Thus a hybrid controller composed of n automata in parallel is chosen to represent the dynamic hybrid system [4]. This modular structure illustrated in Fig. 5.1 makes it possible to better visualize the interactions between the continuous system and the discrete-event system [5].

Urban DC Microgrid
ISBN 978-0-12-803736-2
http://dx.doi.org/10.1016/B978-0-12-803736-2.00005-0

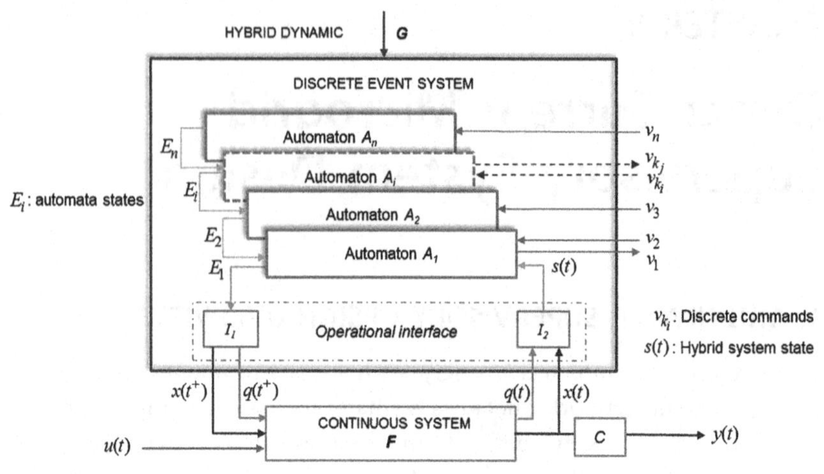

Figure 5.1 Dynamic hybrid system representation based on n automata.

There are three levels:

- Continuous dynamic defined by the function F, which involves the discrete variable $q(t)$, control variable $u(t)$, and system status $x(t)$; the output of the system is obtained from the matrix C.
- Discrete-event system represented by n automata A_i in parallel, each having a finite states number E_i; these automata match the hybrid dynamic that takes into account the discrete control to trigger hybrid phenomena monitored and seen as the set of possible system states.
- Operational interface composed of two functions I_1 and I_2:
  - I_1 affects values at $q(t^+)$ and $x(t^+)$ according to the discrete-event system state $E_i \in E$.
  - I_2 is used to trigger an event $s(t)$ depending on the hybrid state $s(t) = [x(t), q(t)]$.

The G function consists of the discrete-event system and the two functions that reflect the action of events and the triggering of phenomena. Regarding the microgrids, this dynamic hybrid system representation leads one to consider the multisource power system as a continuous dynamic and its supervisory control as the hybrid dynamic that must take into account several events to trigger controlled hybrid actions. Hence, the supervisory structure is a continuous system control composed of several hierarchical levels for which each hybrid state leads to the hybrid dynamic of the overall system. Therefore, as illustrated in Fig. 5.2, four level types are proposed: strategic in terms of end-user and public grid requirements, tactical management involving predictions for consumption and production, energy

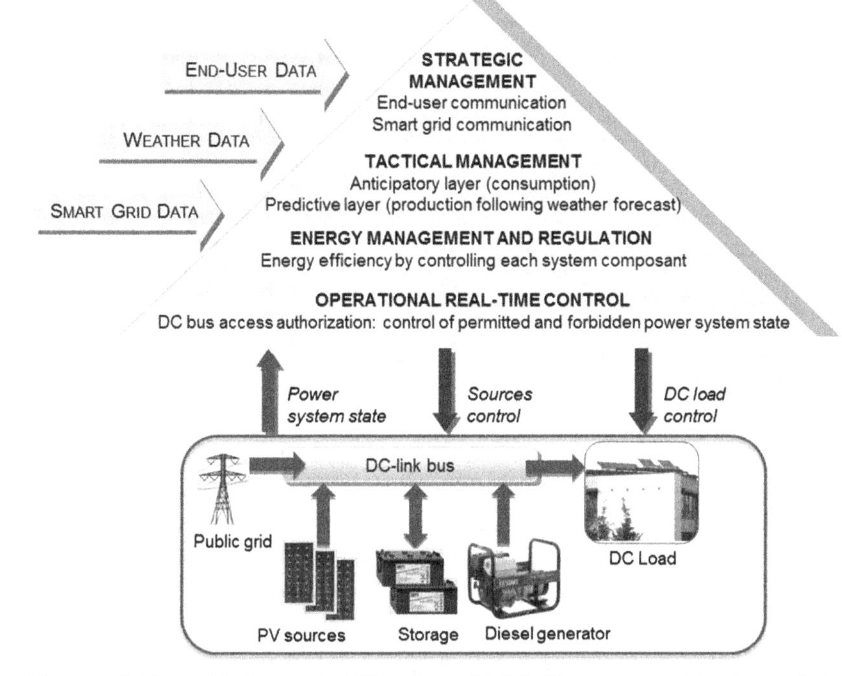

Figure 5.2 Hierarchical control of DC microgrid. *DC*, direct current; *PV*, photovoltaic.

management and regulation that calculate the desired energy optimization, and the essential operational real-time control of continuous system. This hierarchical control is governed first by control orders and second by controlled and autonomous events.

This hierarchical control, as a first synthesis of the controller design, leads to the communication interface between the continuous system and the discrete-event system. This is an operational layer that assigns the values to the power reference and other factors necessary for the control; on the other side it takes into account the hybrid state of the multisource system. The concept automata, in parallel with a hierarchical coupling of identified needed variables, then transforms this modular structure in the supervisory control of the direct current (DC) microgrid as shown in Fig. 5.3.

Finally, the DC microgrid supervisory is proposed as a multilayer structure for local energy management based on information concerning the end-user request, the building appliance forecast, weather forecasts, public grid availability conditions, load shedding criteria, etc. An operational layer is developed to control the DC microgrid power balance taking

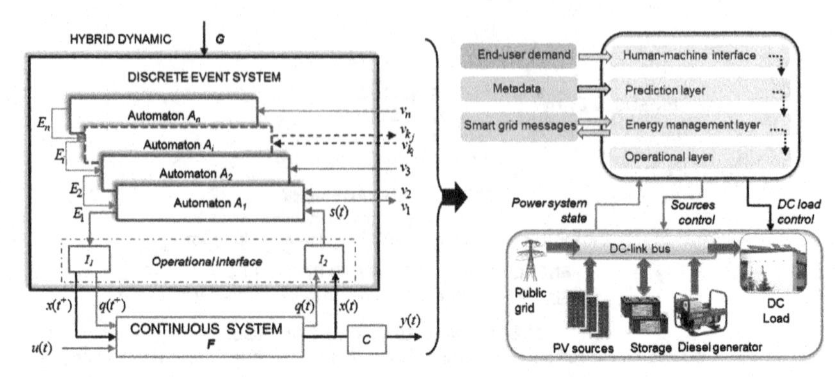

Figure 5.3 Hybrid dynamic system approach and hierarchical supervisory control. *DC*, direct current; *PV*, photovoltaic.

into account multiple constraints. The objective is to provide a continuous supply to the building while energy cost is optimized under constraints such as power limits, electricity prices, energy time of use, and load. The next sections present each hierarchical level of the proposed supervisory control.

2. HUMAN-MACHINE INTERFACE

End users may interact with the DC microgrid supervisory control because of the human–machine interface (HMI) allowing them to adjust, define, and customize the DC microgrid operation criteria such as minimizing energy cost or maximizing the comfort by including/excluding load shedding or limiting the total load shedding amount in the operation. The HMI design is illustrated in Fig. 5.4.

The user can specify the lowest limit of the load shedding coefficient, K_{L_lim}. Aiming to maintain power balance, if the system requires $K_L < K_{L\_\mathrm{lim}}$, then the operation is assumed with the necessary K_L value, but the user will be notified.

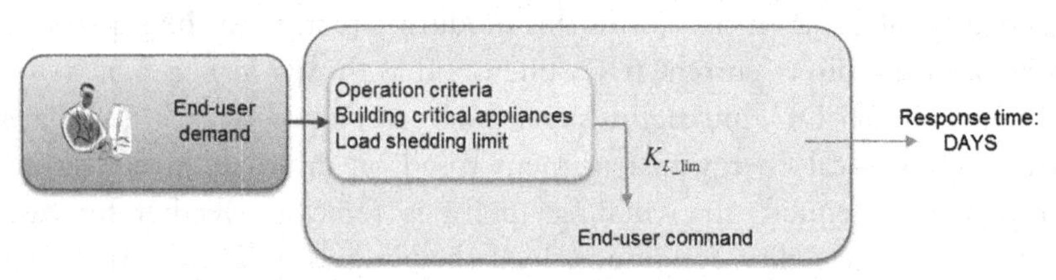

Figure 5.4 Human-machine interface layer design.

Parameters of the system that are used in supervisory control, including the rated power of each element, constraints of each element, load shedding levels, etc., are also assigned in this layer. According to the end-user demand, this layer could modify the corresponding parameter value in the other layers. The end user can also assign different priorities for each building appliance that would be involved in load shedding and define the critical load that requires an uninterruptible power supply.

Thus different operating modes for different end-user types (office building or residential building) could be predefined. Some operating modes could follow a predefined lower limit of K_L. The end users may also choose a criterion to take into account as to minimizing energy cost, including or excluding load shedding if possible. Modifying the optimization criteria or overwriting constraints is also possible in this layer. The HMI layer is assumed to be designed with a graphical interface.

3. PREDICTION LAYER

The prediction layer design is presented in Fig. 5.5. It calculates possible PVA power and load power evolutions for the next hours. It is considered that the solar irradiance [6] and air temperature forecast may be available and integrated over time in a data register. Concerning the load, programming data for the operation of building appliances may be available to calculate power curve forecasts [7]. Therefore these metadata combined with messages from the smart grid and HMI output variable (end-user command) that are considered as discrete inputs enable the predictive layer to calculate the power predictions, photovoltaic array (PVA) power, and load power for the next hours [8].

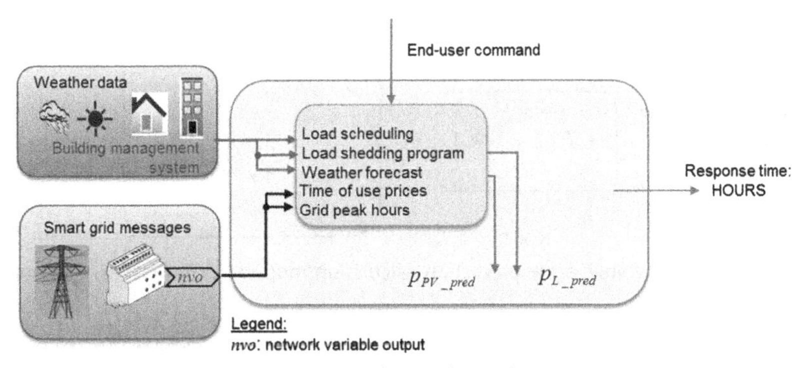

Figure 5.5 Prediction layer design.

3.1 Photovoltaic Power Prediction

For advanced energy management in a DC microgrid, photovoltaic (PV) power prediction is necessary and required by the future smart grid. The weather forecast services related to renewable energy production are able to provide hourly solar irradiance and air temperature forecasting within the smallest error as possible. On the basis of hourly solar irradiance and air temperature forecasting, and on the PV model built with parameter identification or PVA solar irradiance mapping as presented in chapter "Photovoltaic Source Modeling and Control," the PVA predictable power p_{PV_pred} can be calculated with as little error as possible.

To carry on our study, weather prediction data are obtained from Météo France, which provides hourly prediction data of solar irradiance for specific locations. On the basis of these data and on the purely experimental PV model [9] presented in chapter "Photovoltaic Source Modeling and Control," the predictable PVA power p_{PV_pred} is calculated in maximum power point tracking (MPPT) mode [10] following Eqs. [5.2] and [5.3]:

$$p_{PV\_pred} = p_{PV\_MPPT} = f(g, \theta) \quad [5.2]$$

$$\theta = T_{air} + g \cdot \frac{\text{NOCT} - 20}{800} \quad [5.3]$$

with T_{air} the air temperature and NOCT as the nominal operating cell temperature, the value attained by the PV cell with free air circulation all around, under the test conditions of $T_{a\_test} = 20°C$, $G_{test} = 800\ W/m^2$, and wind speed $w_{S\_test} = 1\ m/s$. The NOCT value is specified by the manufacturer and can take a different value for each PV panel. Fig. 5.6 summarizes the PVA power prediction calculation model [11].

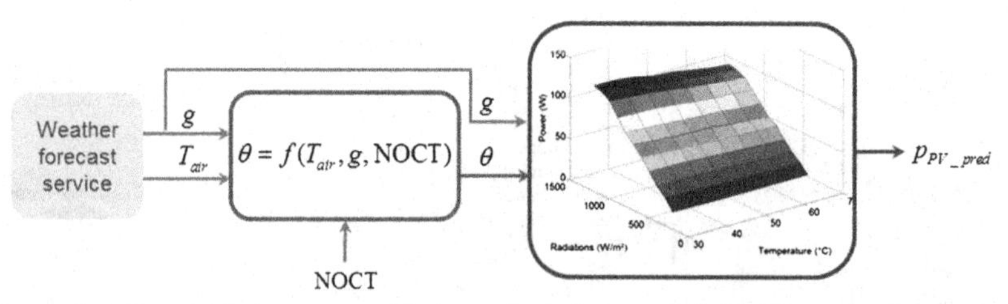

Figure 5.6 Photovoltaic array prediction calculation model. *NOCT*, nominal operating cell temperature.

3.2 Load Power Prediction

The load power supply can be predicted by statistical data and/or by the information of the load scheduling and load shedding program from the building management system and with respect to the end-user demand. The load scheduling, which represents the operating program of building appliances as building energy needs linked to weather, and the partial and full load shedding program are supposed to be known; they are implemented and updated continuously by the building management system. According to the value of K_{L_lim} chosen by the end user, the second variable output that the prediction layer has to calculate is the load power prediction p_{L_pred}.

Every type of tertiary building, such as commerce, office, administration, and education, has a particular profile of energy consumption. Such profiles change mainly according to the occupants' activities and weather conditions—hence the uncertainty of this prediction. Statistical data such as the average power curve are usually available online by the power provider website for its customers (energy tracking service); these data along with the building appliance operation data given by the building management system allow for calculating the predictive power. This power curve must be correlated with the "end-user command" and information from the smart grid concerning the time-of-use prices and grid peak hours. Thus the obtained result may also be confronted with the possible load shedding, indicating the minimum level of power supply.

Concerning the management of building appliances there are very few studies finished [6], but nowadays, taking into account the smart meters, more works are being performed. In this book a specific approach to correctly estimate the predicted load power p_{L_pred} is not developed. Because the experimental validation is based on a load emulator, this prediction is taken into account in the calculations of the energy management layer as a random estimation with an error margin of $\pm 10\%$ with the chosen arbitrary load power curve demanded by the load during the test.

Regarding the load shedding possibilities, for simulation tests arbitrary intervals and values are chosen and imposed. However, continuous load shedding is assigned to highlight the control with respect to power limits in experimental results in chapter "Experimental Evaluation of Urban Direct Current Microgrid." Moreover, the algorithms currently in place can be easily and quickly adapted to the development of automatic calculations of p_{L_pred} and of load shedding limits.

4. ENERGY MANAGEMENT LAYER

The energy management layer shown in Fig. 5.7 interacts with the prediction layer and controls the operational layer by calculating the distribution coefficient K_D following an optimization method [8].

The optimization goal is to obtain the best power distribution between the sources. It is a matter to reduce the energy cost, reduce the grid power peak consumption in grid-connected mode, minimize fuel consumption in the off-grid mode, and avoid load shedding and PVA shedding in both modes. At the same time it has to decide the best contribution of the public grid or diesel generator and the storage. The smart grid message is supposed to provide real-time grid energy tariffs and grid power limits, which assist in reducing peak supply and avoiding undesired injection. Furthermore, because of the smart grid connection, this layer is able to inform the grid operator about the grid supply/injection power prediction.

The inputs of this layer are provided by the prediction layer, p_{PV_pred} and p_{L_pred}, on the one hand and the smart grid message, containing $P_{G\_S\_\lim}$ and $P_{G\_I\_\lim}$ as power limits, time of use energy tariff, on the other hand. In addition, knowledge of the storage limits should be provided.

The main output of this layer is the distribution coefficient K_D, signifying power sharing between the public grid power p_G^* or diesel generator p_{DG}^* and the storage power p_S^*. In addition, for grid-connected mode,

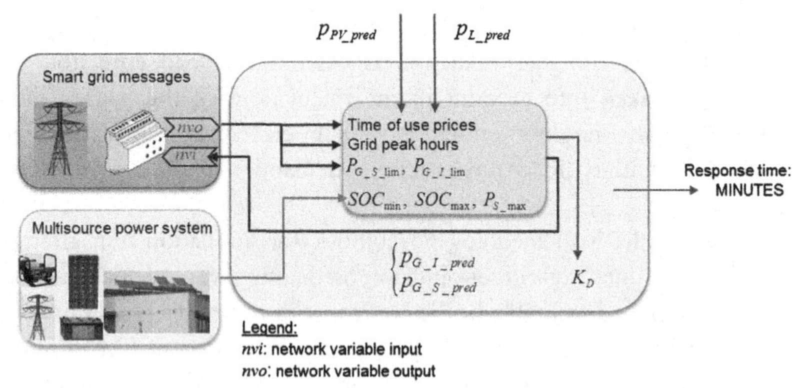

Figure 5.7 Energy management layer design.

optimization results should allow the calculation of the expected power to be injected $p_{G\_I\_pred}$ into the utility grid and of the power required in predictable absorption by the load from the utility grid $p_{G\_S\_pred}$. These two predictions can be transmitted to the smart grid as network variable input messages. This is considered as helpful information for energy production planning for the central grid management operator.

The objective of this layer is to compute off-line but also to be able to calculate online with updated forecasting, the expected optimized operating power flow based on predicted powers previously calculated, and taking into account the grid power limits and element constraints. The optimization focuses on minimizing energy costs; hence, it needs to know in advance the energy tariffs and the time-of-use prices over time provided by the smart grid message. The main result is the calculation of K_D, which is time varying.

4.1 Energy Cost Optimization Problem Formulation

Optimization should take into consideration all power flows in the power system, as well as satisfying the power balance with respect to all imposed constraints. Each source control is different:

- PVA production changes according to weather conditions and load power is governed mainly by end users; therefore they cannot be fully controlled but partially interfered by the supervisory control.
- Storage can be fully controlled within its physical limits.
- Grid power may be fully controlled except during the periods under limited power or very high energy tariffs.
- The diesel generator is assigned to work in duty cycle and also cannot be fully controlled.

Because constraints are different in grid-connected mode and off-grid mode, the optimization problem is separately formulated.

4.1.1 Grid-Connected Mode

During grid-connected mode, the main goals of the proposed DC microgrid are decreasing grid peak consumption, avoiding undesired grid power injection, and reducing the global energy cost [12]. An overview of the power flow in grid-connected mode is shown in Fig. 5.8. PVA, storage, and grid are coupled on the DC bus through their dedicated converters, the conversion efficiency of which is not taken into account. Load demands are powered directly from the DC bus.

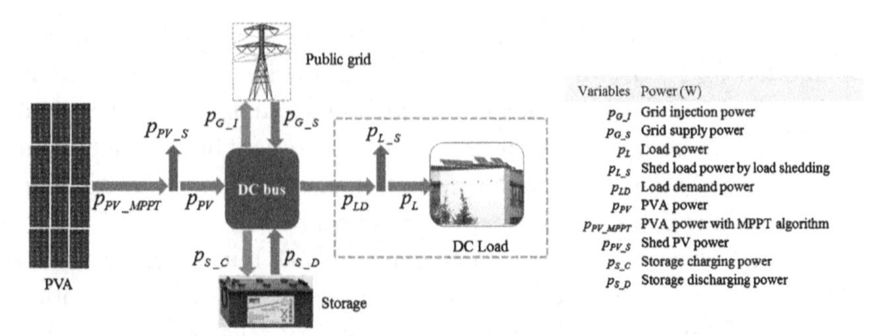

Figure 5.8 Power flow representation using unidirectional parameters for on-grid mode. *DC*, direct current; *PV*, photovoltaic; *PVA*, PV array.

For a fast understanding of the power flow representation, Fig. 5.8 recalls the power notations involved and provides the new notations needed in the mathematical expressions used in this section.

In chapter "Direct Current Microgrid Power System Modeling and Control," the discrete modeling of PVA source has defined the constrained PVA operation mode [13]. Indeed, for excess PV generation, PVA is forced to produce limited power; if the storage reaches its lower limit $SOC = SOC_{\max}$ and if the grid power injection limit is reached, then the PVA constrained power reference is expressed by Eq. [5.4]:

$$p^{*}_{PV\_CONS}(t) = p_L(t) - P_{G\_I\_\mathrm{lim}} \quad [5.4]$$

Thus, in the case of protecting storage from overcharging and respecting grid power injection limits from the smart grid message, the PVA production can be partially shed by the constrained PVA production algorithm. The PVA shed power is noted as p_{PV_S}, and it is considered $p_{PV\_S} = 0$ in the MPPT algorithm. For each time instant t, the PVA production p_{PV} is described by Eq. [5.5]:

$$p_{PV}(t) = p_{PV\_MPPT}(t) - p_{PV\_S}(t) \quad [5.5]$$

The p_{PV_S} values are calculated by optimization; however, the PVA shedding should not induce negative power, therefore it is constrained as in Eq. [5.6]:

$$p_{PV}(t) \geq 0 \quad [5.6]$$

The load should be satisfied according to end-user demand. In the case of insufficient storage and public grid access limits given by the smart grid message, the load power demand p_{LD} cannot be fully met and the load must be partially shed to maintain the operation of the critical load. Therefore if storage is empty $SOC = SOC_{min}$, the load power limit should not be exceeded as in Eq. [5.7].

$$p_{L\_lim} = p_{PV}(t) + P_{G\_S\_lim} \quad [5.7]$$

This limit is controlled by the load shedding coefficient $K_L = p_{L\_lim}/P_{L\_max}$. The proportion in load power that must be shed is noted as p_{L_S}, and the load power p_L follows Eq. [5.8]:

$$p_L(t) = p_{LD}(t) - p_{L\_S}(t) \quad [5.8]$$

The power p_{L_S} is determined by optimization; it also should not induce negative power, therefore it is constrained by Eq. [5.9]:

$$p_L(t) \geq 0 \quad [5.9]$$

Temporary load partial shedding could be a solution to reduce utility grid mismatching or to obtain less energy consumption if agreed by the end user. These operations should be controlled by the supervisory control (eg, minimizing or avoiding load shedding).

The storage is operated by current closed-loop control, and the storage power can be controlled by giving the corresponding current reference [14]. The storage power limit P_{S_max} and its state of charge (SOC) upper and lower limits, SOC_{max} and SOC_{min}, respectively, must be respected to protect the storage from overcharging and overdischarging, as described by Eq. [5.10] with the SOC calculation given by Eq. [5.11]:

$$\begin{aligned} SOC_{min} \leq soc(t) \leq SOC_{max} \\ -P_{S\_max} \leq p_S(t) \leq P_{S\_max} \end{aligned} \quad [5.10]$$

$$SOC(t) = SOC_0 + \frac{1}{3600 \cdot v_S \cdot C_{REF}} \int_{t_0}^{t} \left(p_{S\_C}(t) - p_{S\_D}(t)\right) dt \quad [5.11]$$

where storage power $p_S(t)$ is defined as $p_S(t) = p_{S\_C}(t) - p_{S\_D}(t)$, where $p_{S\_C}(t)$ and $p_{S\_D}(t)$ are the storage charging and discharging power, respectively.

When the SOC limit is not reached, PVA production should not be limited, as in Eq. [5.12]:

$$p_{PV\_S}(t) = 0 \quad \text{if} \quad SOC(t) < SOC_{max} \quad [5.12]$$

PVA power shedding is not permitted when the production can be totally consumed, and the load shedding is not permitted when there is sufficient power supply. These constraints are implied in the optimization objective and are given in explicit form by Eq. [5.13].

$$\begin{aligned}
&\text{if } p_{PV_MPPT}(t_i) > p_{LD}(t_i) \text{ then } p_{L_S}(t_i) = 0 \\
&\text{if } p_{PV_MPPT}(t_i) = p_{LD}(t_i) \text{ then } \begin{cases} p_{L_S}(t_i) = 0 \\ p_{PV_S}(t_i) = 0 \end{cases} \\
&\text{if } p_{PV_MPPT}(t_i) < p_{LD}(t_i) \text{ then } p_{PV_S}(t_i) = 0
\end{aligned} \quad [5.13]$$

Regarding the grid connection, it is also controlled by current closed-loop control [15]. Because limits for grid power supply $P_{G\_S\_\lim}$ and grid power injection $P_{G\_I\_\lim}$ are imposed by smart grid messages, during DC microgrid operation the public grid power should be controlled to satisfy Eq. [5.14]:

$$\begin{aligned}
&0 \le p_{G_I}(t) \le P_{G_I_\lim} \\
&0 \le p_{G_S}(t) \le P_{G_S_\lim}
\end{aligned} \quad [5.14]$$

Considering that all powers are always positive, as sign convention the physical law of power balancing is described by Eq. [5.15]:

$$p_L(t) + p_{G\_I}(t) + p_{S\_C}(t) = p_{G\_S}(t) + p_{S\_D}(t) + p_{PV}(t) \quad [5.15]$$

The stability of the DC bus is performed by the power reference p^* given in chapter "Direct Current Microgrid Power Modeling and Control." Therefore, taking into account the definition of the distribution coefficient K_D and the equalities $p_G(t) = p_{G\_I}(t) - p_{G\_S}(t)$ and $p_S(t) = p_{S\_C}(t) - p_{S\_D}(t)$, the distribution coefficient can be expressed by Eq. [5.16]:

$$K_D(t) = \frac{p_{S\_C}(t) + p_{S\_D}(t)}{p_{S\_C}(t) + p_{S\_D}(t) + p_{G\_I}(t) + p_{G\_S}(t)} \quad [5.16]$$

The objective of the energy management layer is to minimize the total energy cost of system C_{total}, which consists of grid energy cost C_G, storage energy cost C_S, PVA shedding cost C_{PVS}, and load shedding cost C_{LS}, as in Eq. [5.17]:

$$C_{total} = C_G + C_S + C_{PVS} + C_{LS} \quad [5.17]$$

By calculating the energy cost with each time increment Δt, C_G is defined by Eq. [5.18]. According to this definition, the grid power could be bought or sold at the same price.

$$\begin{aligned}
&C_G = \sum_{t_i=t_0}^{t_F} \left[c_G(t_i) \cdot \Delta t \cdot \left(-p_{G_I}(t_i) + p_{G_S}(t_i) \right) \right] \text{ with} \\
&t_i = \{ t_0, t_0 + \Delta t, t_0 + 2\Delta t, \ldots, t_F \}
\end{aligned} \quad [5.18]$$

Although today the PV energy grid injection benefits incentive tariffs, knowing that there is a heavily subsidized part, this study takes into account a single rate for energy purchased or sold, and the grid energy tariff is defined by Eq. [5.19] according to peak hours or normal hours:

$$c_G(t) = \begin{cases} c_{G\_NH} & \text{for } t \in \text{normal hours} \\ c_{G\_PH} & \text{for } t \in \text{peak hours} \end{cases} \qquad [5.19]$$

with c_{G_NH} the normal hour tariff and c_{G_PH} the peak hour tariff.

Storage aging should be considered to give an energy tariff of storage using. However, for this study the cost of storage C_S is defined more simply as expressed by Eq. [5.20], with c_s the storage energy tariff.

$$C_S = \sum_{t_i=t_0}^{t_F} \left[c_S(t_i) \cdot \Delta t \cdot \left(p_{S\_C}(t_i) + p_{S\_D}(t_i) \right) \right] \qquad [5.20]$$

In general, after PV source installation only maintenance involves costs; production does not require extra fees. However, shedding PVA power suggests that the asset is not fully used. Therefore PVA power shedding is penalized in the optimization and the cost of PVA shedding is defined by Eq. [5.21] with c_{PVS} the PVA power shedding tariff:

$$C_{PVS} = \sum_{t_i=t_0}^{t_F} c_{PVS}(t_i) \cdot \Delta t \cdot p_{PV\_S}(t_i) \qquad [5.21]$$

The load power mainly follows the end-user demand; therefore load shedding introduces inconvenience for end users and it must be penalized in optimization. The cost of load shedding is defined by Eq. [5.22], with c_{LS} an arbitrary load power shedding tariff:

$$C_{LS} = \sum_{t_i=t_0}^{t_F} c_{LS}(t_i) \cdot \Delta t \cdot p_{L\_S}(t_i) \qquad [5.22]$$

To limit the power grid fluctuations, grid power changing rate limits are introduced as

$$-\text{Limit} \le p_G(t_i) - p_G(t_{i-1}) \le \text{Limit} \qquad [5.23]$$

Because PVA energy grid injection benefits incentive tariffs, energy grid injection by power grid charged storage is forbidden. Thus the following limits expressed by Eq. [5.24] are imposed to ensure storage energy charge and grid injection only from PVA production.

$$\begin{aligned} p_G(t_i) \ge 0, p_S(t_i) \ge 0 \quad \text{if} \quad p_{PV}(t_i) - p_{LD}(t_i) \ge 0 \\ p_G(t_i) \le 0, p_S(t_i) \le 0 \quad \text{if} \quad p_{PV}(t_i) - p_{LD}(t_i) < 0 \end{aligned} \qquad [5.24]$$

Finally, by considering the discrete time instant t_i, from t_0 to t_F, with the time interval Δt, the problem formulated can be completely mathematically expressed by Eq. [5.25]:

$$\text{Minimize } C_{total} = C_G + C_S + C_{PVS} + C_{LS}$$

$$\text{for } t_i = \{t_0, t_0 + \Delta t, t_0 + 2\Delta t, \ldots, t_F\} \text{ and with respect to :}$$

$$\begin{cases} p_L(t_i) + p_{G\_I}(t_i) + p_{S\_C}(t_i) = p_{G\_S}(t_i) + p_{S\_D}(t_i) + p_{PV}(t_i) \\ p_S(t_i) = p_{S\_C}(t_i) - p_{S\_D}(t_i) \\ p_{PV}(t_i) = p_{PV\_MPPT}(t_i) - p_{PV\_S}(t_i) \\ p_L(t_i) = p_{LD}(t_i) - p_{L\_S}(t_i) \\ \text{if } p_{PV\_MPPT}(t_i) > p_{LD}(t_i) \text{ then } p_{L\_S}(t_i) = 0 \\ \text{if } p_{PV\_MPPT}(t_i) = p_{LD}(t_i) \text{ then } \begin{cases} p_{L\_S}(t_i) = 0 \\ p_{PV\_S}(t_i) = 0 \end{cases} \\ \text{if } p_{PV\_MPPT}(t_i) < p_{LD}(t_i) \text{ then } p_{PV\_S}(t_i) = 0 \\ SOC_{\min} \leq SOC(t_i) \leq SOC_{\max} \\ SOC(t_i) = SOC_0 + \dfrac{1}{3600 \cdot v_S \cdot C_{REF}} \displaystyle\sum_{t_i = t_0}^{t_F} (p_{S\_C}(t_i) - p_{S\_D}(t_i))\Delta t \\ p_{PV}(t_i) \geq 0 \\ p_L(t_i) \geq 0 \\ p_{PV\_S}(t_i) \geq 0 \\ p_{L\_S}(t_i) \geq 0 \\ -P_{S\_\max} \leq p_S(t) \leq P_{S\_\max} \\ 0 \leq p_{G\_I}(t_i) \leq P_{G\_I\_lim} \\ 0 \leq p_{G\_S}(t_i) \leq P_{G\_S\_lim} \\ -\text{Limit} \leq p_G(t_i) - p_G(t_{i-1}) \leq \text{Limit} \\ p_G(t_i) \geq 0, p_S(t_i) \geq 0 \quad \text{if } p_{PV}(t_i) - p_{LD}(t_i) \geq 0 \\ p_G(t_i) < 0, p_S(t_i) < 0 \quad \text{if } p_{PV}(t_i) - p_{LD}(t_i) < 0 \\ \qquad p_{PV\_S}(t) = 0 \quad \text{if } SOC(t) < SOC_{\max} \end{cases} \quad [5.25]$$

4.1.2 Off-Grid Mode

The isolated DC microgrid has to supply the DC load while reducing global energy cost. The diesel generator works in bang-bang mode as presented in chapter "Direct Current Microgrid Power System Modeling and Control;" thus only the storage can be continuously controlled. The power flow diagram used in power management and optimization for off-grid operation mode is shown in Fig. 5.9 [16].

The total energy cost C_{total} is given by Eq. [5.26]:

$$C_{total} = C_{DG} + C_S + C_{PVS} + C_{LS} \qquad [5.26]$$

where C_{DG} is the diesel generator energy cost for which the definition is expressed for each time duration Δt and $t_i = \{t_0, t_0 + \Delta t, t_0 + 2\Delta t, ..., t_F\}$. In Eq. [5.27] c_{DG} is the diesel generator energy tariff:

$$C_{DG} = \sum_{t_i=t_0}^{t_F} c_{DG}(t_i) \cdot \Delta t \cdot p_{DG}(t_i) \qquad [5.27]$$

Concerning the storage, PVA, and load, the same constraints as in on-grid mode are implied in this optimization objective. The diesel generator operating mode follows the model given in chapter "Direct Current Microgrid Power System Modeling and Control." In addition, for the off-grid mode, to ensure continuous operation the SOC lower limit at the end of one operation can be assigned as given in Eq. [5.28]:

$$SOC(t_F) \geq SOC_F \qquad [5.28]$$

with SOC_F a constant for the desired SOC value at the final time.

Because the optimization goal is to minimize fuel consumption with respect to SOC, to reduce load shedding and PV source shedding the off-grid mode optimization problem is formulated by Eq. [5.29] [17–19].

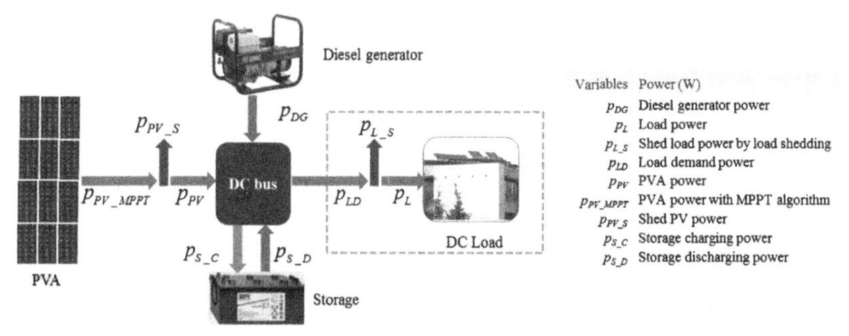

Figure 5.9 Power flow representation using unidirectional parameters for off-grid mode. *DC*, direct current; *MPPT*, maximum power point tracking; *PV*, photovoltaic; *PVA*, photovoltaic array.

$$
\begin{aligned}
&\text{Minimize } C_{total} = C_{DG} + C_S + C_{PVS} + C_{LS}\\
&\text{with respect to :}\\
&\left\{
\begin{array}{l}
p_{PV}(t_i) + p_{DG}(t_i) + p_{S\_D}(t_i) = p_{S\_C}(t_i) + p_L(t_i)\\
p_S(t_i) = p_{S\_C}(t_i) - p_{S\_D}(t_i)\\
p_{PV}(t_i) = p_{PV\_MPPT}(t_i) - p_{PV\_S}(t_i)\\
p_L(t_i) = p_{LD}(t_i) - p_{L\_S}(t_i)\\
\text{if } p_{PV\_MPPT}(t_i) > p_{LD}(t_i) \text{ then } p_{L\_S}(t_i) = 0\\
\text{if } p_{PV\_MPPT}(t_i) = p_{LD}(t_i) \text{ then } \left\{\begin{array}{l} p_{L\_S}(t_i) = 0\\ p_{PV\_S}(t_i) = 0\end{array}\right.\\
\text{if } p_{PV\_MPPT}(t_i) < p_{LD}(t_i) \text{ then } p_{PV\_S}(t_i) = 0\\
SOC_{\min} \leq SOC(t_i) \leq SOC_{\max}\\
SOC(t_i) = SOC_0 + \dfrac{1}{3600 \cdot v_S \cdot C_{REF}} \displaystyle\sum_{t_i=t_0}^{t_F} p_S(t_i)\Delta t\\
SOC(t_F) \geq SOC_F\\
p_{PV}(t_i) \geq 0\\
p_L(t_i) \geq 0\\
p_{PV\_S}(t_i) \geq 0\\
p_{L\_S}(t_i) \geq 0\\
0 \leq p_{DG}(t_i) \leq P_{DG\_P}\\
-P_{S\_MAX} \leq p_S(t_i) \leq P_{S\_MAX}\\
p_{DG}(t_i) = \lambda \cdot P_{DG\_P} \quad \text{with} \quad \lambda \in \{0, 1\}\\
t_i = \{t_0, t_0 + \Delta t, t_0 + 2\Delta t, \ldots, t_F\}
\end{array}
\right.\\
&\left\{
\begin{array}{l}
p_{DG}(t_i) = p_{DG}(t_{i-1}) \quad \text{if} \quad rem(t_i/dt_{BG}) \neq 0\\
t_i = \{t_0 + \Delta t, t_0 + 2\Delta t, \ldots, t_F\}
\end{array}
\right.
\end{aligned}
\qquad [5.29]
$$

4.2 Solving the Problem

Linear programming techniques are common approaches to solve optimization problems that can be expressed in the standard form given by Eq. [5.30]:

$$\mathrm{Max}(\mathrm{Min})Z = c_1x_1 + c_2x_2 + \cdots\cdots + c_nx_n$$
$$\text{subject to} \begin{cases} a_{11}x_1 + a_{12}x_2 + \cdots + a_{1n}x_n \leq (=, \geq) b_1 \\ a_{21}x_1 + a_{22}x_2 + \cdots + a_{2n}x_n \leq (=, \geq) b_2 \\ \quad \vdots \qquad \vdots \qquad \vdots \\ a_{m1}x_1 + a_{m2}x_2 + \cdots + a_{mn}x_n \leq (=, \geq) b_m \\ x_1, x_2, \cdots, x_n \geq 0 \ (\leq 0, \text{freedom}) \end{cases} \quad [5.30]$$

The Simplex algorithm, proposed by Dantzig [20], is widely used to solve such problems very efficiently when all variables are real. It can be verified that the mathematical formulation given by Eqs. [5.25] and [5.29] follows the standard form expressed by Eq. [5.30] except for the last constraint, which does not take a linear form.

However, this constraint can be easily linearized by introducing a variable array x_i for each time point t_i. Different from the other variables of the formulation that take a continuous real value, variable x_i has to take its value in $\{0,1\}$ as a binary variable. The following constraints expressed by Eq. [5.31] are used to ensure that $x_i = 0$ if the $SOC_{\max}$ is not reached at time t_i and $x_i = 1$ in the other case:

$$x_i \leq \frac{SOC(t_i) + \varepsilon}{SOC_{\max}} \quad [5.31]$$

where ε is a small real constant (0.005 in this application) allowing an acceptable margin around the value of $SOC_{\max}$. As soon as the quantity $SOC(t_i) + \varepsilon$ is greater than $SOC_{\max}$, variable x_i can take the value 1. Finally, the constraint given by Eq. [5.32] is used to ensure that the constraint given by Eq. [5.12] can be fully respected:

$$p_{PV\_S}(t_i) \leq x_i \cdot (p_{PV}(t_i) - p_{LD}(t_i)) \quad [5.32]$$

Eq. [5.32] expresses the fact that $p_{PV\_S}(t_i) = 0$ when $x_i = 0$ (ie, if the $SOC_{\max}$ is not reached at time t_i) and $p_{PV\_S}(t_i)$ can take a value no greater than $(p_{PV}(t_i) - p_{LD}(t_i))$ when $x_i = 1$ (ie, if the $SOC_{\max}$ is reached at time t_i).

Note that the x_i variable data type is an integer (more precisely binary), which is why the previous formulation is now said to be a mixed integer linear program. The direct consequence is that the very efficient Simplex algorithm cannot be used directly anymore because it works only with real variables. Instead, more advanced techniques, often using Simplex as subroutines (eg, cutting plane methods, branch and bound algorithms, branch and cut, or branch and price methods) are used to solve these problems.

Nevertheless, numbers of mixed integer linear programming solvers implementing these optimization techniques exist, such as CPLEX [21], LP_SOLVE, and GUROBI. In this study the formulated optimization problems are solved using the IBM ILOG CPLEX solver, which is a powerful tool for solving different types of optimization problems. However, any other mixed integer linear programming solver also can be used. Although mixed integer linear programs are commonly more difficult to solve than linear programs because of the integer variables, our problem can be solved very efficiently by such a solver, even with a huge set of time points.

To express the optimization problem in the solver syntax and call the solver solution algorithm, a procedure written in C++ is used. This procedure outputs the optimum power flow in the form of a file corresponding to the definition of the control parameter K_D, which is subsequently transmitted to the operational layer. This procedure is illustrated in Fig. 5.10 [12].

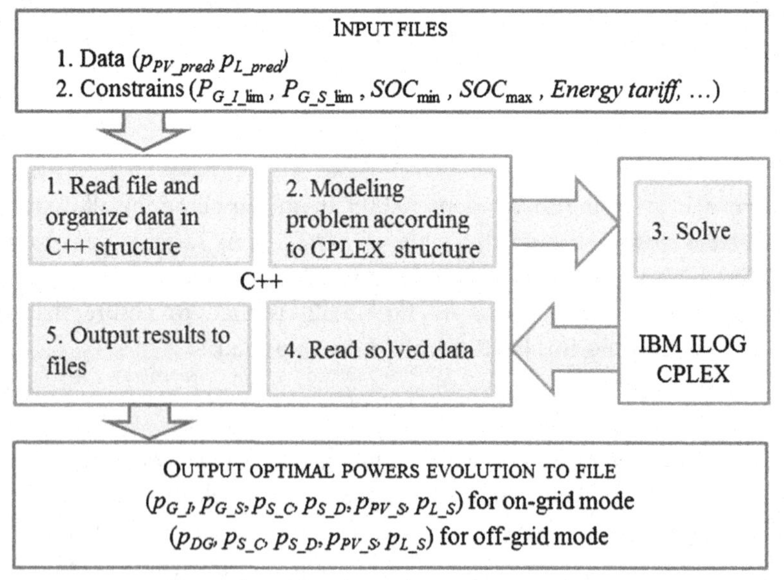

Figure 5.10 Flowchart of optimization solving.

Therefore, for the DC microgrid grid-connected mode, the result obtained by the output of this procedure is used to calculate the distribution coefficient K_D following Eq. [5.16]. Thus K_D becomes a time-variant sequence depending on the time duration taken into account in the optimization calculation.

For the off-grid mode the energy management gives the optimized power flow as the time series evolution of p_{DG}, p_{S_D}, p_{S_C}, p_L, and p_{PV}, which is translated into distribution coefficient K_D. Because the diesel generator is considered at two states, $p_{DG} = 0$ and $p_{DG} = P_{DG\_P}$, K_D is defined as the switching signal for diesel generator operation whereas storage power is assigned in control strategy to balance power.

4.3 Interface for Operation Layer

The optimization calculates optimal power flow for the grid-connected mode and off-grid mode. The prediction layer provides the prediction of possible PVA power p_{PV_pred} and load power evolutions p_{L_pred} for the next hours. According to the optimization formulation, the energy management layer gives the optimized power flow of the sources and for the load. The optimized power flow is then translated into an interface parameter, which is the distribution coefficient K_D that is used for both grid-connected and off-grid mode, to control the microgrid to reproduce the optimization results in real operation.

With different operating modes K_D is differently defined.

In grid-connected mode, K_D represents the best power distribution between the grid and the storage [22]. By the prediction metadata, the energy management layer estimates the optimum control power flow for the operation of next hours. The estimated optimum control power flow is then translated into control parameter K_D to run the operation layer. The optimum K_D is calculated by Eq. [5.16]. Thus the optimum K_D sequence represents optimum power flow in one parameter. K_D is the interface parameter between the energy management layer and the operational layer. The advantage of using K_D lies in coupling easily power balancing and energy management so that robust power balancing and optimization can be achieved at the same time. The power balancing strategy is designed to satisfy all constraints and is parameterized with K_D. The energy management layer gives only K_D through low-speed communication to control the operation instead of updating all power references in real time. Optimization can be reperformed online, updating the K_D sequence without interrupting power balancing, which takes full advantage of the latest prevision and real-time system status. Moreover, because the operational layer can keep power balancing with any K_D value, the operation is robust to withstand prediction errors [23].

In off-grid mode the two controllable sources are storage and the diesel generator. Storage power can continuously change. The diesel generator is assigned with two states: on or off. In this case storage power is assigned in control strategy to balance power, and K_D is defined as a switch signal for the diesel generator as in Eq. [5.33]:

$$K_D(t) = boolean(p_{DG}(t) > 0) \quad [5.33]$$

Thus when $K_D = 1$ the diesel generator starts and outputs the rated power. For $K_D = 0$ the diesel generator has no specific order; therefore it is the power balancing control algorithm that determines turning on or off the diesel generator [16]. For the two operating modes K_D is the time-variant sequence that contains sufficient information to represent the optimized power flow.

5. OPERATION LAYER

According to the energy management layer output $K_D(t)$, grid power limits, and power system states, the operation layer presented in Fig. 5.11 aims at balancing power in the power system while meeting all constraints presented above for each operating mode [8].

This layer aims at balancing power in the power system while respecting the energy management performed in the upper layer. Having defined $K_D(t)$ as the interface between the energy management layer and the operational layer, adequate strategies in the operational layer must be developed. The control strategy should satisfy the following requirements. On one hand, given the same predicted operating condition as in optimization, the control strategy is required to reproduce optimized power flow in real operating conditions. On the other hand, the operation strategy must ensure robustness and withstand uncertainties introduced in predicted power due to forecast data.

The operational layer calculates and outputs the power references and the load shedding coefficient. Thus the operational layer controls in real time the power balancing by applying the $K_D(t)$ calculated offline, but it must still be able to provide self-correcting actions. Indeed, the actual operating conditions lead to a reference power p^* different from that

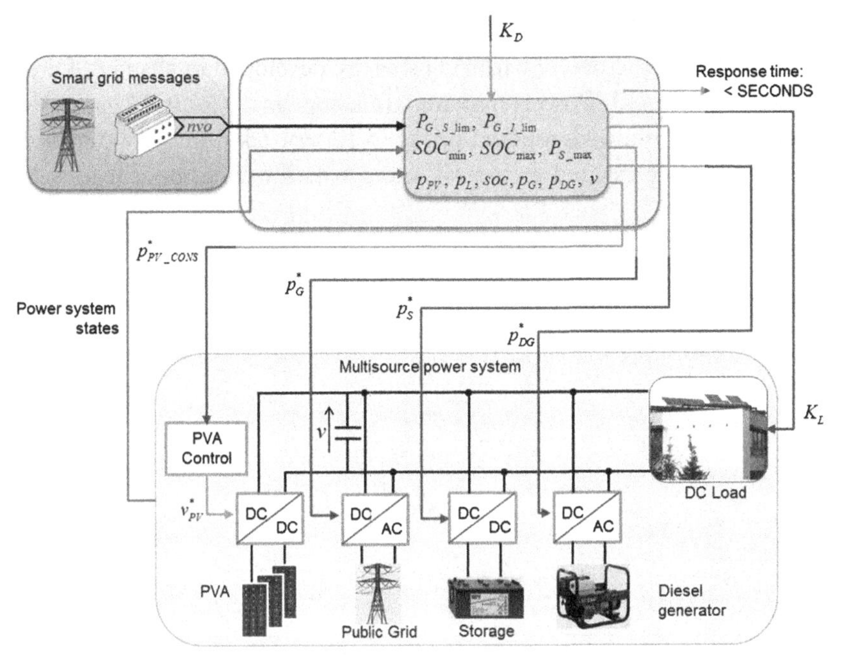

Figure 5.11 Operational layer design. *AC*, alternating current; *DC*, direct current; *PVA*, photovoltaic array.

calculated offline during the optimization step. Thus the power references $p^*_{PV\_CONS}$, p^*_G, p^*_S, and p^*_{DG} calculated with optimum $K_D(t)$ are most of the time different from those optimized. In real-time operation, self-correction is applied according to three degrees of flexibility. It is the grid power or the diesel generator power that provide the largest share of flexibility, but the limitation of the PVA production and load shedding are also used where appropriate.

To ensure the correct operating of the power system, at least two algorithms are needed and must be implemented: one for the PVA control, as described in chapter "Photovoltaic Source Modeling and Control," and the second to control the power balancing, the description of which is given in this section. In theory a third algorithm to control the load, by appliance shedding, should be required; however, as mentioned already, this is not developed in this book. On the basis of the control strategy developed in the power system study, this section gives the operating algorithm for grid-connected mode and off-grid mode.

5.1 Control Algorithm for Grid-Connected Mode

On the basis of the power balancing strategy developed in chapter "Direct Current Microgrid Power System Modeling and Control," a power balancing algorithm is proposed in Fig. 5.12 that takes into consideration the constraints presented previously and allows for continuous load shedding [12].

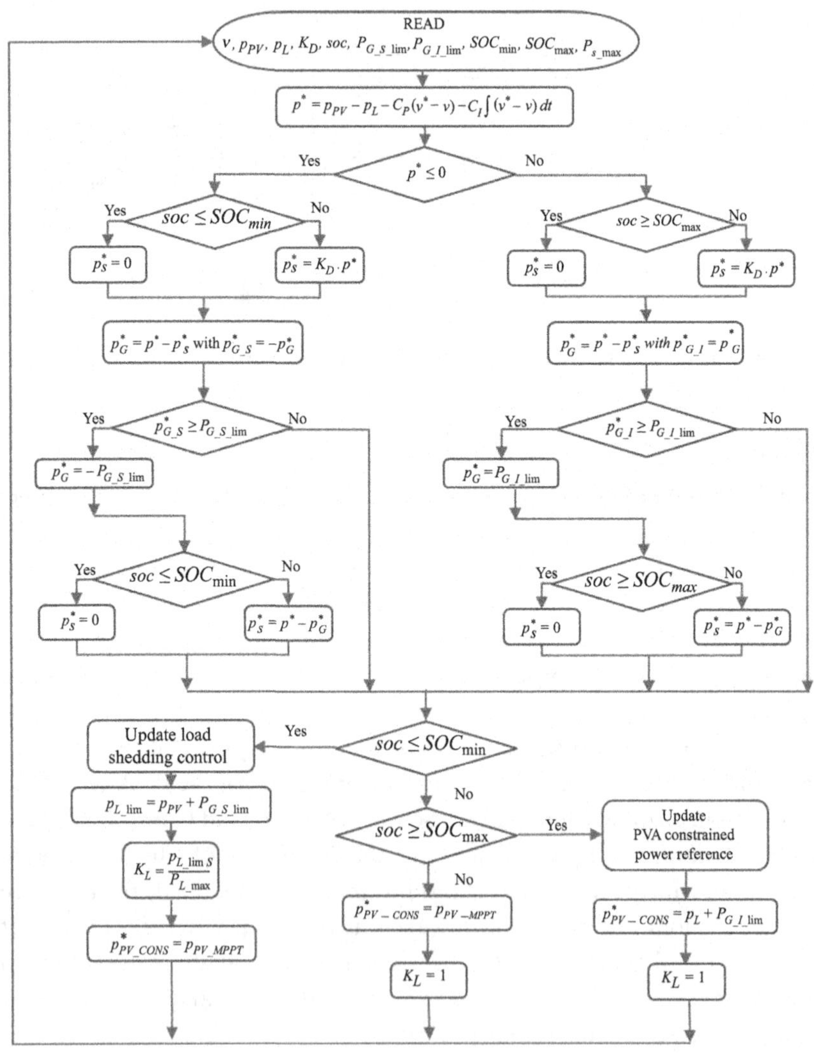

Figure 5.12 Flowchart of grid-connected operation algorithm.

In this algorithm storage power control reference p_S^* is updated once more if grid power reaches its limit, signifying that the storage, if available, is used first in power balancing before performing load shedding or PVA constrained production. Power balancing can be maintained with any K_D value. The power sharing proportion defined by K_D can be satisfied except for extreme cases. Concerning self-correcting ability in power balancing, the grid power represents the most important degree of flexibility, but PVA constrained production and load shedding are also performed if necessary.

5.2 Control Algorithm for Off-Grid Mode

Regarding grid-connected mode, in off-grid mode the power balancing reveals more difficulty because of the lack of one degree of flexibility. This is explained by the diesel generator bang-bang operating mode; thus only the storage can be continuously controlled. The control algorithm for the DC microgrid in off-grid operation mode is presented in Fig. 5.13 [16].

As mentioned earlier, in off-grid mode it could happen that the PVA outputs rated power whereas the diesel generator is turned on and load consumption is low. Charging storage with high power will shorten storage life; therefore at least one source needs to be limited or cut off. The priority is defined as shedding PVA first and then cut off the diesel generator if $SOC_{\max}$ is reached.

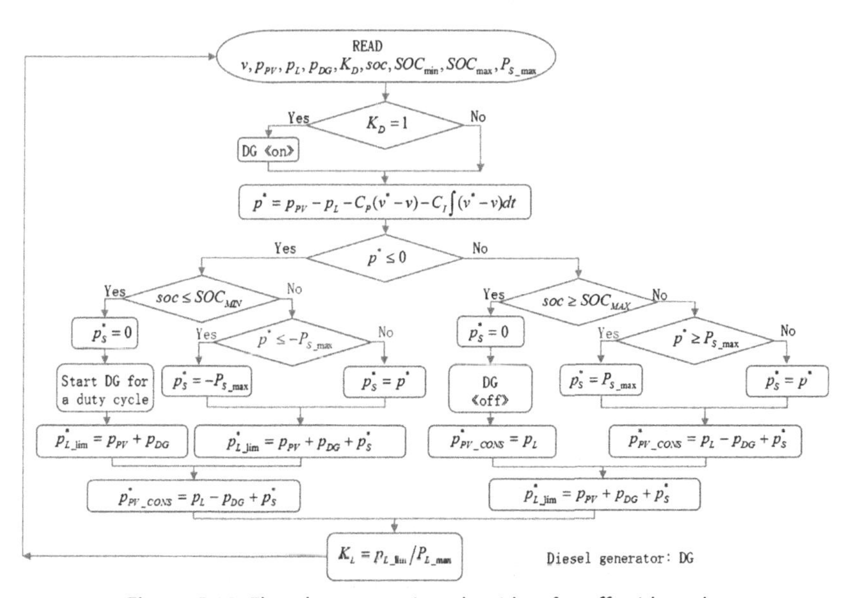

Figure 5.13 Flowchart operation algorithm for off-grid mode.

6. EVALUATION OF THE SUPERVISORY SYSTEM BY SIMULATION

The supervisory control design proposed in Fig. 5.14 is able to optimize predicted power flow and to balance instantaneous power based on unique interface parameter $K_D(t)$ [23].

The supervisory control takes into consideration prediction data and various constraints, such as grid power limits and storage capacity, which can be fully respected for optimization and real operation. Each layer provides an independent function; thus the structure is flexible and can be implemented in several microcontrollers or computers so that real-time power balancing control and complicated optimization can be executed at the same time without affecting each other. The multilayer structure simplifies implementation of such a complex control strategy. Thus the control of the power balancing is separated from the energy management layer, yet they are linked through one interface parameter $K_D(t)$. On one hand, the energy management layer is able to optimize the microgrid energy cost through predictive data and thus obtain the predicted optimized power flow, which is then translated into the $K_D(t)$ sequence. On the other

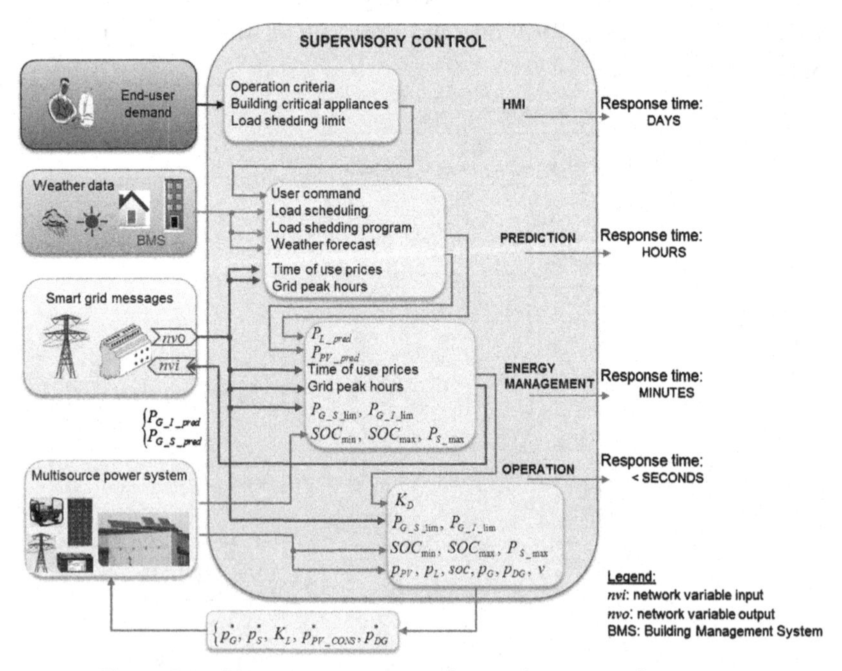

Figure 5.14 Direct current microgrid supervisory control overview.

hand, the power balancing control in the operational layer is an independent function that can work with any $K_D(t)$ value. The distribution coefficient $K_D(t)$ is a single interface parameter, but it represents the power flow from different sources. Hence the communication of $K_D(t)$ does not need high-speed communication between layers.

The microgrid supervisory control is tested by simulation with numerical power values based on the experimental DC microgrid platform [8]. In the following subsections, simulation results are given for both microgrid operating modes (ie, grid-connected and off-grid mode).

Concerning the energy tariffs used in simulation tests some clarification is needed. Taking into consideration criteria such as peak shaving, avoiding undesired injection, making full use of available storage capacity, and avoiding possible load shedding and PVA shedding together, it is difficult to refer to real energy tariffs. In this study, tariff and cost function are rather a technique for optimizing multiple criteria at the same time than only calculating energy invoices. On the other hand, concerning PVA shedding, load shedding, and storage, their energy tariff calculation is complex and depends on the chosen technology. This is why the energy tariffs used in this study are somewhat arbitrary; however, the numerical values of energy tariffs given here are chosen to highlight the logic of management strategy that seems to be an energy trend for the next 20 years. Nevertheless, only a relative energy tariff can change the optimization results, which means that the optimized power flow remains the same if the energy tariff absolute values are changed but the energy tariff for different sources remain in the same order as given by Eq. [5.34]:

$$\begin{aligned} c_{PVS(LS)} \gg c_{G\_PH} > c_{G\_NH} > c_S \quad & \text{for the on-grid mode} \\ c_{DG} > c_{PVS(LS)} \gg c_S \quad & \text{for the off-grid mode} \end{aligned} \qquad [5.34]$$

Regarding the grid energy tariff c_G, for normal hours c_{G_NH} an average energy tariff, close to that offered by most providers, is considered. In contrast, for peak hours, a very penalizing purchase tariff c_{G_PH} is chosen, which is not proposed today. The reason for this assumption lies in providing demand response, for most critical hours, to perform power peak shaving.

6.1 Simulation Results for Grid-Connected Mode

The DC microgrid supervisory control for the grid-connected mode is simulated for an operation based on the weather conditions of April 23,

2011, in Compiegne, France [8]. The objective of this study is more to validate a comprehensive approach rather than purely numerical results. For this reason we do not give the numerical values of various system components studied, but power values are based on the existing experimental multisource system platform. The simulation results are obtained under a MATLAB-Simulink environment. All of the implemented automatic controls are satisfactorily working. The proportional integral controller, the gains of which are used in power balancing, provides a wide control bandwidth adapting to the simulation step. However, synthesis for parameter tuning is not realized in this work.

The PVA prediction data are calculated from the real measurement data to assign the day-ahead prediction data uncertainty error as random. The measured PVA MPPT power, the values of which have been recorded by the PVA experimental platform, is shown Fig. 5.15a, in which the green curve shows the real-time PVA power evolution and the gray bars are the hourly average PVA power.

Fig. 5.15b gives the considered PVA power prediction data (in red (black in print versions)), which are hourly data having $\pm 10\%$ random error with an hourly average of the measurement data (in green (gray in print versions)).

Load power prediction data are supposed to be given by the load management system, which implies additional uncertainties. In this study a simple arbitrary load power evolution is considered. The difference between the load power and the load power prediction is shown in Fig. 5.16.

To perform an optimized operation for the next hours, the prediction layer is supposed to give to the energy management layer the calculated PVA predicted power and the load predicted power hourly evolutions.

Peak hours during the day are arbitrarily assumed to be 11:00–13:00 and 16:00–18:00. Table 5.1 provides the parameters used for optimization and the power balancing control strategy.

These parameters are selected according to system configuration with an aim to involve as much system behavior as possible during the tests, such as storage events (full, empty), load shedding, and PV source power limiting.

Taking into consideration these power predictions, the optimized power flow and $K_D(t)$ evolution, supposed to be calculated a few hours before the operation of April 23, are shown in Fig. 5.17a and b, respectively. Note that the prediction powers are also shown in the same graphic with the optimized powers: the optimized powers are noted with the subscript OPT.

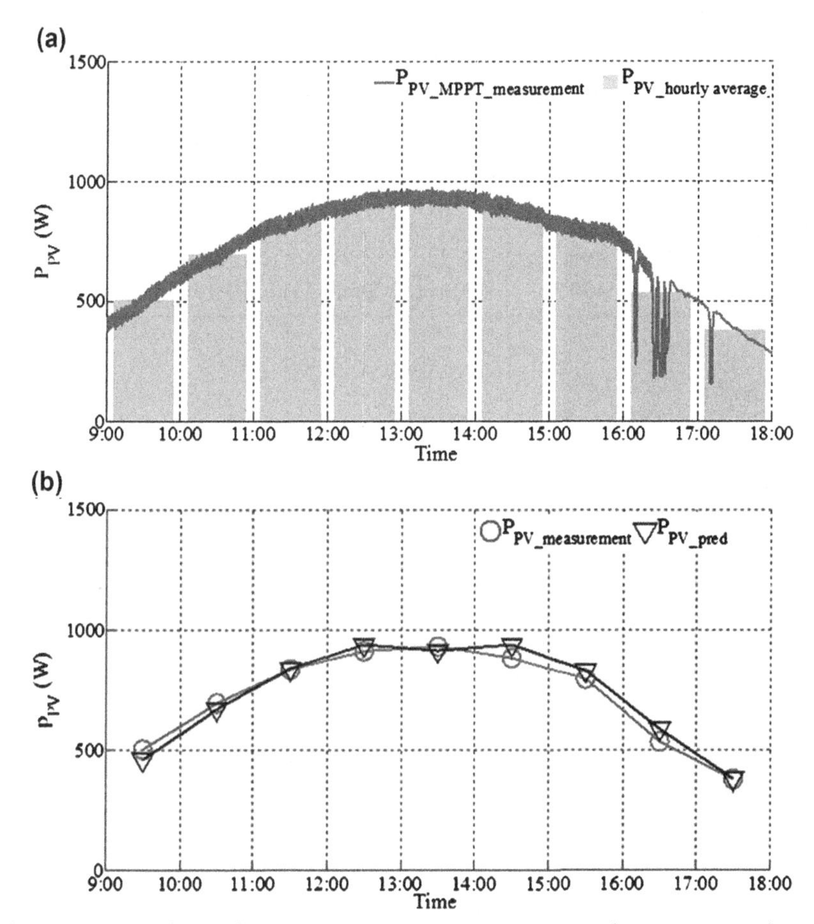

Figure 5.15 (a) Photovoltaic array maximum power point tracking power evolution and (b) hourly photovoltaic power measurement and prediction.

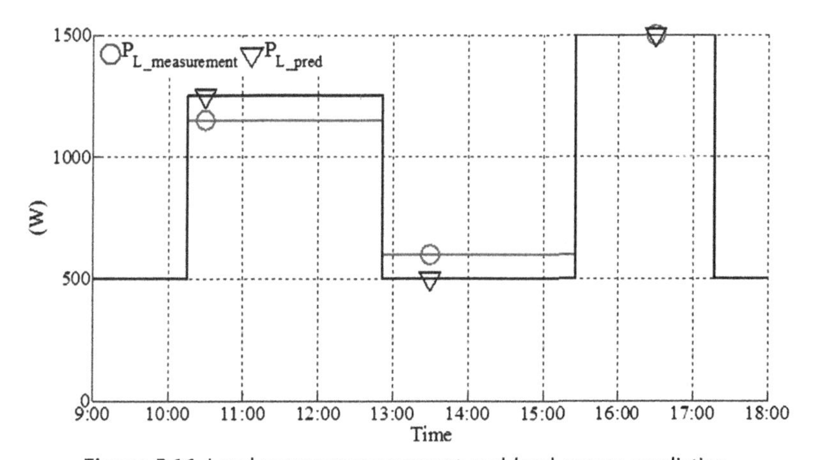

Figure 5.16 Load power measurement and load power prediction.

Table 5.1 Optimization and simulation parameter values

$P_{G\_S\_lim}$	600 W	SOC_{min}	45%
$P_{G\_I\_lim}$	700 W	SOC_{max}	55%
P_{L_max}	1500 W	c_{G_NH}	0.1 €/kWh
v^*	400 V	c_{G_PH}	0.7 €/kWh
P_{S_max}	1500 W	c_S	0.05 €/kWh
P_{PV_MPPT} at STC	1000 W	c_{PVS}	1 €/kWh
SOC_0	50%	c_{LS}	1 €/kWh
C_{REF}	130 Ah	Grid peak hours	11:00–13:00; 16:00–18:00

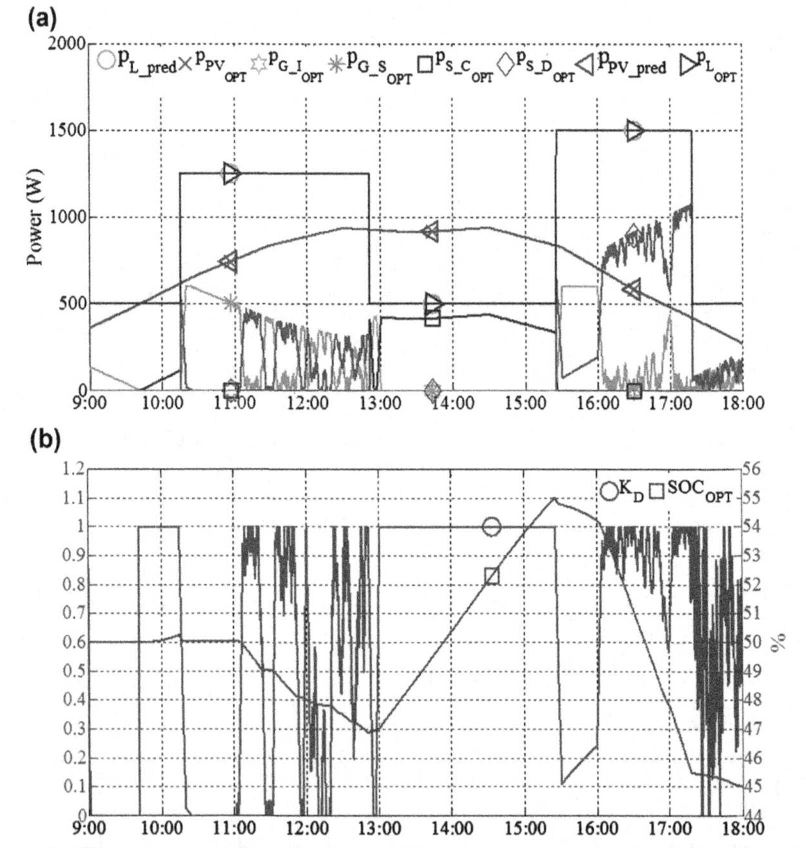

Figure 5.17 (a) Optimized power flow and (b) $K_D(t)$ and state of charge evolution for on-grid mode. *SOC*, state of charge.

On the basis of the prediction information, the energy management layer calculates the optimization problem by CPLEX and gives the optimum power flow evolution as presented in Fig. 5.17a. The corresponding $K_D(t)$ sequence is calculated by Eq. [5.16] from the optimum power flow evolution, as presented in Fig. 5.17b. For performing a day optimization during 9 h, the data resolution is chosen at 10 s/point (ie, 3240 points for each power curve). The optimization program execution time is within 10 s for a computer with a CORE i5 processor. The optimum total energy cost for this period estimated by CPLEX is 0.517 €.

Fig. 5.17 shows the optimization calculation results. It can be observed that for peak hours the PVA production is not sufficient to supply the load; therefore the storage and the grid have to supply the load for the remaining part of the power. The considered optimization problem is formulated to minimize the global energy cost for the whole period from 9:00 to 18:00 while respecting all constraints. Therefore, for the peak hours the grid energy tariff is very high and largely superior to storage energy tariff, and it seems normal that the storage has to be proposed to supply the load. During the peak hours the *SOC* decreases continually, as shown in Fig. 5.17b. Hence, the storage is used to supply the load as much as possible with respect to its *SOC* lower limit and its power limit. However, because the storage energy is not enough, the grid power is also used to supply the load. This is why during peak hours the storage power and grid power are proposed to share the necessary power to supply the load in an optimized manner while respecting all constraints. This sharing power is proposed in an intermittent manner by the used solver.

6.1.1 Power Flow Simulation Controlled by $K_D(t)$

The calculated $K_D(t)$ optimum evolution is transmitted to the operational layer to run the power system following the conditions given on April 23, 2011 (meteorological and load). The operation power flow, as a real situation, simulated by MATLAB, is shown in Fig. 5.18a. During this day operation, grid and storage share power for supplying energy or for receiving energy at the same time. In the first off-peak hours (ie, 9:00–11:00) the power grid mainly supplies the load for reserving storage for peak hour supply.

During the first peak hours (ie, 11:00–13:00) the load is supplied by storage and the grid, and the sharing proportion is determined by optimization calculation, which also aims to reserve storage for supplying during the second period of peak hours. Just before 13:00 the surplus of the

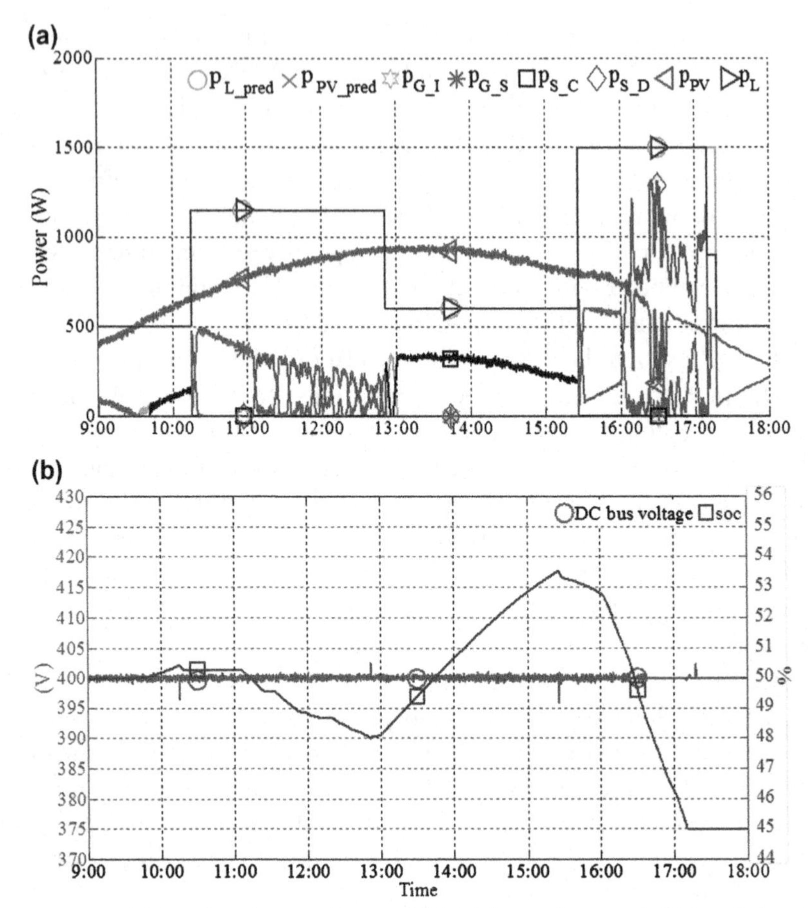

Figure 5.18 (a) Simulated power flow and (b) DC bus voltage and *SOC* evolution for optimum $K_D(t)$ for on-grid mode. *DC*, direct current; *SOC*, state of charge.

PVA production is injected into the grid to make the maximum profit. Aiming to reduce the energy cost by avoiding the power grid supplying the load during peak hours, in the second peak hour period (16:00–18:00) the storage is mainly used for supplying the load. During 13:00–15:00 the storage is charged with the excess PVA production, and then it is discharged for supplying the load in the second peak hour. Grid power injection limit and supply limit are respected.

Short-time load shedding can be seen in the operation after 17:00, when the storage is empty. The load shedding is performed based on instantaneous power information. To avoid load shedding fluctuations due to the PVA power fluctuations, it is also possible to impose duration for load shedding in CPLEX optimization; therefore optimized load shedding information could be given to the operational layer to override the operation layer load shedding control.

The total energy cost for the concerned period is 0.512 €, which is close to the optimization total cost. The *SOC* evolution with the optimum $K_D(t)$ and the DC bus voltage are illustrated in Fig. 5.18b. The DC bus voltage fluctuations are negligible compared with the value of 400 V, signifying the power is well balanced.

Comparing the results presented in Figs. 5.17 and 5.18, it can be seen that the simulated power flow is slightly different from the optimization because of the uncertainties of solar irradiance prediction and load power prediction. During the solar irradiance fluctuations between 16:00 and 17:00, the storage provided more power. However, in this case the storage is still able to be the main load supply during the second period of peak hours, but a slight load shedding occurs when the *SOC* reaches its low limit.

6.1.2 Power Flow Simulation Controlled by Constant K_D

To further analyze, a simulation case for a constant K_D is presented in Fig. 5.19. $K_D = 0.5885$ is chosen as the constant value, which is the average value of optimum $K_D(t)$ evolution shown in Fig. 5.17b.

In this case the obtained total energy cost is 0.652 €; the difference with optimization is larger compared with using optimal $K_D(t)$, and longer load shedding can be seen during this operation. The *SOC* evolution, illustrated in Fig. 5.19b, is very different from the optimum *SOC* evolution shown in Fig. 5.18b. Nevertheless, even if the optimization effect is affected, the power balancing is robust. Regarding the DC bus voltage illustrated in Fig. 5.19b, it can be seen that the DC bus voltage remains stable with very slight fluctuations, signifying that the power is well balanced.

6.1.3 Comparison and Discussion

Table 5.2 shows the total energy cost of the DC microgrid C_{total} for the considered period (ie, 9:00–18:00) and occurrences of load shedding for these three cases: (1) optimized operation by energy management layer with ±10% uncertainties prediction data, (2) simulated operation in the case of a real PVA production with the calculated optimum $K_D(t)$, and (3) simulated operation with constant $K_D = 0.5885$. It can be seen that the simulated cost for the optimum $K_D(t)$ is close to the prediction cost, and the error is within 1%, which is due to prediction uncertainties.

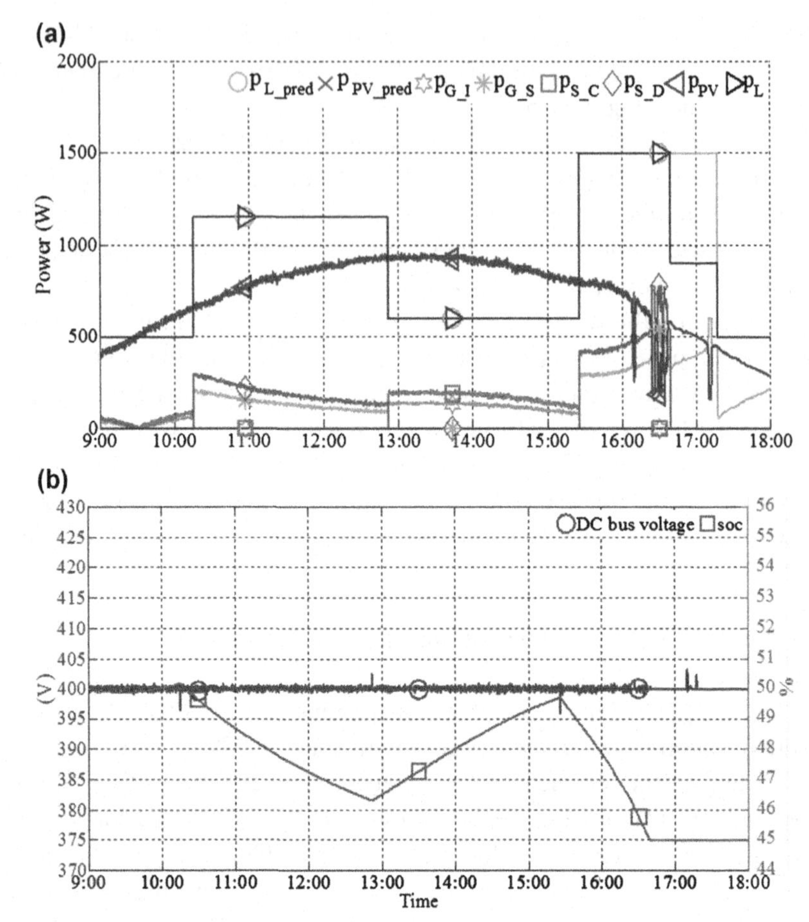

Figure 5.19 (a) Simulated power flow and (b) DC bus voltage and *SOC* evolution for constant $K_D(t) = 0.5885$ for on-grid mode. *DC*, direct current; *SOC*, state of charge.

Table 5.2 Comparison of different simulations cases for on-grid mode

Case	K_D	C_{total} (€)	Load shedding
Optimization	Optimum $K_D(t)$ as given in Fig. 5.17b	0.517	No shedding
Simulation	Optimum $K_D(t)$ as given in Fig. 5.17b	0.512	7 min
Simulation	$K_D = 0.5885$	0.652	37 min

For the constant K_D case, the cost is 26% more than the prediction case; moreover, longer load shedding can be seen in this case. Even with some uncertainties the optimized $K_D(t)$ operates the microgrid with respect to the utility grid requirements and storage capacity. This comparison validates the presented simulation case for the proposed supervisory control of the proposed DC microgrid.

As previously mentioned, the communication of $K_D(t)$ does not need high-speed communication between layers. Therefore the supervisory control provides the possibility of reperforming optimization and updating the $K_D(t)$ sequence during the actual operation without interrupting the power balancing. Thus hourly or more frequent optimization that updates the sequence with the latest prediction and power system status is expected to provide better energy performance of the supervisory control.

However, the limit of the supervisory control is that optimizing effectiveness is affected on the prediction precision. Prediction uncertainties do not influence power balancing but the optimal energy cost is affected. Future research should focus on enhancing optimization performance, especially facing low prediction precision. An optimization technique that is able to optimize power flow with consideration to uncertainties of the prediction combined with a rule-based algorithm in the operation layer that corrects $K_D(t)$ in real time with respect to power system status can be developed as one solution. In addition, a second storage can be installed as a backup for correcting the errors between optimized power and real operation.

6.2 Simulation Results for Off-Grid Mode

The supervisor control for the off-grid mode is simulated for the same operating conditions on April 23, 2011, in Compiegne, France. The test conditions are the same as in the grid-connected mode. Table 5.3 gives the specific parameters used for optimization and the power balancing control strategy for off-grid mode.

Table 5.3 Optimization and experimental parameter values for off-grid mode

P_{DG_P}	1500 W	C_{REF}	130 Ah
dt_{DG}	1200 s	$SOC_{\min}$	45%
$P_{L\_\max}$	1500 W	$SOC_{\max}$	55%
v^*	400 V	c_{DG}	1.1 €/kWh
$P_{S\_\max}$	1500 W	c_S	0.01 €/kWh
P_{PV_MPPT} at standard test conditions	1000 W	c_{PVS}	1 €/kWh
$SOC_0 = SOC_F$	50%	c_{LS}	1 €/kWh

The optimized power flow is given in Fig. 5.20a whereas the *SOC* and $K_D(t)$ evolution are given in Fig. 5.20b. For better visibility of the power curves, p_{DG} is presented as a negative value, although the value is positive in equations. This exception is used for figures of off-grid mode throughout the remainder of this book.

In the off-grid case, the final *SOC*, SOC_F, is imposed as 50% with the aim to maintain continuous operation for several days. It can be seen in power flow optimization that the diesel generator is started for several duty cycles and the storage is operated within its power limits of 1500 W and *SOC* limits between 45% and 55%. The final *SOC* value is greater than 50%. The optimized diesel generator working cycle is translated as the

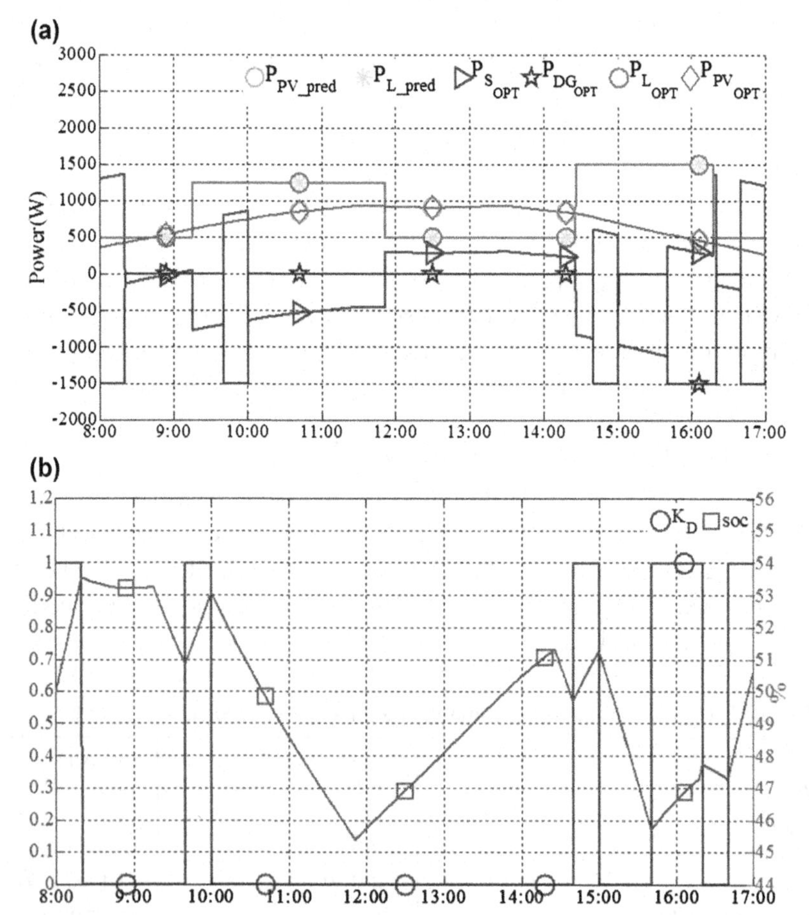

Figure 5.20 (a) Optimized power flow and (b) $K_D(t)$ and *SOC* evolution for off-grid mode. *SOC*, state of charge.

control parameter $K_D(t)$ to control the on/off state of the diesel generator in power balancing operation.

6.2.1 Power Flow Simulation Controlled by $K_D(t)$

On the basis of the $K_D(t)$ value, the operation in off-grid mode is simulated with MATLAB-Simulink and Stateflow. The obtained power evolution and *SOC* evolution are shown in Fig. 5.21a and b, respectively.

It can be seen even with uncertainty that the simulated power flow is in accordance with the optimization power flow and the diesel generator is started with the optimization order. However, the final *SOC* value is less than the SOC_F.

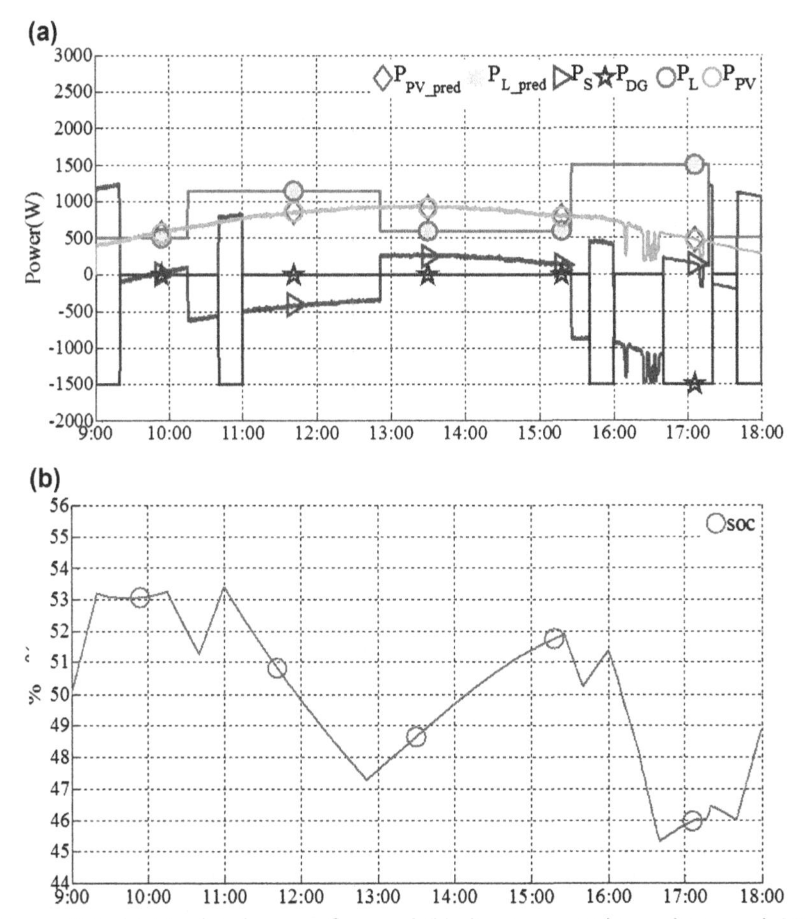

Figure 5.21 (a) Simulated power flow and (b) direct current bus voltage and *SOC* evolution for optimum $K_D(t)$ for off-grid mode. *SOC*, state of charge.

Table 5.4 Comparison of different cases for off-grid mode simulation test

Case	C_{total} (€)	Load shedding	Constrained photovoltaic array power
Optimization	3.295	No	No
Simulation	3.278	No	No

6.2.2 Comparison and Discussion

Table 5.4 summarizes the test of off-grid mode. The optimization estimated that the overall cost of the operation is 3.295 €. The cost of the simulated case is 3.278 €. The cost difference is approximately 0.5%. The simulation cost is less than the optimization cost, but the difference is not significant. This is due to prevision uncertainties between optimization and simulation on one hand and on the other hand that the storage used is more than the expected SOC_F, which is not penalized by the cost function.

It can be seen that the simulated cost with the optimum $K_D(t)$ is close to the optimization cost and that the optimized power flow can be reproduced through a simple interface in real operating conditions even with uncertainties. This comparison validates the presented simulation case for the proposed supervisory control of a multisource power system.

7. CONCLUSIONS

DC microgrid control combining power balancing, optimization, and smart grid interaction is proposed in this chapter. The research issue of implementing optimization in real-time operation is particularly addressed.

On the basis of the definition and representation of the dynamic hybrid system, the supervisory system has been designed to control and monitor the multisource power system, which represents the continuous dynamics of the microgrid. To take into account multiple discrete events, a supervisory control structure has been proposed in the form of four parallel controllers, which led to a multilayer system. This hierarchical control of the DC microgrid aims at managing the balance of the instantaneous power in the microgrid on the basis of energy cost optimization with constraints such as storage limits, public grid power limitations, and energy tariffs, which are variable in time.

Therefore, taking into account the forecast of PVA production and load power demand, the four-layer supervisory system performs optimization and implements the result in instantaneous power balancing through a simple interface. It also handles constraints such as storage capability; grid power limitations from the smart grid; grid time-of-use pricing; grid peak hours in grid-connected mode; and the diesel generator minimum fuel consumption, working duty cycle and storage lifetime in off-grid mode. The optimization is based on mixed integer linear programming and it is solved by CPLEX. Simulation results, even with uncertainties of prediction and arbitrary energy tariffs, show that the proposed supervisory control design is able to perform efficiently and with cost-effective power flow in real-time operation with respect to imposed constraints. Load shedding and PVA constrained power ensure power balancing in any case. The simulation shows that the optimization gives better energy performance while minimizing load shedding and PVA production limitation, and the operational layer respects all constraints of power system elements. On the other side, the optimization efficiency is based on prediction precision, which may limit the final performance. The designed operational layer can work with any $K_D(t)$ value, therefore the prediction errors and nonoptimum $K_D(t)$ does not affect the power balance. However, if uncertainties are higher than $\pm 10\%$, then the energy cost and the load shedding duration could be severely affected. The cost is close, with errors within 1%. It is proved that the proposed supervisory control is feasible to apply optimized power flow in real operating conditions and withstand prevision uncertainties. Moreover, in grid-connected mode grid power limits from the smart grid message are fully respected with the robust control.

To summarize, the feasibility of the proposed DC microgrid supervisory control structure, which combines grid interaction and energy management with power balancing, is proved by simulation results. Although the microgrid only refers to a building scale and involves only a few sources, the idea of parameterized power balancing and interfacing with optimization, as well as smart grid interaction, can be generalized; thus it can be used as a solution for advanced energy management for other microgrids to optimize local power flow and improve future PV penetration.

The next chapter presents experimental results based on real conditions.

REFERENCES

[1] Antsaklis PJ, Stiver JA, Lemmon MD. Lecture notes in computer science, hybrid system. Springer-Verlag; 1993.

[2] Wang BC, Sechilariu M, Locment F. Power flow Petri Net modelling for building integrated multi-source power system with smart grid interaction. Math Comput Simul 2013;91:119–33.

[3] El Bagdouri M, Cebron B, Sechilariu M, Burger J. Variational formalism applied to control of autonomous switching systems. Control Cybern 2004;33(4):535–49.

[4] Henzinger TA. The theory of hybrid automata. In: The 11th IEEE symposium on logic in computer science LICS; 1996.

[5] Cebron B, Sechilariu M, Burger J. Comparison of two calculation methods of the optimal switching instants of a hybrid dynamic system. In: IEEE international conference on systems. Man and Cybernetics SMC; 1999.

[6] Lorenz E, Hurka J, Heinemann D, Beyer HG. Irradiance forecasting for the power prediction of grid-connected photovoltaic systems. IEEE J Sel Top Appl Earth Obs Remote Sens 2009;2:2−10.
[7] Amjady N, Keynia F, Zareipour H. Short-term load forecast of microgrids by a new bilevel prediction strategy. IEEE Trans Smart Grid 2010;1:286−94.
[8] Sechilariu M, Wang BC, Locment F. Supervision control for optimal energy cost management in DC microgrid: design and simulation. Int J Electr Power Energy Syst 2014;58:140−9.
[9] Houssamo I, Wang B, Sechilariu M, Locment F, Friedrich G. A simple experimental prediction model of photovoltaic power for DC microgrid. In: IEEE international symposium on industrial electronics ISIE; 2012.
[10] Houssamo I, Locment F, Sechilariu M. Experimental analysis of impact of MPPT methods on energy efficiency for photovoltaic power systems. Int J Electr Power Energy Syst 2013;46:98−107.
[11] Denoix T, Sechilariu M, Locment F. Experimental comparison of photovoltaic panel operating cell temperature models. In: The 40th annual conference of the IEEE industrial electronics society IECON; 2014.
[12] Sechilariu M, Wang BC, Locment F, Jouglet A. DC microgrid power flow optimization by multi-layer supervision control. Design and experimental validation. Energy Convers Manag 2014;82:1−10.
[13] Wang BC, Houssamo I, Sechilariu M, Locment F. A simple PV constrained production control strategy. In: IEEE international symposium on industrial electronics ISIE; 2012.
[14] Locment F, Sechilariu M, Houssamo I. DC load and batteries control limitations for photovoltaic systems. Experimental validation. IEEE Trans Power Electron 2012; 27(9):4030−8.
[15] Sechilariu M, Wang BC, Locment F. Building integrated photovoltaic system with energy storage and smart grid communication. IEEE Trans Ind Electron 2013; 60(4):1607−18 [Special Issue on Distributed Generation and Micro-grids].
[16] Sechilariu M, Locment F, Wang BC. Photovoltaic electricity for sustainable building. Efficiency and energy cost reduction for isolated DC microgrid. Energies 2015;8(8):7945−67 [Special Issue on Solar Photovoltaics Trilemma: Efficiency, Stability and Cost].
[17] Sechilariu M, Wang BC, Locment F. Power management and optimization for isolated DC microgrid. In: The 22nd IEEE international symposium on power electronics, electrical drives, automation and motion SPEEDAM; 2014.
[18] Trigueiro Dos Santos L, Sechilariu M, Locment F. Prediction-based optimization for islanded microgrid resources scheduling and management. In: IEEE international symposium on industrial electronics ISIE; 2015.
[19] Trigueiro Dos Santos L, Sechilariu M, Locment F. Day-ahead microgrid optimal self-scheduling. Comparison between three methods applied to isolated DC microgrid. In: The 40th annual conference of the IEEE industrial electronics society IECON; 2014.
[20] Dantzig GB. Maximization of a linear function of variables subject to linear inequalities. Activity analysis of production and allocation (Cowless Commission Monograph). Wiley; 1950.
[21] IBM ILOG CPLEX Optimizer. http://www.ibm.com.
[22] Locment F, Sechilariu M, Houssamo I. Multi-source power generation system in semi-isolated and safety grid configuration for buildings. In: The 15th IEEE Mediterranean electrotechnical conference MELECON; 2010.
[23] Wang BC, Sechilariu M, Locment F. Intelligent DC microgrid with smart grid communications: control strategy consideration and design. IEEE Trans Smart Grid 2012;3(4):2148−56 [Special Issue on Intelligent Buildings and Home Energy Management in a Smart Grid Environment].

CHAPTER 6

Experimental Evaluation of Urban Direct Current Microgrid

1. INTRODUCTION

Through simulation tests in the last chapter, it validated that with ±10% random errors between prediction and simulated operating condition, the optimized power flow can be reproduced in operation through a simple interface. In this chapter the proposed supervisory control is validated through experimental tests to identify the merits and shortages in real-time operating conditions. The experimental tests are performed with an experimental platform, the description of which is given thereafter.

The experimental platform shown in Fig. 6.1 is installed in the laboratory of the EA 7284 AVENUES research team (Pierre Guillaumat Center of Universite de Technologie de Compiegne, France). The image of the photovoltaic (PV) sources shows that the PV panels are installed on the roof of the Pierre Guillaumat Center.

The PV array (PVA) consists of 16 PV panels. The storage uses a set of eight 12-V/130-Ah lead-acid batteries in series (Sonnenschein). The grid connection is emulated by reversible alternating current (AC) voltage

Figure 6.1 Image of main components of the experimental platform. *DC*, direct current; *IGBTs*, insulated gate bipolar transistor; *PC*, personal computer; *PV*, photovoltaic.

Urban DC Microgrid
ISBN 978-0-12-803736-2
http://dx.doi.org/10.1016/B978-0-12-803736-2.00006-2

sources and the direct current (DC) load is emulated by programmable DC electronic load (PEL). Indeed, regarding the public grid connection, given the strict conditions required by Electricité de France, the public grid is emulated by a linear amplifier of four quadrants of 3 kVA (POWER +). Finally, the DC load is emulated by PEL, which thermally dissipates the power it receives—2.6 kW (Chroma). The real-time controller is dSPACE 1006.

The PVA, storage, public grid, and diesel generator are coupled to the DC bus through a six-leg power converter. The proposed power converter, referenced as SKM100GB063D, is based on an insulated gate bipolar transistor (IGBT) with a commutation frequency at 20 kHz controlled by a driver SKHI 22A (all from home SEMIKRON); they are all protected by the "Protistor" type of security device. To ensure compatibility among the different elements, a set of inductors and capacitors are added.

The specifications of the main devices were given in Table 4.6 in chapter "Direct Current Microgrid Power Modeling and Control"; nevertheless, because Table 6.1 presents the components of the experimental platform, it could be useful to give them again. Concerning

Table 6.1 Components of the experimental platform

Element	Parameter	Device
PVA (16 PV panels in series, with given STC parameters)	$I_{MPP} = 7.14$ A, $V_{MPP} = 280$ V	Solar-Fabrik SF-130/2-125
Storage (8 serial battery units)	96 V/130 Ah	Sonnenschein solar S12/130 A
Public grid emulator	3 kVA	Puissance+; Bidirectional linear amplifier
Diesel generator emulator	3 kVA	Puissance+; bidirectional linear amplifier
Programmable DC electronic load	2.6 kW	Chroma 63202
Controller board		dSPACE 1006
Power electronic converter	600 V/100 A	SEMIKRON SKM100GB063D
Pyranometer		CT-RM 1010BS2000TDC24SKt
Solar radiation sensor		Pyranometer 6450
Temperature sensor		COF115 PT100
Acquisition data system		YOKOGAWA SL 1000
Driver		SKHI 22A

DC, direct current; *PV*, photovoltaic; *PVA*, PV array; *STC*, standard test condition.

the PVA, current and voltage are given for the maximum power point at the standard test conditions.

Concerning losses, note that the conversion efficiency is not considered and the measurements also contain errors due to power electronic device switching the transient noise and temperature drift of components in the current sensing board. Thus it may occur that the load shedding and PVA constrained power calculation given by the operational layer algorithm can result in grid power reaching or even slightly surpassing its limits. Therefore security margins are added to shed a little more load or PV power than calculation to maintain the grid power within its limit. That is also why in the case of load shedding and PVA constrained power the grid power is a little less than the given power limits in the following experimental results.

2. CONSIDERATIONS ON MULTILAYER SUPERVISORY COMMUNICATION

The supervisory control system, as described in chapter "Direct Current Microgrid Supervisory System Design", is the interface between the multisource power system and the end user, metadata, and smart grid messages. The necessary communication lines used in experimental tests are shown in Fig. 6.2.

The end user can define the low limit of the load shedding coefficient K_{L_lim}. Considering the difficulty in setting all DC microgrid system parameters to be able to reveal the main phenomena that lead to validation, in experimental tests K_{L_lim} is chosen low; therefore it has no impact on the DC microgrid operation.

Concerning PVA power prediction, it is calculated by the solar irradiance prediction data provided by the national weather forecast service Météo France. The uncertainties of PVA power prediction are related to each forecast precision.

The load power profile is arbitrarily assigned for the same reason to clearly show as many events involved in the supervisory control as possible. The load power profile is shown in Fig. 6.3. The difference between the predicted load power demand and the used experimental load power demand is assigned to be at the most 200 W, which results in relative prediction errors ranging from 13.3% to 28.6%.

The optimization algorithm result, based on power predictions, is obtained before the real-time DC microgrid operation.

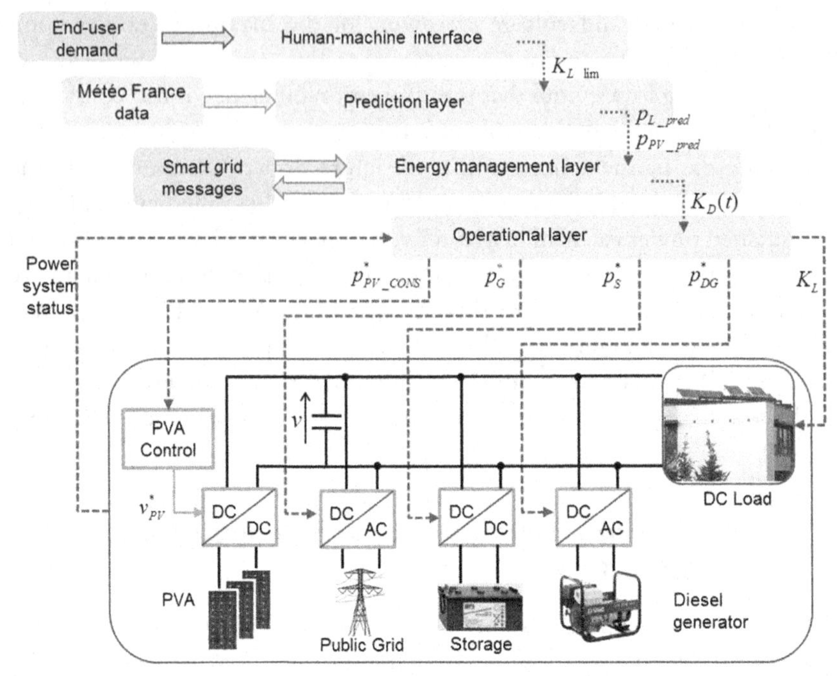

Figure 6.2 Communication lines. *AC*, alternating current; *DC*, direct current; *PVA*, photovoltaic array.

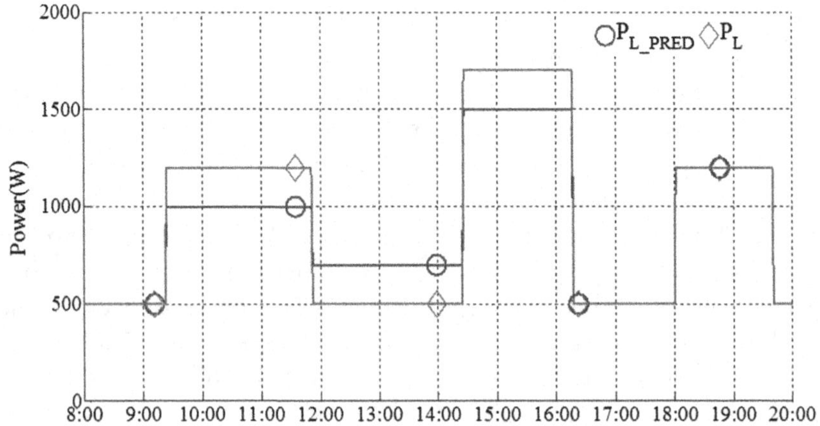

Figure 6.3 Load power prediction and actual load power during the experimental test.

The operational layer performs the power balancing algorithm by applying the energy management result $K_D(t)$ and the PVA constrained power algorithm. Thus it has to calculate and transmit the power references p_G^*, p_S^*, and p_{DG}^* and the load shedding coefficient K_L because of the first algorithm and the PVA maximum power point tracking (MPPT) or constrained power reference $p_{PV\_CONS}^*$ with the second algorithm. At the same time, the multisource power system is able to communicate its status as input in the power balancing algorithm.

3. CONSIDERATIONS ON POWER CONTROL ALGORITHMS IMPLEMENTATION

Considering the external interface, corresponding controllers, smart grid messages, and the two control algorithms, the overall control diagram for the operational layer implemented in the experimental platform is presented in Fig. 6.4 [1,2].

This diagram only provides the overall control for the grid-connected mode; however, in off-grid mode the diagram is very similar. To make it work in real-time, it is compiled by dSPACE, and then the system operates using the ControlDesk from dSPACE in real time as a dialogue interface. All controller coefficients are referred to as C_I and C_P, the integral gain and

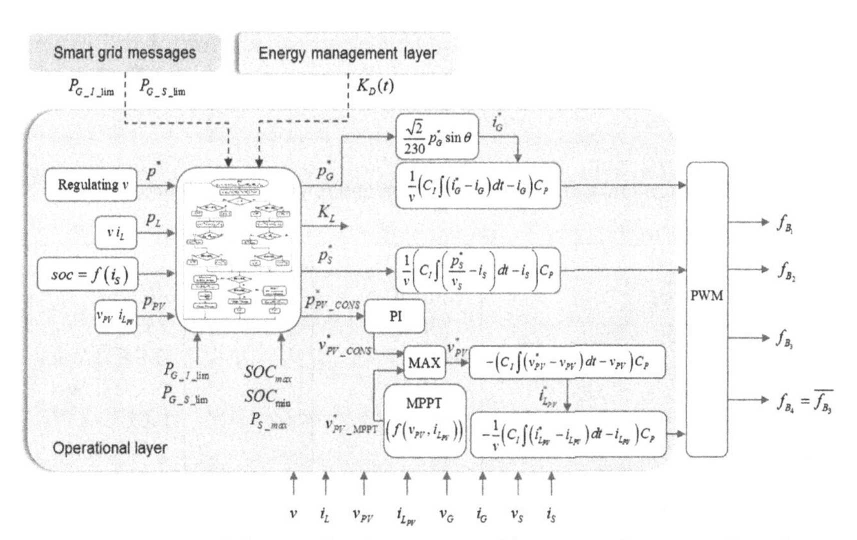

Figure 6.4 Control diagram for the operational layer in grid-connected mode.

the proportional gain, respectively. For different controllers, C_I and C_P are not the same value; they are calculated so that each controller can properly work. Finally, the controllers output duty cycle to the corresponding converters for pulse width modulation (PWM) control, which is a classic technique for power electronic control. All controls are linear controls, operating with proportional integral (PI) correctors, with PWM at 20 kHz. Finally, the PWM signals are routed to the driver cards, which independently control the IGBT components. To simplify the control, in this study all PI correctors have disturbance compensation. Thus note that the synthesis of PI correctors is not performed with high accuracy.

4. DIRECT CURRENT MICROGRID OPERATING IN GRID-CONNECTED MODE

For grid-connected tests, the parameters given in Table 6.2 are used for optimization and power balancing control strategy.

These parameters are selected according to experimental platform configuration aiming at involving as much system behavior as possible during the tests, such as storage events (full, empty), load shedding, and PVA source constrained power. The energy tariff values are chosen completely arbitrarily, but to enhance the control strategy it is mandatory to respect the inequality $c_{PVS(LS)} \gg c_{G\_PH} > c_{G\_NH} > c_S$. For technical reasons, and after verification of the AC versus DC efficiency, in this study, the 400-V DC bus voltage is adopted. The electrical scheme of the experimental platform for grid-connected operation mode is presented in Fig. 6.5.

Table 6.2 Optimization and experimental parameters values for grid-connected operation

$P_{G\_S\_lim}$	1000 W	C_{REF}	130 Ah
$P_{G\_I\_lim}$	1000 W	SOC_{min}	45%
P_{L_max}	1700 W	SOC_{max}	55%
v^*	400 V	c_{G_NH}	0.1 €/kWh
P_{S_max}	1200 W	c_{G_PH}	0.7 €/kWh
P_{PV_MPPT} at STC	2000 W	c_S	0.01 €/kWh
SOC_0	50%	c_{PVS}	1 €/kWh
Grid peak hours	10:00–12:00, 17:00–19:00	c_{LS}	1 €/kWh
Nominal operating cell temperature	48°C		

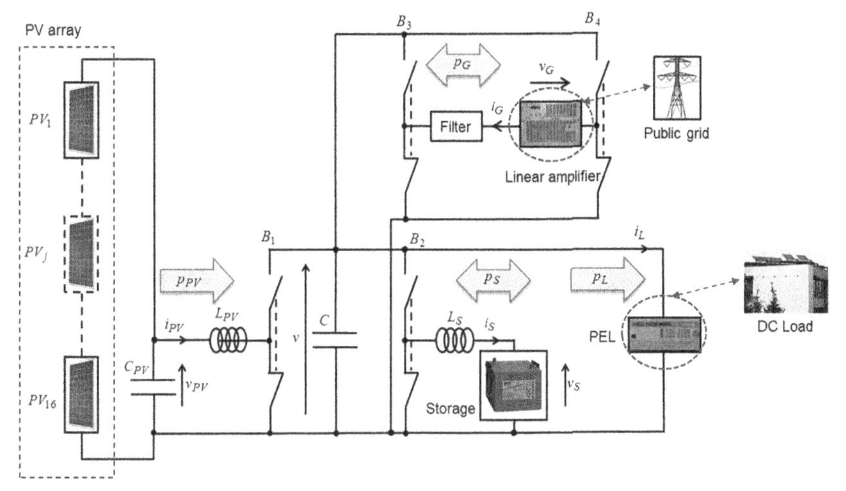

Figure 6.5 Multisource power system electrical schema for on-grid mode. *DC*, direct current; *PEL*, programmable DC electronic load; *PV*, photovoltaic.

In the following sections a series of three tests is given for validating the proposed supervisory control in grid-connected operating mode.

4.1 Experimental Test Description for Grid-Connected Mode

The experimental results are strongly influenced by the solar irradiance evolution and the induced prediction errors. Hence, to analyze the proposed DC microgrid validation, many tests were operated, which have permitted choosing three typical cases. These case studies correspond to three types of solar irradiance evolution: high irradiance almost without fluctuations (test 1), high irradiance with strong fluctuations (test 2), and low irradiance with strong fluctuations (test 3) [3].

4.1.1 Test 1 for Grid-Connected Mode

Test 1 was performed for operation on August 21, 2013. The load prediction uncertainties were shown in Fig. 6.3. Concerning PVA prediction uncertainties, the PVA power raw data prediction, corrected prediction to correspond to the PV panel tilt, and measurement are shown in Fig. 6.6.

Compared with the raw data prediction, the corrected prediction is lower at the beginning and the end of the day and higher in the middle of the day as the Sun position changes. However, solar irradiance uncertainties show that the prediction corresponds better to the measurement during the first few hours, especially before 11:00.

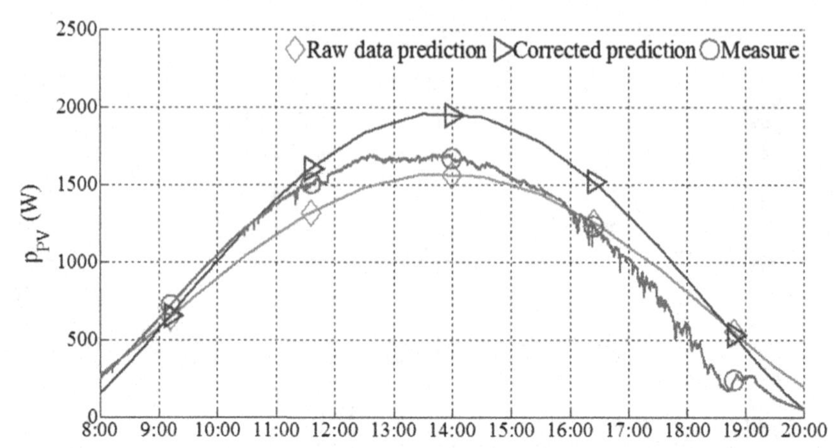

Figure 6.6 Photovoltaic array (PVA) power prediction and actual PVA power measure for on-grid test 1.

According to the prediction information, the energy management layer solves the optimization problem by CPLEX and gives the optimum power flow evolution, $K_D(t)$, and optimum SOC as presented in Fig. 6.7.

The power flow optimization effect can be observed. During peak hours, the excess power is optimized to be injected into the public grid to increase benefits (ie, 10:10–12:00, 17:00–18:00). In the case of supply during peak hours, the storage supplies the load for avoiding an expensive public grid peak supply (ie, 10:00–10:10, 18:00–19:00). In grid peak hours, the supervisory control acts similar to a simple rule-based control that simply injects excess power into the grid and supplies the load by storage. That is because this is the optimal way to reduce energy cost. Meanwhile the optimization keeps storage available for the grid peak hour supply.

The difference between optimization and rule-based control remains in off-peak grid hours (ie, normal hour energy tariff). For example, storage charging is preferred in off-peak grid hours for energy cost reduction. However, the storage charging is not necessary to be immediately started at the beginning of off-peak grid hours. Because the storage is fully charged too early, the excess PVA power may not be fully absorbed due to the public grid power injection limit, which can result in PVA constrained power that waste energy and increase energy cost. Another issue is that if load power demand is low during peak hours, it is also possible for the storage to supply the load during off-peak hours. The storage charging or discharging proportion can hardly be determined by the rule-based control, but the prediction-based supervision control can easily determine these

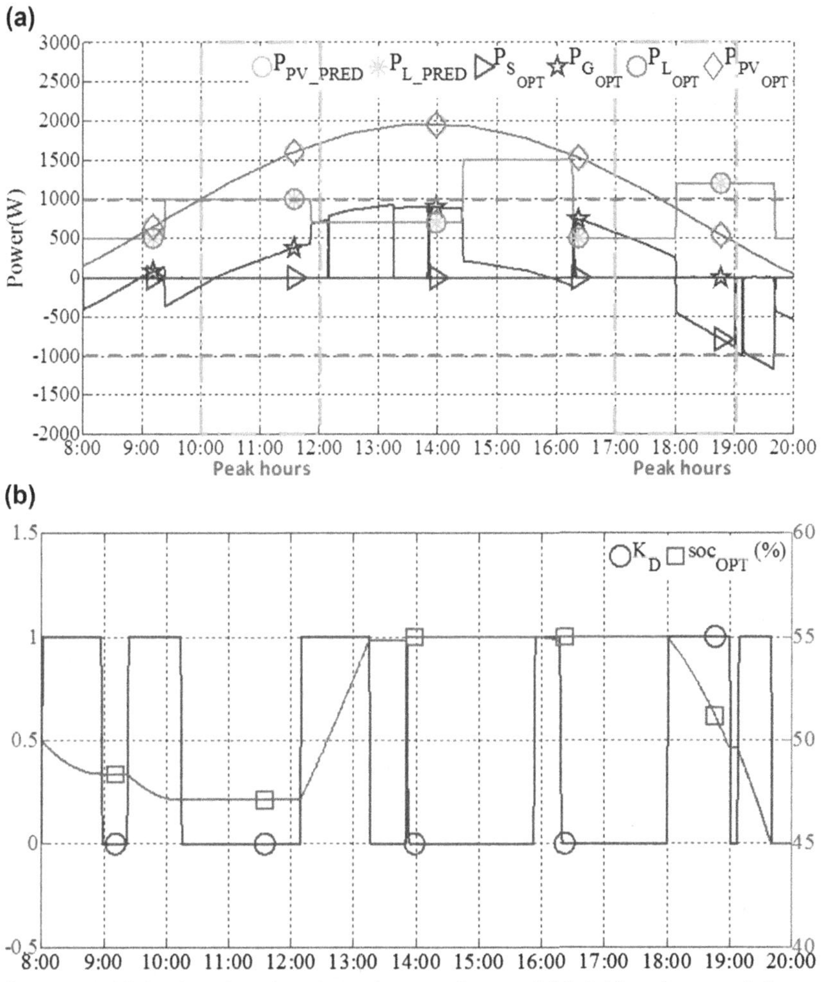

Figure 6.7 (a) Predicted and optimized power flow and (b) $K_D(t)$ and state of charge (*SOC*) evolution given by optimization for on-grid test 1.

aspects from an overall viewpoint. That is why the storage is optimized to supply power during off-peak hours (ie, 8:00–9:00, 9:20–10:00), and PV excess power is injected into the grid (ie, 9:00–9:20). The optimization result cannot be replaced by rule-based control because the operation varies according to conditions for minimizing the energy cost.

The operation is performed based on the optimum $K_D(t)$ sequence and the on-grid power balancing control algorithm given in chapter "Direct Current Microgrid Supervisory System Design". The PVA operating real

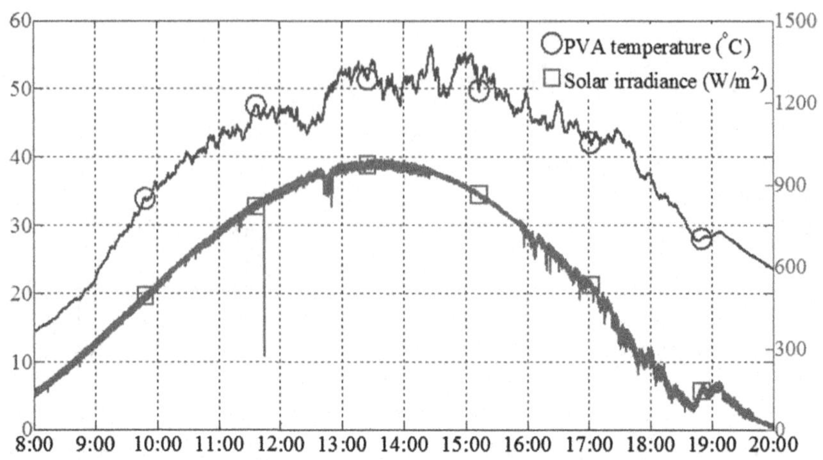

Figure 6.8 Photovoltaic array (PVA) cell temperature and solar irradiance on August 21, 2013.

conditions are shown in Fig. 6.8. The experimental real-time power flow is shown in Fig. 6.9a. The predicted load power is noted as load demand power by P_{L_D} and the actual load power is noted P_L. The experimental SOC and DC bus voltage evolution are shown in Fig. 6.9b.

During this test the public grid and storage supply or absorb power according to $K_D(t)$ except for the moment when SOC limits are reached in the period of 18:50–20:00. Because of uncertainties from load prediction and PVA power prediction, the experimental power flow cannot evolve as similar to the optimized power flow. Because the PVA produces less power than predicted, the obvious result is that storage is not fully charged during the day and reaches its $SOC_{\min}$ earlier than expected, resulting in load shedding in the period of 18:50–19:40, according to power balancing control as public grid supply is limited. When the public grid is able to supply within its limits, the load is supplied by the public grid without load shedding, (ie, 19:40–20:00).

The experimental power flow corresponds to the optimized power flow from 8:00 to 9:20. From 9:20 to 11:50, the PV power prediction error is not significant whereas the load demands 200 W more power than predicted. This results in more storage discharge than the prediction between 9:20 and 10:10. In the period of 10:10–12:05, it is expected that the public grid absorbs excess PVA power. Because of the error of load prediction, the public grid supply during peak hours is involved at 10:10 until 11:00 and less power is injected in the grid from 11:00 to 12:05.

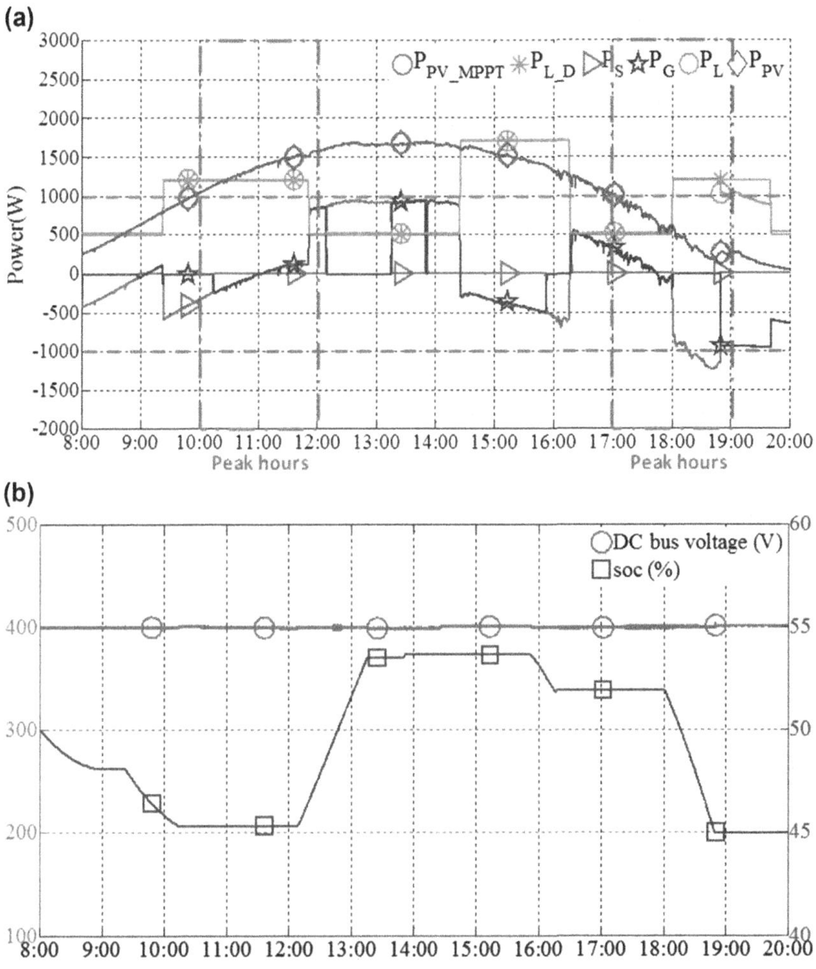

Figure 6.9 (a) Experimental real-time power flow and (b) experimental DC bus voltage and *SOC* evolution for on-grid test 1. *DC*, direct current; *SOC*, state of charge.

From 12:05 to 14:20, PVA power and load demand are less than their predictions. As a result, the public grid power and storage power correspond to the optimization result.

From 14:20 to 16:15, the load demand is higher than prediction and the PVA power is less than the prediction, which is the period that introduced the largest error between the optimized and actual power flow: the public grid supplies power at 14:20 until 15:50 instead of power injection expected by optimization, and the storage supplies more power than expected from 15:50 to 16:15.

During 16:20−18:00, the PVA power is less than prediction; therefore less power is injected into the public grid. From 18:00 to 20:00, the storage capacity is less than the optimization result and it supplies more power than expected; therefore it is empty at 18:50 instead of 19:40. Then the public grid supplies the load from 18:50 to 20:00 with load shedding as mentioned earlier.

Note that prediction errors can affect the optimization performance. However, compared with prediction, if the actual production and consumption are with almost the same errors in the same direction (ie, the difference between the actual production and consumption corresponds to the difference given by prediction), then the actual power flow of storage and the public grid can correspond to the optimization result (eg, the period of 12:05−14:20). Although those prediction errors can affect the optimization performance, the prediction errors do not affect power balancing. It is shown in Fig. 6.9b that the DC bus voltage is steady, signifying the power is well balanced during the operation.

The actual power sharing ratio between storage and the grid is shown in Fig. 6.10. Comparing the actual power sharing with the $K_D(t)$, it can be noted that the $K_D(t)$ resulting from optimization is respected for most of the time, except for the short period by the end when the storage is empty. At that moment the operational layer performs load shedding to ensure power balancing. The slight difference is due to measurement errors.

Table 6.3 compares the energy costs between optimization and the experiment. The total energy cost C_{total} is calculated for three cases: (1)

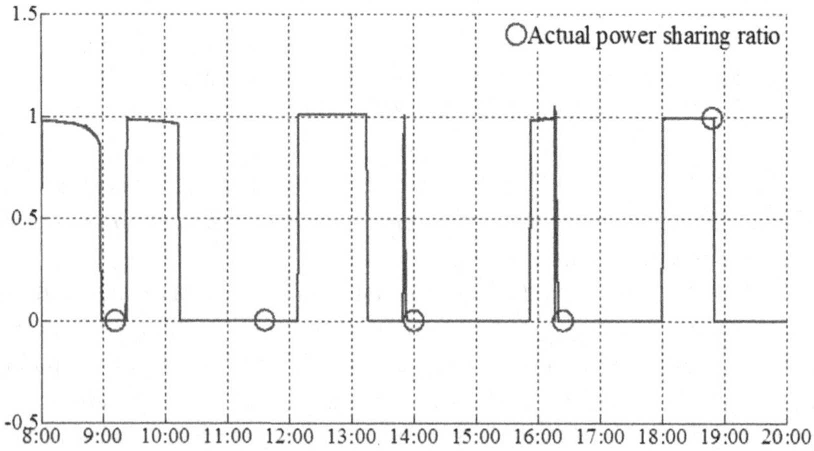

Figure 6.10 Actual power sharing ratio between storage and grid for on-grid test 1.

Table 6.3 Comparison of different energy cost calculation for on-grid test 1

Case operation	**Energy cost C_{total} (€)**
Estimated optimum energy cost following power predictions	−0.777
Actual energy cost (experiment)	0.225
Optimum energy cost for real conditions calculated after operation	−0.247

estimated optimum energy cost following power predictions; (2) the experimental cost, which is the actual energy cost; and (3) the optimum energy cost for real conditions calculated after operation.

The estimated optimum energy cost following power predictions is negative because during this day local renewable production is sold to the grid and it can bring benefits that can compensate for the energy consumption cost.

Because of uncertainties the actual energy cost is higher than the estimated cost by optimization before the operation. The load shedding action is the main contribution to this cost error. The other error is introduced by excess public grid and storage supply involved by uncertainties. Thus, aiming at a reasonable comparison, at the end of the test an optimization for real operating conditions is performed, which gives the ideal experiment cost. In this last case the obtained cost is the total optimum energy cost that would have been possible to achieve following the real solar irradiance and the real load power.

4.1.2 Test 2 for Grid-Connected Mode

Test 2 was performed for operation on August 9, 2013. The PVA power prediction uncertainties are shown in Fig. 6.11. Because the prediction is given in the form of hourly data, the fluctuations presented in the real PVA power are not taken into account. In addition, the concave character of prediction around 14:30 is delayed around 15:45 in the actual condition.

The optimized power flow is shown in Fig. 6.12a. On the basis of the optimum power flow evolution, the optimum $K_D(t)$ sequence is calculated as shown in Fig. 6.12b.

During peak hours the DC microgrid injects the excess PVA production into the public grid and supplies the load by storage. In off-peak hours storage and public grid power sharing are optimized. In this case neither load shedding nor PVA constrained power is performed.

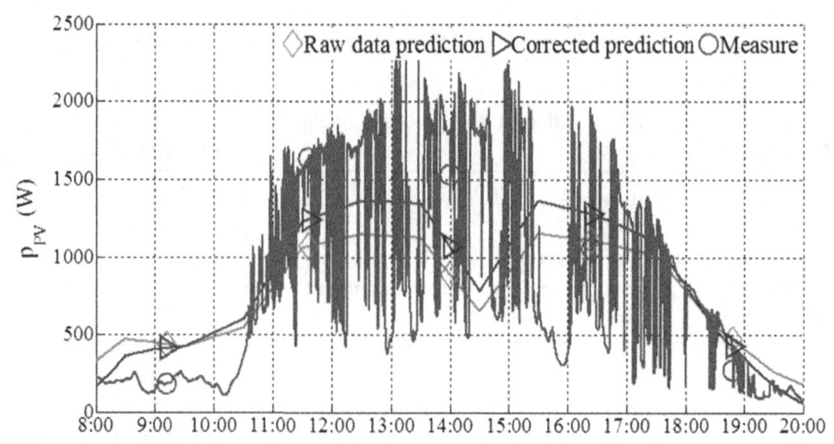

Figure 6.11 Photovoltaic array (PVA) power prediction and actual PVA power measure for on-grid test 2.

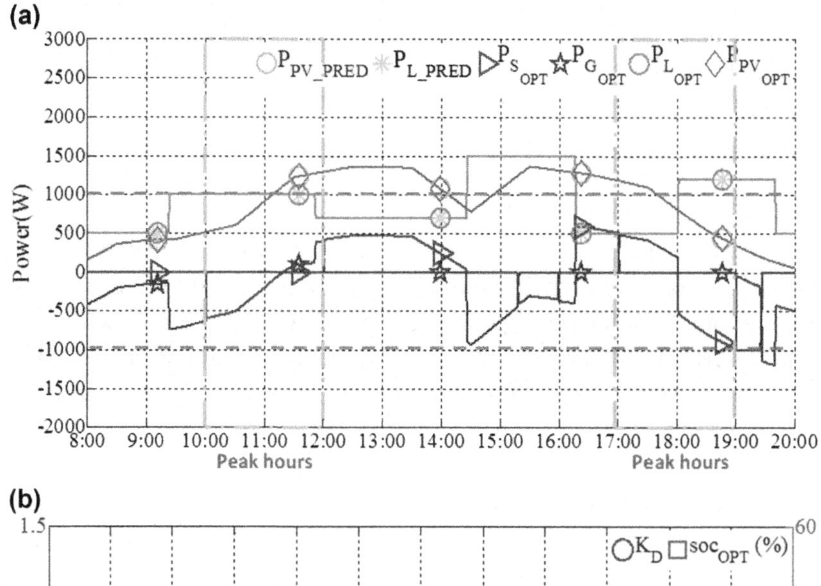

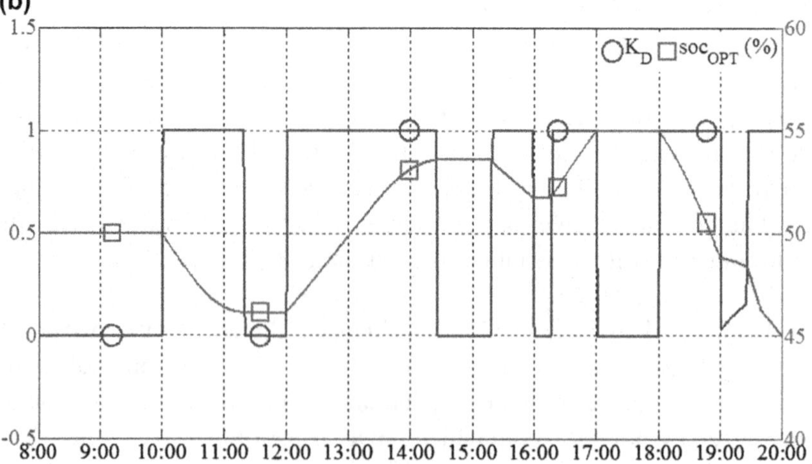

Figure 6.12 (a) Predicted and optimized power flow and (b) $K_D(t)$ and *SOC* evolution given by optimization for on-grid test 2. *SOC*, state of charge.

The real operating conditions for PVA are shown in Fig. 6.13. The experimental power evolution obtained with $K_D(t)$ and operational layer control strategy are shown in Fig. 6.14a. Experimental SOC and DC bus voltage evolution are shown in Fig. 6.14b.

During this day of operation the public grid and storage supply or absorb power according to $K_D(t)$, except for the period when SOC limits are reached or public grid power limits are reached.

From 8:00 to 9:20, public grid supplies the load as commanded by $K_D(t)$. From 9:20 to 10:00 the load demands more power than predicted that the grid alone is not able to supply. Because the storage is available for supply (ie, $SOC > SOC_{min}$), storage supplies power together with the public grid, which is the self-correcting action operated by the operational layer control strategy.

During 10:00–10:20, the storage supplies the load according to $K_D(t)$. The storage reaches the SOC_{min} at 10:20 because it has supplied power before 10:00 and supplies more power to the load because of prediction uncertainties. For the period of 10:20–11:20 the public grid supplies the load for most of the time without regard to $K_D(t)$ by self-correcting actions given by the operation strategy. Because the public grid supply power is limited, load shedding is performed by 10:25–10:35 to maintain power balancing. During the PVA production fluctuations from 10:55 to 11:20, the storage is slightly charged and supplies again for a short while.

Public grid is used for power balancing at 11:20 until 12:00, as commanded by $K_D(t)$. Although the supply of public grid is involved instead of a total injection into the public grid, the injected energy is greater than the energy to supply the DC bus during the fluctuations.

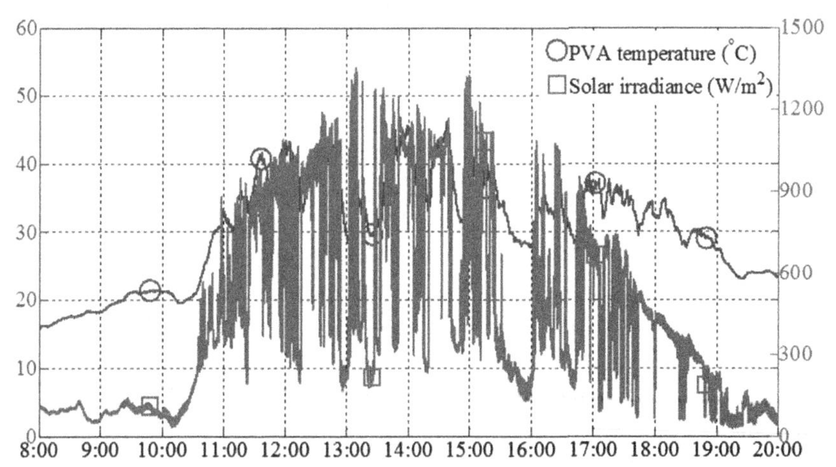

Figure 6.13 PVA cell temperature and solar irradiance on August 9, 2013. *PVA*, photovoltaic array.

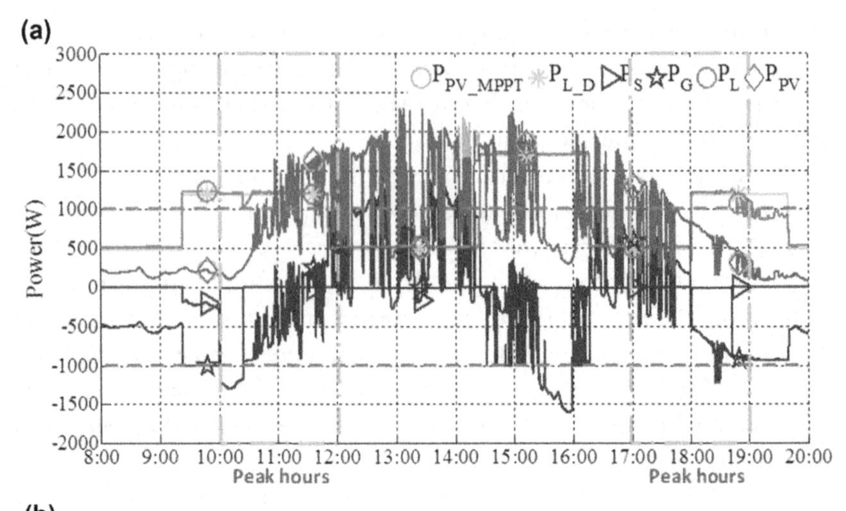

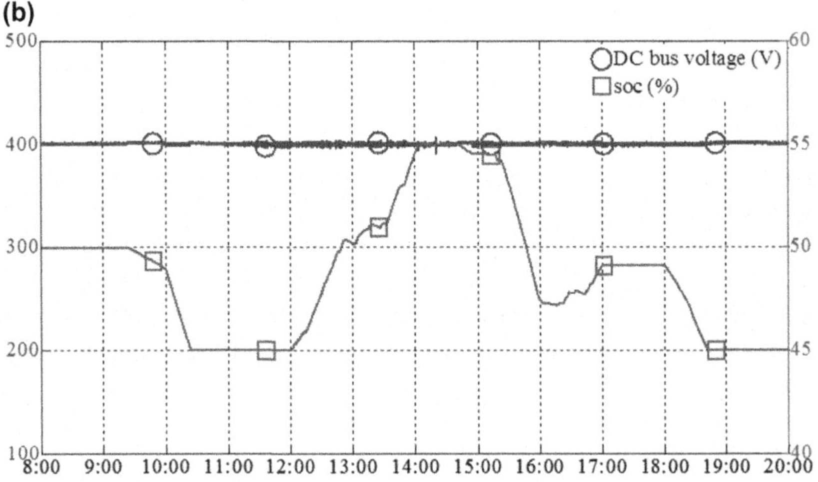

Figure 6.14 (a) Experimental real-time power flow and (b) experimental DC bus voltage and *SOC* evolution for on-grid test 2. *DC*, direct current; *SOC*, state of charge.

From 12:00 to 14:20 storage is charged following the command by $K_D(t)$. Because load consumes less than prediction, the storage is fully charged at 14:00. From 14:00 to 14:20 the excess PVA production is injected into the grid. Because the surplus of PVA production exceeds the public grid power injection limit, the surplus of PVA production cannot be absorbed; therefore the PVA constrained power is performed as self-correcting action by the operational layer algorithm. During this period the actual PVA power is less than the estimated PVA MPPT production according to the real weather conditions.

At 14:20 until 15:15 the public grid is commanded by $K_D(t)$ to supply the load. However, when the PVA production is too low and the grid supply limit is reached, the storage supplies power as a result of self-correcting. The storage follows the $K_D(t)$ command to supply power at 15:15 until 16:00. From 16:00 to 16:15, the public grid supplies the load and storage is involved when the public grid supply is limited. During 16:15–17:00 storage is charged following the $K_D(t)$ command. During 17:00–18:00 the power is mainly injected into the public grid as commanded by $K_D(t)$. From 18:00 to 20:00 storage supplies first, but $SOC_{\min}$ is reached before expected and then public grid supplies power. When public grid power supply limit is reached, load shedding is performed to maintain power balancing.

The steady DC bus voltage in Fig. 6.14b signifies that the power is well balanced during the test. Although prediction involves uncertainties and the weather conditions result in fluctuating PVA production, the power balancing is maintained with respect to all rigid constraints such as public grid power limits and *SOC* limits. The self-correcting actions' effect on power balancing can be fully observed and load shedding and PVA constrained power appear only for short durations.

The actual power sharing ratio between storage and the public grid is shown in Fig. 6.15. Because of fluctuations and prediction uncertainties, $K_D(t)$ cannot be fully respected. However, the mean actual power sharing ratio value still represents the optimization information.

Table 6.4 compares different energy cost calculations for on-grid test 2. Because of uncertainties, the experimental cost is larger than the estimated

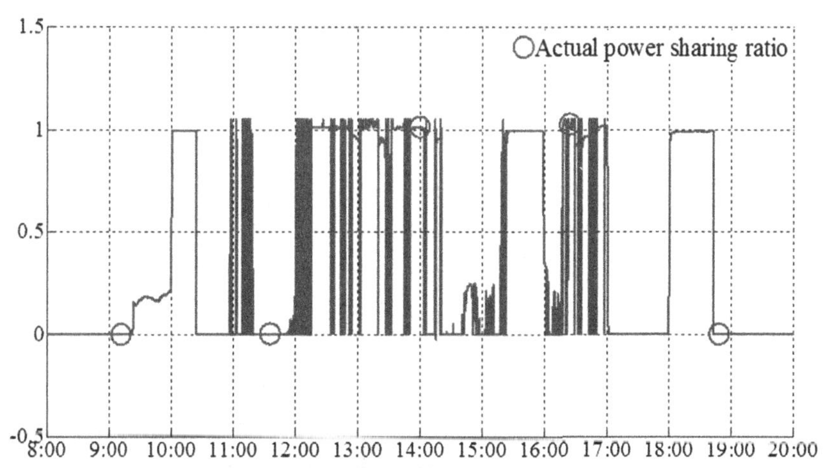

Figure 6.15 The actual power sharing ratio between storage and grid for on-grid test 2.

Table 6.4 Comparison of different energy cost calculations for on-grid test 2

Case operation	Energy cost C_{total} (€)
Estimated optimum energy cost following power predictions	−0.149
Actual energy cost (experiment)	0.929
Optimum energy cost for real conditions calculated after operation	0.357

optimum energy cost following power predictions, but it is closer to the optimum energy cost for real conditions calculated after operation.

In this case load shedding and limited PVA power represent the main part of the cost error; the other part is introduced by prediction uncertainties that result in public grid and storage of extra power supplies during peak hours and off-peak hours.

4.1.3 Test 3 for Grid-Connected Mode

Test 3 was performed for operation on July 30, 2013. The weather condition is cloudy with generally low solar irradiance. The PVA power prediction uncertainties are shown in Fig. 6.16.

The optimized power flow by CPLEX is shown in Fig. 6.17a. The optimum $K_D(t)$ sequence, which is applied for the experimental operation of July 30, 2013, and the optimum *SOC* are shown in Fig. 6.17b.

In this test during the first peak hours of 10:00−12:00 it can be seen that the optimization result proposes to inject the excess PVA production into the public grid and to supply the load by storage. During the second peak

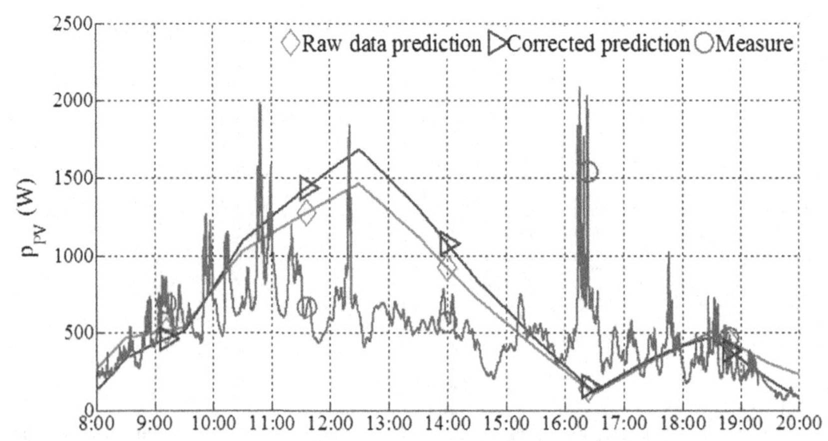

Figure 6.16 Photovoltaic array (PVA) power prediction and actual PVA power measure for on-grid test 3.

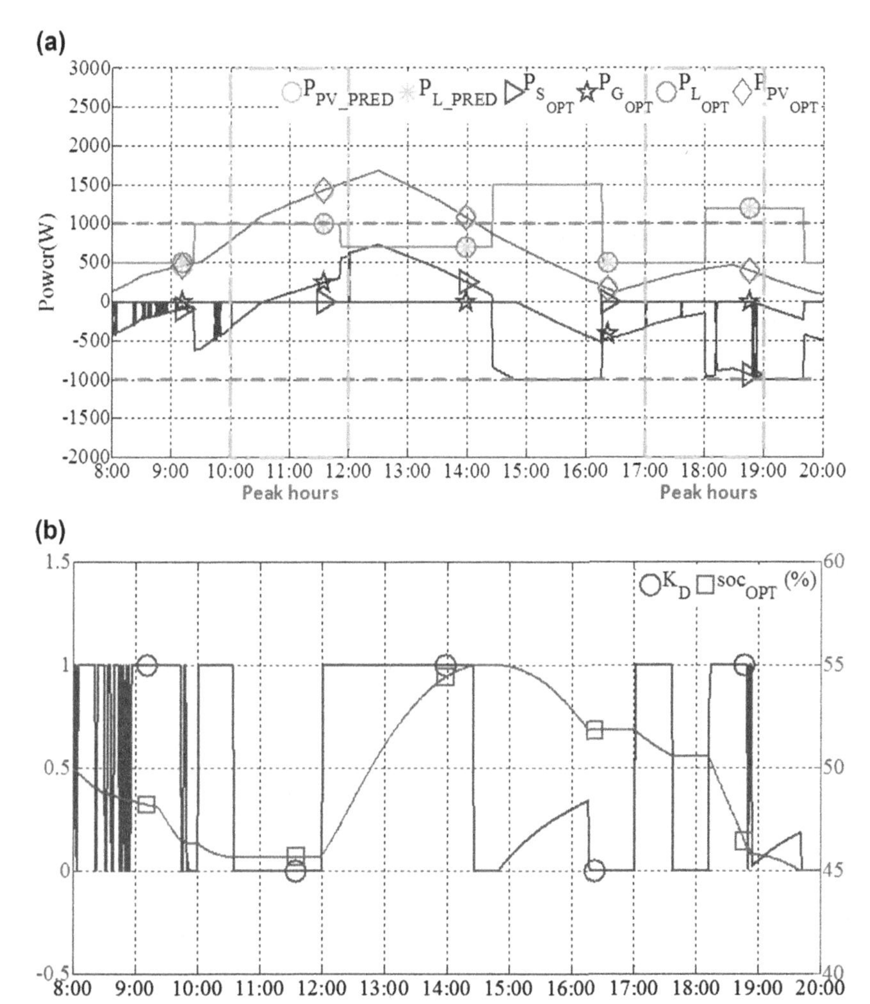

Figure 6.17 (a) Predicted and optimized power flow and (b) $K_D(t)$ and *SOC* evolution given by optimization for on-grid test 3. *SOC*, state of charge.

hours of 17:00–19:00, because the storage is not able to be fully charged before the second peak hours, the public grid alone supplies from 17:40 to 18:10 and together with the storage from 18:50 to 19:00. In off-peak hours the optimization result relates an optimized sharing between storage and public grid powers. It can be seen that the optimization tries to maximally charge the storage. The storage is also used to prevent load shedding when the public grid supplies at its power limit. With the optimization neither load shedding nor PV power limiting is performed.

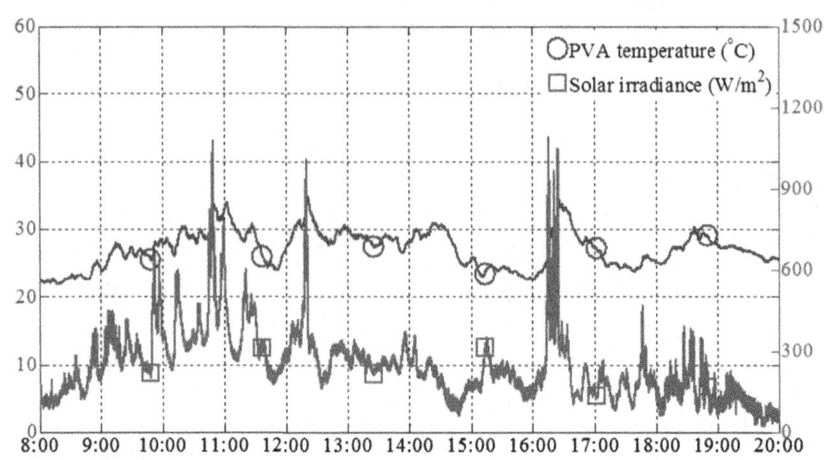

Figure 6.18 PVA cell temperature and solar irradiance on July 30, 2013. *PVA*, photovoltaic array.

The operating real conditions for PVA are shown in Fig. 6.18. The experimental power evolution is shown in Fig. 6.19a. Experimental *SOC* and DC bus voltage evolution are shown in Fig. 6.19b.

Because of uncertainties, the storage reached SOC_{min} before 10:00 and does not participate in the supply during the first peak hours. The storage is slightly charged around 12:10 and contributes too little in the subsequent power balancing. Because of the prediction uncertainties, the load power is supplied by the public grid with respect to its power limits. When the grid power reaches the limit, load shedding is performed to maintain power balancing during 14:20−16:15 and 17:55−19:40.

In this test the error between actual PVA production and prediction is significant. As a result, the storage is only used for a short period and does not participate in public grid peak shaving. However, the power can be balanced successfully by the operational layer control strategy.

The actual power sharing ratio between storage and the public grid is shown in Fig. 6.20. Because of prediction uncertainties, $K_D(t)$ can only be respected at the beginning and the optimization result does not suit the real conditions. However, the operation layer can maintain robust power balancing.

Table 6.5 summarizes the different energy cost calculations for on-grid test 3. Because of prediction uncertainties and low solar irradiance, the experimental cost is much greater than the estimated optimum energy cost following power predictions, but it is always much closer to the optimum energy cost for real conditions calculated after operation.

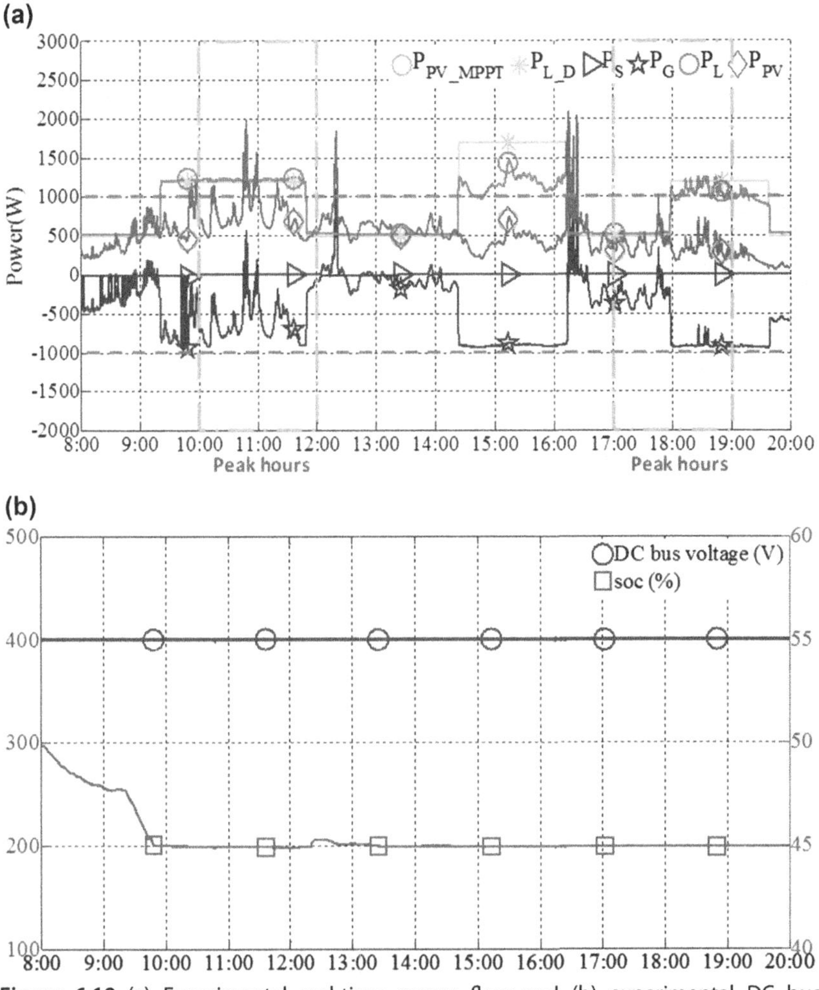

Figure 6.19 (a) Experimental real-time power flow and (b) experimental DC bus voltage and *SOC* evolution for on-grid test 3. *DC*, direct current; *SOC*, state of charge.

Because of uncertainties, the storage cannot be optimized to participate during the peak hours to supply the load. Hence, the energy cost is largely increased by the public grid peak supply and load shedding.

4.2 Results Analysis and Discussions for Grid-Connected Mode

Three tests for grid-connected operating mode are provided. As a general result, the proposed supervisory control is experimentally validated and it is shown that it can work under different weather conditions while providing a

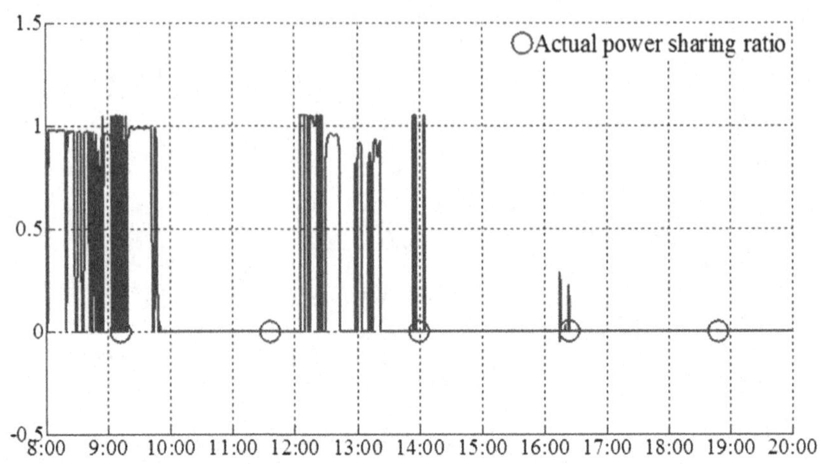

Figure 6.20 Actual power sharing ratio between storage and the grid for on-grid test 3.

Table 6.5 Comparison of different energy cost calculations for on-grid test 3

Case operation	Energy cost C_{total} (€)
Estimated optimum energy cost following power predictions	0.386
Actual energy cost (experiment)	3.219
Optimum energy cost for real conditions calculated after operation	2.165

robust power balancing control strategy and interfacing with optimization. The difficulty of implementing optimization in real operation while respecting rigid constraints is solved by the proposed multilayer supervisory control, which parameterizes the power balancing strategy that provides an interface with optimization. Table 6.6 summarizes the comparison of the cost of energies between the operated three tests for the grid-connected mode.

For each test the effect of the optimization result can be seen in actual power flow with a certain degree of uncertainties. For the first two tests it can be seen that even with uncertainties the storage can participate during the peak hours for the power balancing aiming at peak shaving and avoiding undesired injection. The load shedding contributed to 24.4% of the cost error for the first test whereas load shedding and PVA constrained power formed 29.4% of the cost error for the second test.

Table 6.6 Energy cost comparison among the operated three tests for on-grid mode

Test	Case operation	Total energy cost (€)	Load shedding cost (€)	Photovoltaic array constrained power cost (€)
1.	Optimization	−0.777	0	0
	Experimentation	0.225	0.244	0
	Optimization for real conditions	−0.247	0	0
2.	Optimization	−0.149	0	0
	Experimentation	0.929	0.266	0.052
	Optimization for real conditions	0.357	0	0
3.	Optimization	0.386	0	0
	Experimentation	3.219	1.300	0
	Optimization for real conditions	2.165	0.257	0

For the third case, when the storage does not participate in peak hour power balancing, the energy cost can be increased largely by two aspects: grid peak supply and load shedding. Indeed, the load shedding formed 40.4% of the experimental cost and the public grid peak hour supply cost (ie, 1.626 €) formed 50.5% of the experimental cost. Because of uncertainties the storage cannot be optimized to participate in the peak hour to supply. Hence, the energy cost is largely increased by grid peak supply and load shedding.

The actual power flow can be better optimized when the prediction data provide more precision. Even with uncertainties, the experiment cost can be controlled near ideal experiment cost, which validated the effectiveness of the proposed supervisory control.

Regarding the public grid, no matter how the power evolves in the DC microgrid, the public grid power can be maintained within its limits. Because of the smart grid messages, it can be easy to regulate these limits and to increase or to decrease the supplied power and injected power, which facilitate the demand-side management for the public grid operation. It is also noted that if the prediction error for production and the error for consumption are similar and in the same direction, the power flow for the sources is less affected. These experimental results validate the feasibility of the proposed control structure of the studied DC microgrid.

5. DIRECT CURRENT MICROGRID OPERATING IN OFF-GRID MODE

For off-grid mode tests, the parameters that are used for optimization and power balancing control strategy are given in Table 6.7. These parameters are selected according to system configuration with an aim to involve as much system behavior as possible during the tests, such as storage events (full, empty), load shedding, and PVA power limiting.

These parameters are selected according to experimental platform configuration aiming at involving as much system behavior as possible during the tests, such as storage events (full, empty), load shedding, PVA source constrained power, and diesel generator operation. The energy tariff values are chosen completely arbitrarily, but to enhance the control strategy it is mandatory to respect the inequality $c_{DG} > c_{PVS(LS)} \gg c_S$. The DC bus voltage is always 400 V DC. For the same reason as in grid-connected mode, security margins are added to shed a little more load or PV power than calculated to maintain the storage power within its limit. In addition to these, once the PVA power limiting is triggered, it lasts for a period for avoiding storage charging and discharging oscillations. The electrical scheme of the experimental platform for grid-connected operation mode is presented in Fig. 6.21.

In off-grid mode, the diesel generator is considered as a single-phase AC source and is emulated by the same linear amplifier. However, the difference is that a diesel generator works only for supplying power and is assigned a working duty cycle at rated power. There are two reasons for assigning diesel generator duty cycle. The first reason is that the highest diesel generator efficiency is when it outputs its rated power. In contrast, the fuel is wasted when the diesel generator makes always spinning and outputs low power, or when it is frequently started and stopped. The second reason is that storage can be used as a buffer for power balancing while the diesel generator works in duty cycle.

Table 6.7 Optimization and experimental parameters values for off-grid operation

P_{DG_P}	1500 W	C_{REF}	130 Ah
dt_{DG}	1200 s	SOC_{min}	45%
P_{L_max}	1700 W	SOC_{max}	55%
v^*	400 V	c_{DG}	1.1 €/kWh
P_{S_max}	1200 W	c_S	0.01 €/kWh
P_{PV_MPPT} at standard test conditions	2000 W	c_{PVS}	1 €/kWh
$SOC_0 = SOC_F$	50%	c_{LS}	1 €/kWh
Nominal operating cell temperature	48°C		

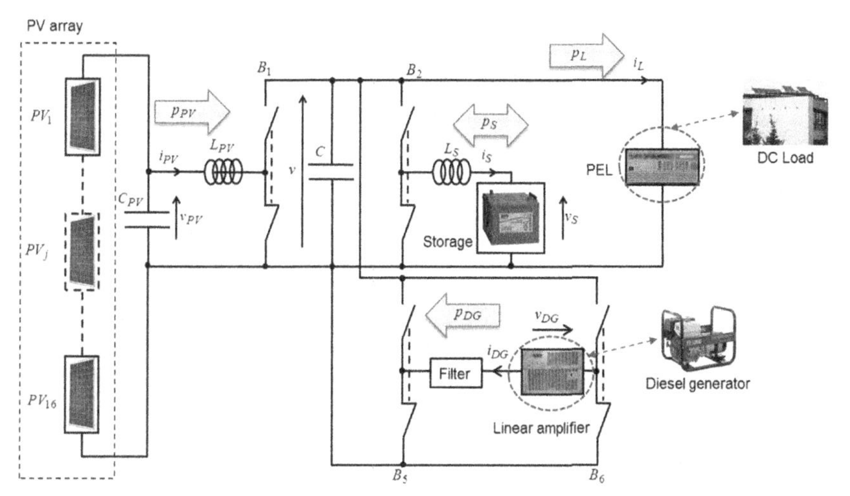

Figure 6.21 Multisource power system electrical schema for off-grid mode. *DC*, direct current; *PEL*, programmable DC electronic load; *PV*, photovoltaic.

5.1 Experimental Test Description for Off-Grid Mode

Regarding grid-connected operating mode, three experimental tests are presented. Depending on the meteorological day profile, the case studies retained are high solar irradiance almost without fluctuations (test 1), high solar irradiance with strong fluctuations (test 2), and mixed high irradiance with strong fluctuations and low irradiance without fluctuations (test 3) [4,5].

5.1.1 Test 1 for Off-Grid Mode

Test 1 was performed for operation on September 4, 2013. The load prediction uncertainties are the same as shown in Fig. 6.3. A few minutes ahead of the test, based on the solar irradiance hourly forecast data and PVA model, the PVA power prediction is calculated and corrected to correspond to the PV panel tilt. The PVA power uncertainties are shown in Fig. 6.22.

On the basis of these predictions, the microgrid optimizes the power flow by CPLEX as shown in Fig. 6.23a. Concerning the power flow curves, note that for storage, negative power means supplying the load whereas positive means receiving power. For graphical clarity the diesel generator power is represented by negative values. The proposed optimization shows that the storage is used for power balancing and the diesel generator is started by duty cycles to maintain continuous supply for the load and ensure that at the end of the operation the storage capacity is

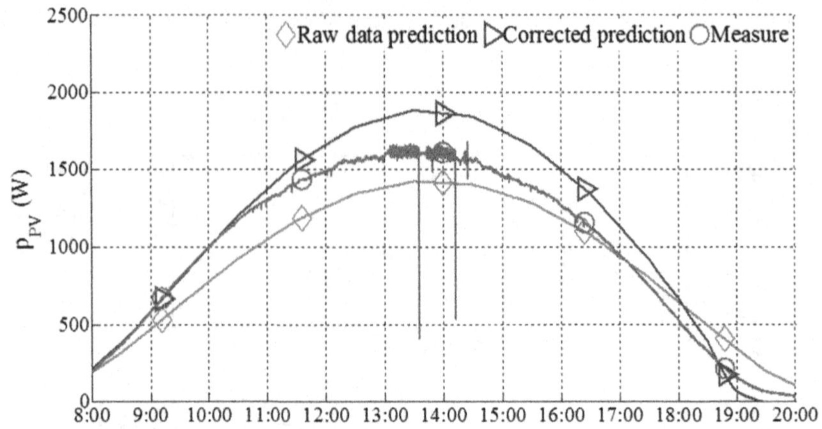

Figure 6.22 Photovoltaic (PVA) power prediction and actual PVA power measure for off-grid test 1.

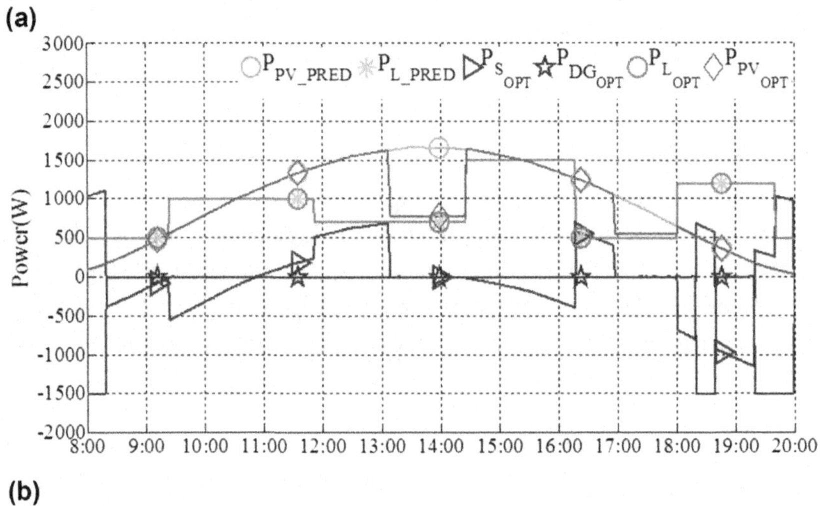

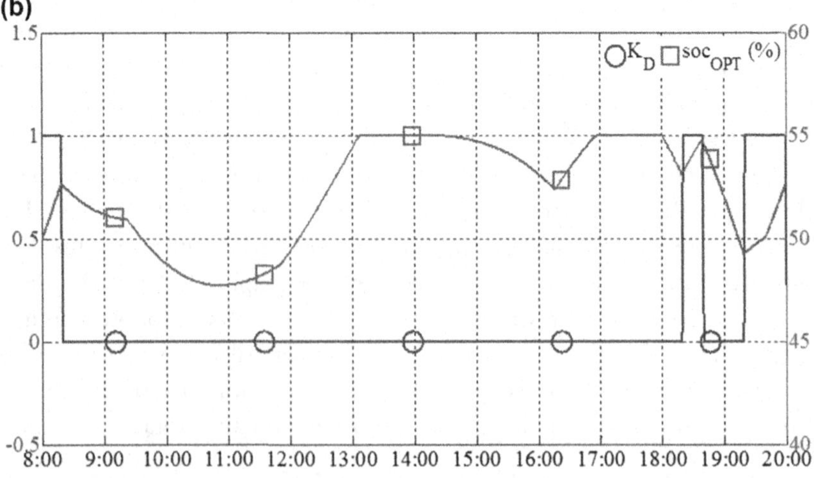

Figure 6.23 (a) Predicted and optimized power flow and (b) $K_D(t)$ and *SOC* evolution given by optimization for off-grid test 1. *SOC*, state of charge.

above a preferred level of SOC_F. When the storage reaches the SOC_{max}, the only way to maintain power balancing is to limit PVA power production; this is why the PVA power is limited in the period of 13:10–14:20. The optimum $K_D(t)$ time series sequence, calculated as the on-off signal for the diesel generator, and the estimated *SOC* evolution given by the optimization are shown in Fig. 6.23b.

The operation is performed based on the optimum $K_D(t)$ sequence and the off-grid power balancing control algorithm given in chapter "Direct Current Microgrid Supervisory System Design". The PVA operating real conditions are shown in Fig. 6.24. The experimental real-time power flow is shown in Fig. 6.25a. The experimental *SOC* and DC bus voltage evolution are shown in Fig. 6.25b.

In this test the diesel generator is started by the duty cycle as commanded by $K_D(t)$ for 8:00–8:20, 18:20–18:40, and 19:20–20:00. No load shedding is performed. During 8:00–13:05 the power is balanced following $K_D(t)$. Despite uncertainties in load and PV source power prediction, the storage is fully charged around 13:05, as expected by optimization. During 13:05–14:20 there is no other possibility that can absorb PVA production, and it is hard to calculate the PVA constrained power reference to produce the accurate power needed by the load. As a solution the control strategy slightly overlimits the PVA production, inducing storage discharging with low power. After the *SOC* is reduced by a certain amount, the PVA is recovered to produce MPPT power again until the SOC_{max} is reached

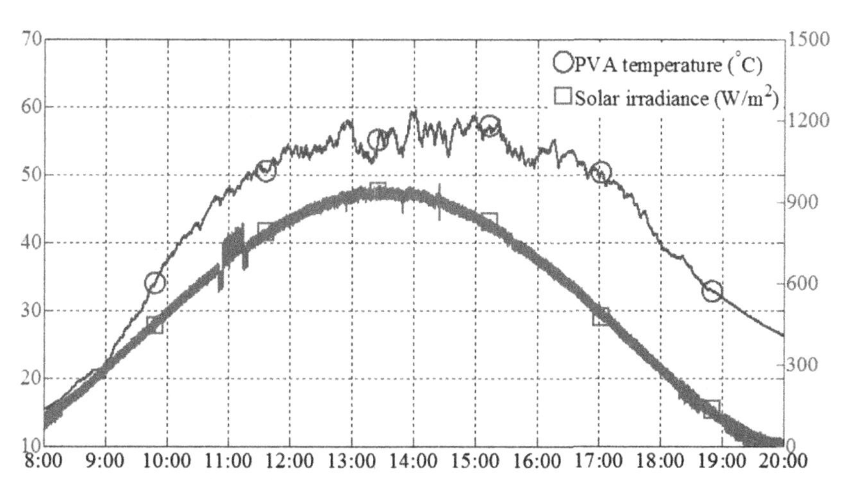

Figure 6.24 PVA cell temperature and solar irradiance on September 4, 2013. *PVA*, photovoltaic array.

(a)

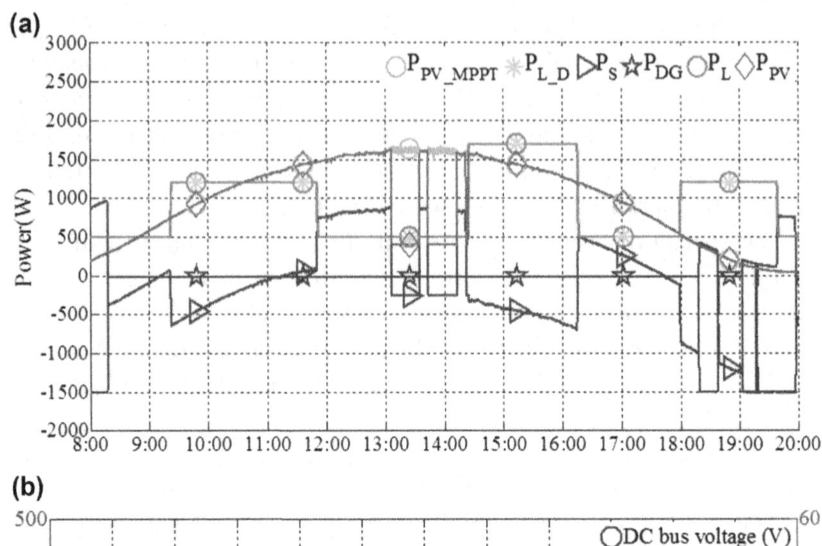

(b)

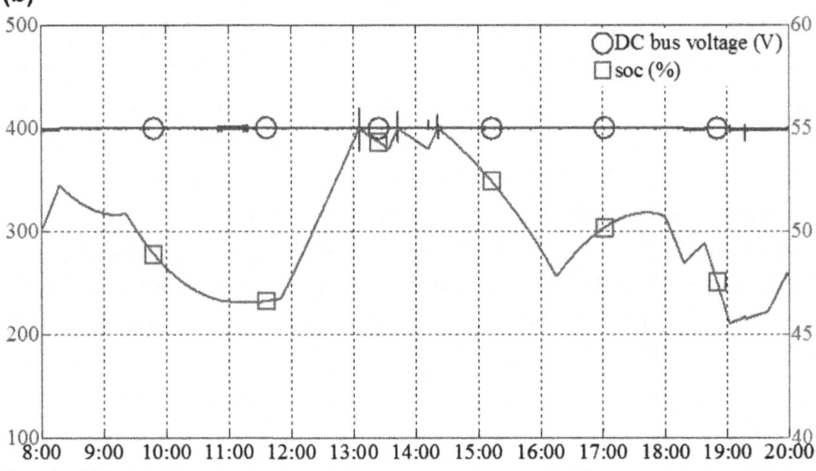

Figure 6.25 (a) Experimental real-time power flow and (b) experimental DC bus voltage and *SOC* evolution for off-grid test 1. *DC*, direct current; *SOC*, state of charge.

again. That is why the PVA power oscillation between MPPT mode and power limiting mode can be observed.

During 14:20–20:00 two differences relative to the optimization can be noted: firstly, the PVA constrained power during 16:55–18:00 is not performed in actual power flow because actual higher load power than PV production caused $SOC < SOC_{max}$; secondly, the diesel generator is started during 19:00–19:20, which is controlled by the power balancing control algorithm that starts the diesel generator when SOC approaches SOC_{min}.

As presented in Fig. 6.25b, because of uncertainties the final *SOC* value is less than 50%. The power is well balanced during the operation, as shown also in Fig. 6.25b by the steady DC bus voltage. The DC bus voltage fluctuates approximately 5% at the instants of limited PVA power and starting diesel generator control, which is related to corresponding control dynamics and is acceptable. The DC bus voltage pulse is generally coming from two sides: on the one hand, the voltages and currents are filtered and the powers are calculated by the filtered signals; on the other hand, power references to stabilize the DC bus voltage are compensated by the filtered power, which is also filtered therefore it has a delay in time. This problem, the development of which is not considered here, could be solved by using less filtered signal or improving the performance of correctors to cancel the compensation.

The actual diesel generator on/off signal is shown in Fig. 6.26. If one compares the $K_D(t)$ with the actual diesel generator on/off operation, it can be seen that the optimization command $K_D(t)$ is respected for most of the time, except for the additional duty cycle at last, which is the self-correcting action by operation layer.

Table 6.8 shows the total energy cost C_{total} based on the energy tariff (€/kWh) given in Table 6.7 and taking into account the operating time period of 12 h (ie, from 8:00 to 20:00).

The total energy cost C_{total} is calculated for three cases: (1) estimated optimum energy cost following power predictions; (2) the experimental cost, which is the actual energy cost; and (3) the optimum energy cost for

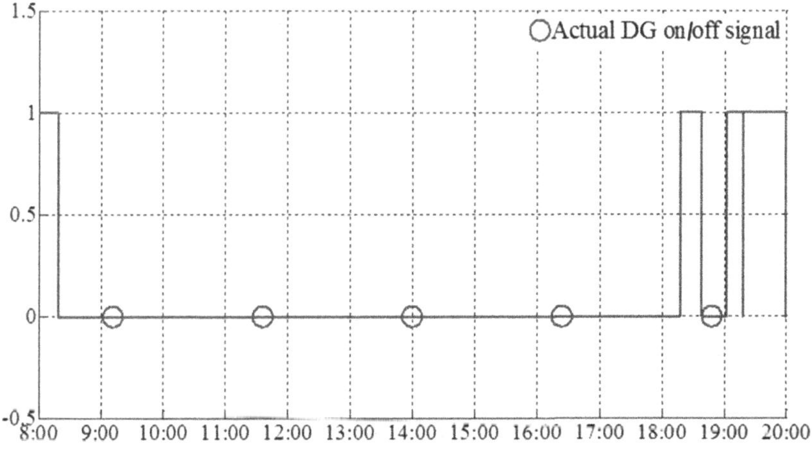

Figure 6.26 Actual diesel generator on/off signal for off-grid test 1. *DG*, diesel generator.

Table 6.8 Comparison of different energy cost calculations for off-grid test 1

Case operation	Energy cost C_{total} (€)
Estimated optimum energy cost following power predictions	3.629
Actual energy cost (experiment)	3.658
Optimum energy cost for real conditions calculated after operation	3.380

real conditions calculated after operation. Thus the actual energy cost is higher than the estimated cost by optimization before the operation; it is due to uncertainties. Aiming at a fair comparison, the total energy cost is calculated again at the end of the test based on real test conditions and using the optimization problem formulation. In this last case the obtained cost is the total optimum energy cost that would have been possible to achieve following the real solar irradiation and the real load power. The difference between the actual total energy cost and the total optimum energy cost is approximately 8% for this experimental test. Given this small difference, the optimization problem formulation could be considered as demonstrated.

5.1.2 Test 2 for Off-Grid Mode

Test 2 was performed for operation on August 27, 2013. PVA power prediction uncertainties are shown in Fig. 6.27. It can be noted that the measure is less than the corrected prediction; for the periods of 8:00–10:00 and 15:30–20:00 the weather is heavily clouded, which is not predicted.

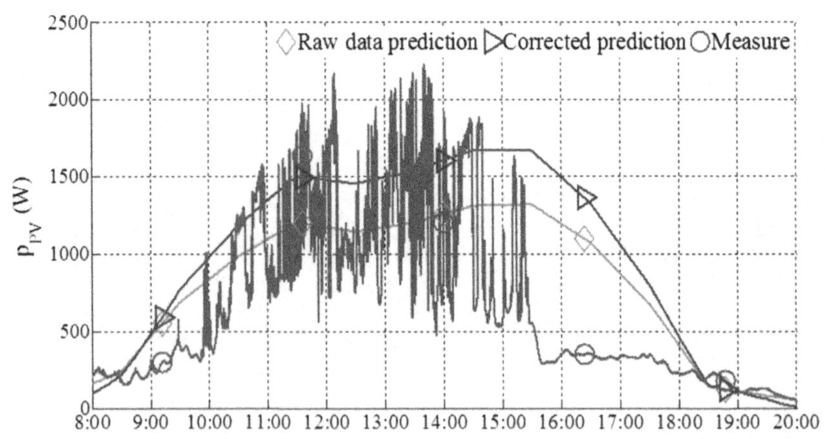

Figure 6.27 Photovoltaic array (PVA) power prediction and actual PVA power measure for off-grid test 2.

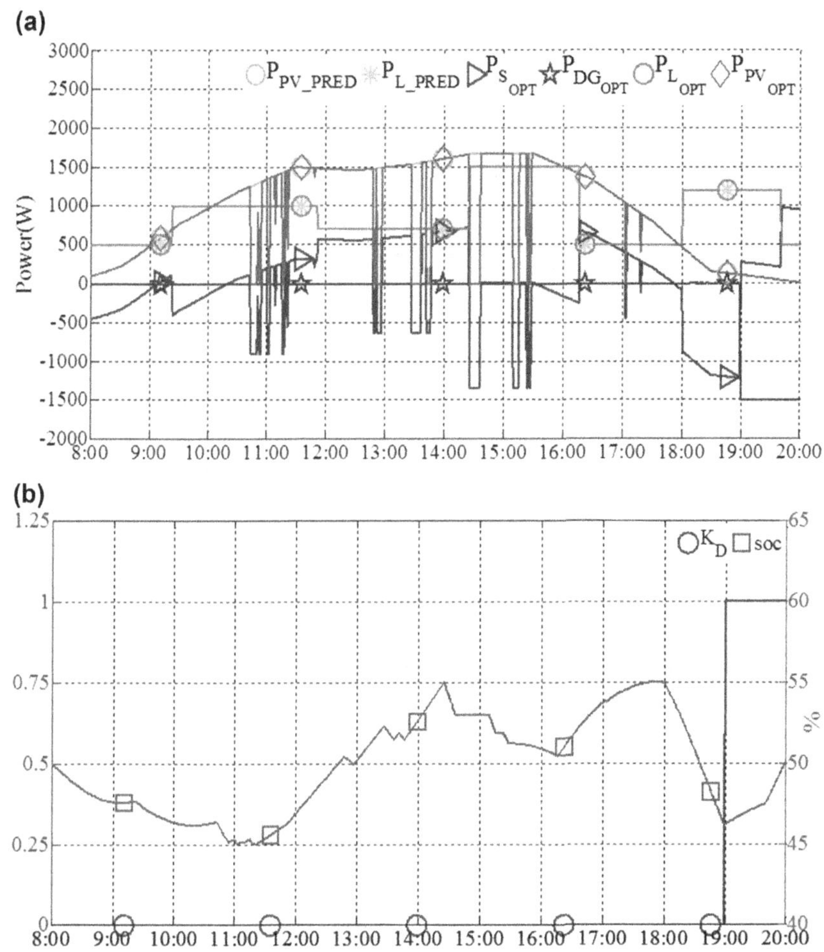

Figure 6.28 (a) Predicted and optimized power flow and (b) $K_D(t)$ and *SOC* evolution given by optimization for off-grid test 2. *SOC*, state of charge.

The optimized power flow by CPLEX is shown in Fig. 6.28a. As mentioned earlier, when the storage reaches $SOC_{\max}$ the only way to maintain power balancing is to limit PVA power production; therefore PVA constrained power must be performed. In this case the limited PVA production is distributed randomly during 10:45–17:20. On the basis of optimum power flow evolution, the optimum $K_D(t)$ sequence is calculated for the experimental operation, which is the on/off signal for the diesel generator, as shown in Fig. 6.28b.

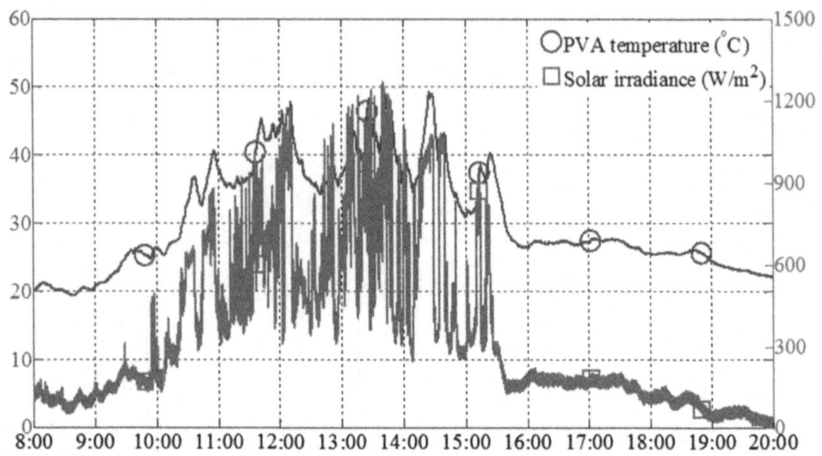

Figure 6.29 PVA cell temperature and solar irradiance on August 27, 2013. *PVA*, photovoltaic array.

The experimental operation real conditions for PVA are shown in Fig. 6.29. On the basis of the power balancing algorithm, which implies the use of the control parameter $K_D(t)$, the experimental real-time power flow is shown in Fig. 6.30a. The experimental *SOC* and DC bus voltage evolution are shown in Fig. 6.30b.

During this day operation storage is used for regulating the power balance. The diesel generator is started with interface value $K_D = 1$ for 19:00–20:00 as commanded by optimization or when the *SOC* approaches SOC_{min} (ie, 9:10–9:30, 9:45–10:05, 10:15–10:35, 11:30–11:50, 15:40–16:40, 17:40–18:00, 18:15–18:35, and 18:37–18:57), sequences that are controlled by power balancing strategy.

Storage power is limited to avoid high power charging by high PVA power plus diesel generator generation that can shorten the storage lifetime. Therefore, in the case when redundant power exceeds the storage power limit, the PVA production is limited to protect storage (ie, 10:20–10:35 and 11:30–11:50). When storage reaches SOC_{max}, PVA constrained power is performed from 13:20 to 13:50. To avoid oscillation the limited PVA power is recovered with hysteresis as mentioned in test 1.

Load shedding is performed as another control degree of flexibility in the case that when storage is empty and diesel generator plus PVA power cannot supply the load demand (ie, 16:00–16:15). In this case the prediction uncertainties are significant; therefore more diesel generator production is performed. Nevertheless, the power is able to be maintained as

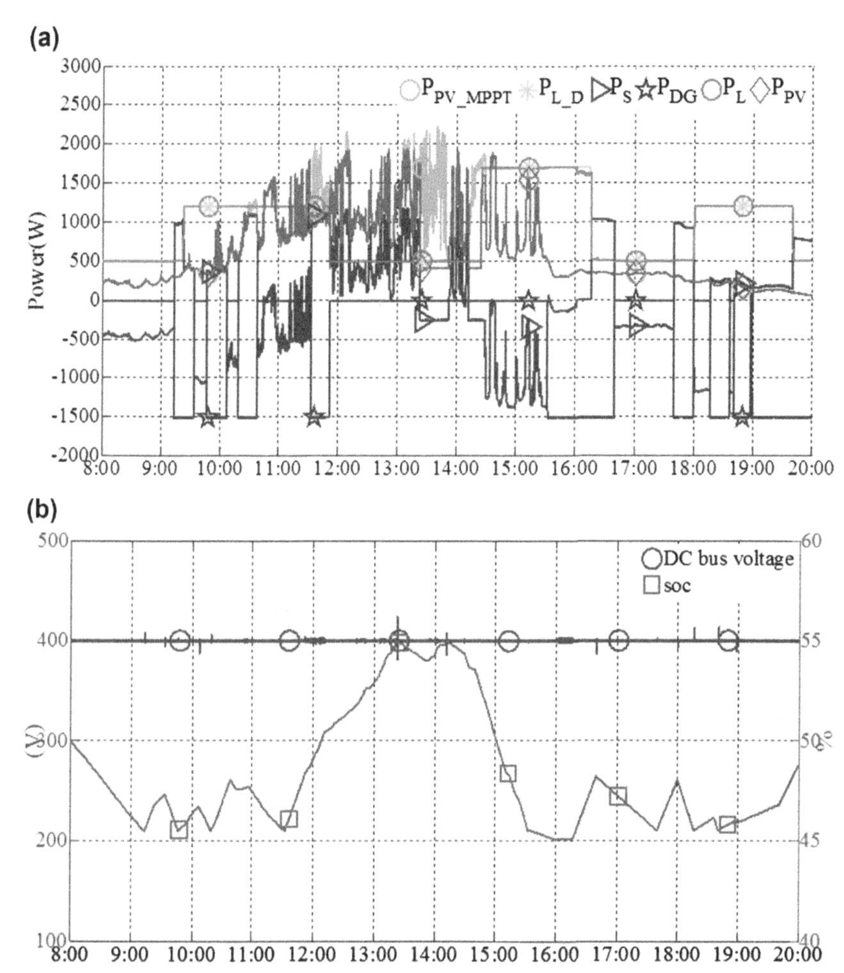

Figure 6.30 (a) Experimental real-time power flow and (b) experimental DC bus voltage and *SOC* evolution for off-grid test 2. *DC*, direct current; *SOC*, state of charge.

indicated by steady DC bus voltage in Fig. 6.30b. Because of prediction uncertainties, it can be seen that the biofuel generator is started more than expected to maintain continuous load supply.

The actual diesel generator on/off signal is shown in Fig. 6.31. Because of prediction uncertainties, it can be seen that the diesel generator is started more than expected to maintain continuous load supply.

Table 6.9 compares the energy cost among optimization, experiment, and optimum for real conditions. The experimental cost is much greater than the estimated cost by optimization.

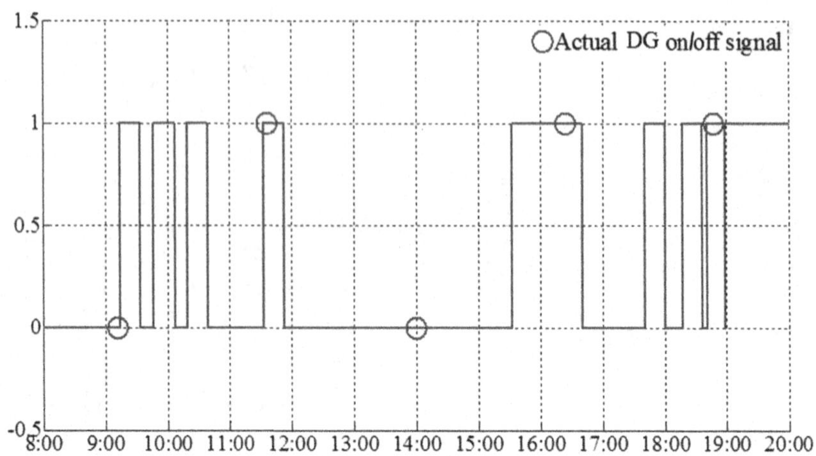

Figure 6.31 Actual diesel generator on/off signal for off-grid test 2. *DG*, diesel generator.

Table 6.9 Comparison of different energy cost calculations for off-grid test 2

Case operation	Energy cost C_{total} (€)
Estimated optimum energy cost following power predictions	3.259
Actual energy cost (experiment)	7.807
Optimum energy cost for real conditions calculated after operation	7.596

The difference is obvious because more biofuel generator production is involved to ensure power balance whereas in real conditions the PVA production is much less than expected by prediction. However, the experimental cost is close to the optimum energy cost for real conditions calculated after the experiment.

5.1.3 Test 3 for Off-Grid Mode

Test 3 was performed for operation on September 6, 2013. The PVA power prediction uncertainties are shown in Fig. 6.32. The prediction can correspond to the two peak periods of production.

The optimized power flow by CPLEX is shown in Fig. 6.33a. Because the irradiance is relatively low, no PVA constrained power is performed. On the basis of optimum power flow evolution, the optimum $K_D(t)$ sequence is given in Fig. 6.33b.

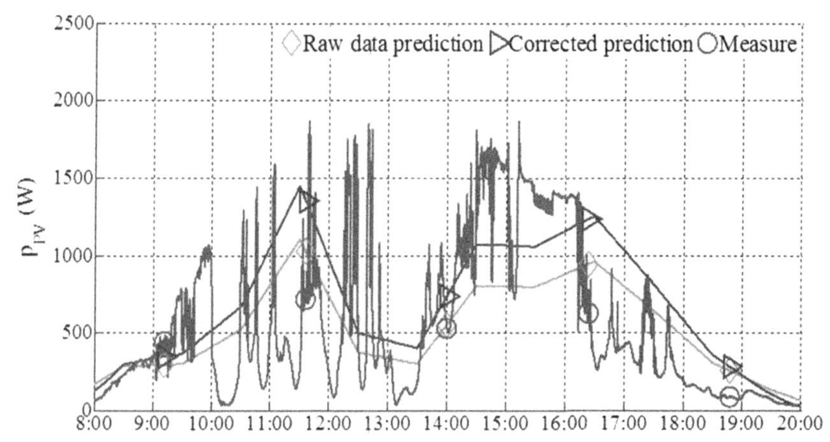

Figure 6.32 Photovoltaic array (PVA) power prediction and actual PVA power measure for off-grid test 3.

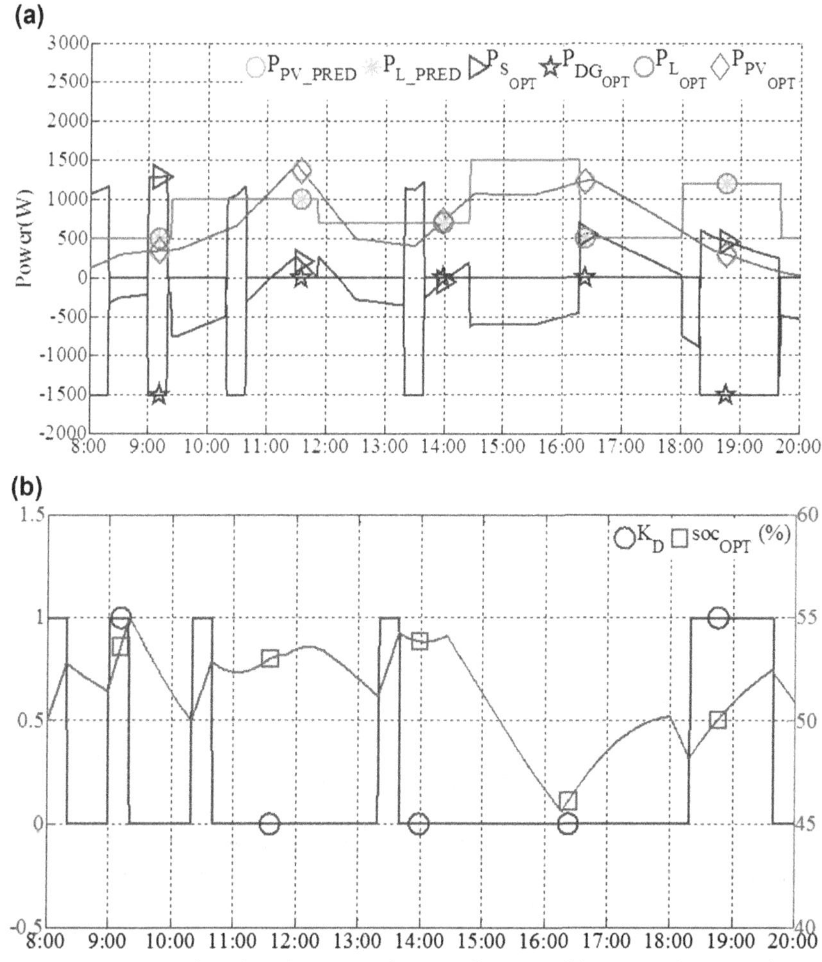

Figure 6.33 (a) Predicted and optimized power flow and (b) $K_D(t)$ and *SOC* evolution given by optimization for off-grid test 3. *SOC*, state of charge.

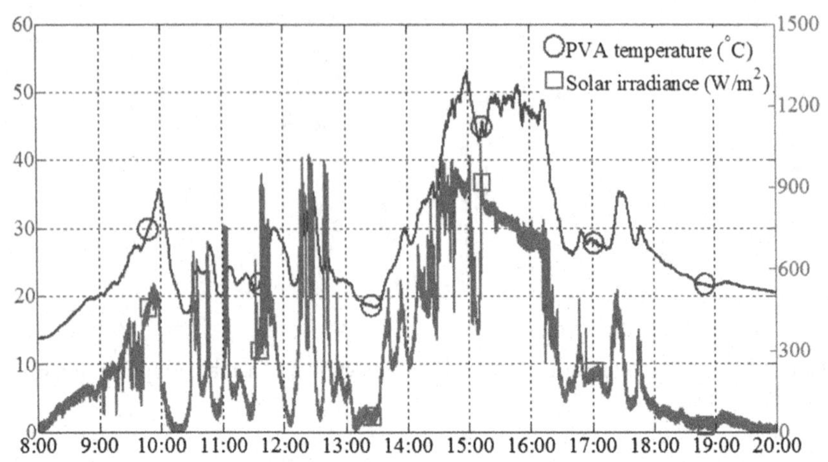

Figure 6.34 PVA cell temperature and solar irradiance on September 6, 2013. *PVA*, photovoltaic array.

The experimental operation real conditions for PVA are shown in Fig. 6.34. The obtained experimental power flow is shown in Fig. 6.35a. Experimental *SOC* and DC bus voltage evolution are shown in Fig. 6.35b.

Identical to test 2, storage power is limited to avoid high power charging by high PVA power plus diesel generator production that can shorten the storage lifetime. Therefore, in the case of storage injection power tending to exceed storage power limit P_{S_max}, the PVA production is limited to protect storage (around 9:10, 10:30, 11:30, 12:30, 15:40, and 17:20). The uncertainties make the actual instantaneous power evolution different as predicted. Nevertheless, the power balancing is maintained and load can be supplied without load shedding. The power balancing is indicated by steady DC bus voltage in Fig. 6.35b.

The actual diesel generator on/off signal is shown in Fig. 6.36. Because of prediction uncertainties, it can be seen that the diesel generator is started more than expected to maintain continuous load supply.

The energy cost calculations are given in Table 6.10, which validates again that in real conditions the experimental cost is close to the optimum energy cost for real conditions calculated after operation.

5.2 Results Analysis and Discussions for Off-Grid Mode

The optimization objective is to minimize fuel consumption and maintain a certain storage capacity at the end of the operation. The experimental tests validate the proposed control strategy, which can work under different

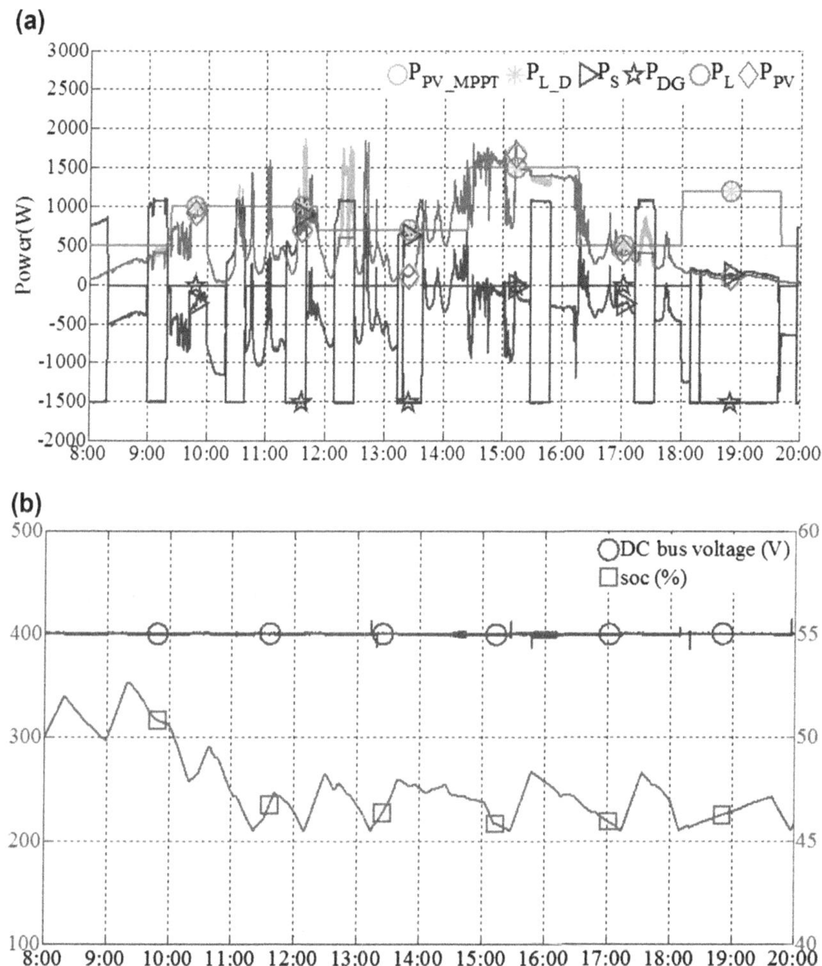

Figure 6.35 (a) Experimental real-time power flow and (b) experimental DC bus voltage and *SOC* evolution for off-grid test 3. *DC*, direct current; *SOC*, state of charge.

weather conditions for providing robust power balancing while taking into account optimization results. Because $K_D(t)$ is defined as a switch for diesel generator operation, its values are given for each optimization case study. Regarding the experimental tests, when the diesel generator starts and outputs the rated power the actual diesel generator on/off signal is 1; when the diesel generator has no specific order the actual diesel generator on/off signal is 0. It can be noted that for experimental test, the actual on/off signal for the diesel generator is not always identical to $K_D(t)$ obtained by the

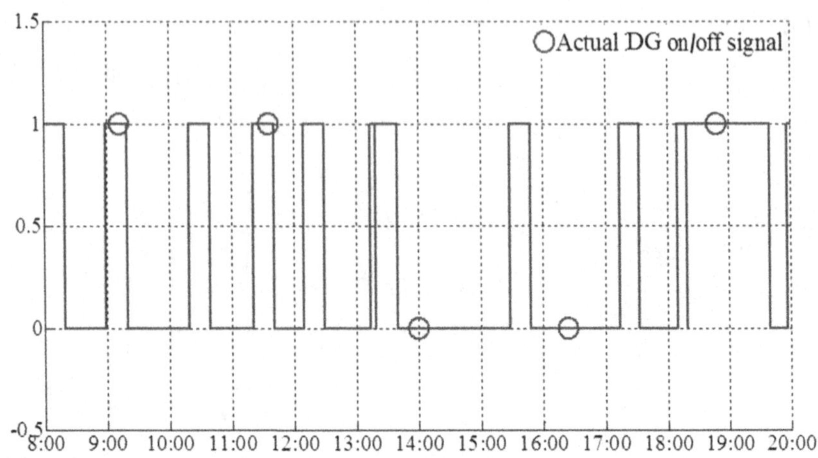

Figure 6.36 Actual diesel generator on/off signal for off-grid test 3. *DG*, diesel generator.

Table 6.10 Comparison of different energy cost calculations for off-grid test 3

Case operation	Energy cost C_{total} (€)
Estimated optimum energy cost following power predictions	4.366
Actual energy cost (experiment)	6.689
Optimum energy cost for real conditions calculated after operation	6.017

optimization calculation. This shows that the proposed control algorithm is robust and able to maintain the power balancing with self-correcting ability and not only to follow the predicted optimization operation.

Table 6.11 summarizes the comparison of the cost of energies among the operated three tests.

Experimental results show that the proposed microgrid structure is able to implement optimization in real power control and ensures self-correcting capability for off-grid mode. The power flow can be controlled to approach the optimum cost when the prediction error is within certain limits. Even if the prediction is imprecise, the power balancing can be maintained with respect to rigid constraints.

It is obvious that the optimization effectiveness depends on prediction precision. In the first test, despite the diesel generator duty cycles assigned by optimization, the diesel generator is started for one more duty cycle by

Table 6.11 Energy cost comparison among the operated three tests for off-grid mode

Test	Case operation	Total energy cost (€)	Load shedding cost (€)	Photovoltaic array constrained power cost (€)	Diesel generator production cost (€)
1.	Optimization	3.629	0	1.463	2.166
	Experimentation	3.658	0	1.121	2.423
	Optimization for real conditions	3.380	0	1.180	2.2
2.	Optimization	3.259	0	1.575	1.684
	Experimentation	7.807	0.211	0.835	6.761
	Optimization for real conditions	7.596	0	1.030	6.566
3.	Optimization	4.366	0	0	4.366
	Experimentation	6.689	0	0.175	6.514
	Optimization for real conditions	6.017	0	0.001	6.016

the operation control algorithm; therefore the diesel generator production is more than that calculated by optimization by 11.9%. In addition, for this first test the PVA constrained power was less performed than the optimization result by 23%. During the second test, the diesel generator is started for more duty cycles, but the final SOC is close to 50%. In the third test the final *SOC* is far from 50%. Load shedding occurs only in test 2, although it was not expected in the optimization. This is due to a relatively large gap between the solar irradiance prediction and reality on the one hand and the imposed limits to storage on the other hand. Regarding the PVA shedding there are differences between optimization and experimentation in all three tests; however, these facts do not affect the total energy costs.

In general, in addition to the prediction precision other factors that affect the effectiveness exist, such as converter efficiency and control security margin. Because the general prediction data are provided for a large area of approximately 20 km^2, for a single location the prediction precision may not be satisfactory, especially for cloudy weather conditions. Improvement can be done by adding local forecasting techniques such as a sky camera and local weather measurement and forecasting station. On the other hand, for these tests the storage is used for only 10% of its capacity, which corresponds to the energy of 1.25 kWh; however, this condition is imposed just to show more control events during a day test. A larger storage capacity could certainly be more resistant to the prediction errors.

The experimental costs, which are the actual energy costs, are different from the estimated cost by anterior optimization; this is due to forecast uncertainties. In posterior optimization for a real conditions case the obtained costs are the ideal optimum energy costs that could be reached following the real solar irradiation and the real load power. Indeed, this calculation is performed by completely eliminating uncertainties. Therefore the total energy cost, as result of this operation, can be considered the indicator of optimum operating. This indicator would be helpful to analyze the system behavior, but its calculation is not possible until the end of the operation. The results presented in Table 6.11 highlight the role of uncertainties in DC microgrid control to obtain the minimum energy cost. The gap between the indicator and the actual outcome depends heavily on the gap between the meteorological forecast data and actual measurements.

Even with uncertainties, the experimental cost can be controlled close to optimization for the real conditions cost, which is the ideal experimental cost. Thus it can be considered that these results validated the effectiveness of the proposed optimization and power management control. The power balancing can be maintained and rigid constraints such as storage power limit and storage capacity limit are fully respected.

6. CONCLUSIONS

In this chapter experimental tests are performed to validate the proposed supervisory control for the grid-connected and off-grid mode. These experimental tests show the feasibility of the DC microgrid overall control, which can work under different weather conditions for providing a robust power balancing control strategy and interfacing with the optimization result. Nevertheless, energy cost optimization effectiveness depends on the prediction precision.

Furthermore, following the obtained experimental results, the feasibility of implementing optimization in real operation, while respecting the rigid constraints, is validated. In addition, the proposed supervisory control that parameterized the power balancing strategy that provides an interface for optimization is also technically verified. Even with uncertainties the experiment cost can be controlled near the ideal experiment cost.

The supervisory control handles various constraints at the same time, such as storage capability, public grid power limitations, grid time-of-use pricing, grid peak hour, storage power limits, diesel generator working duty cycle, etc. With the same constraints a rule-based control strategy could be more complicated than optimization-based control. Combined with a robust power balance strategy, the energy management layer can optimize the power flow to make full use of produced energy and reduce load shedding as well as PVA constrained power. It can also help to reduce public grid supply, to avoid undesired public grid power injection in on-grid mode, and to minimize fuel consumption in off-grid mode.

The optimization effectiveness is mainly affected by the prediction precision but also by other existing factors such as converter efficiency and security margin. Prediction uncertainties do not influence power balancing, but the optimal energy cost is affected. Future work may focus on the impact of reducing uncertainties. One approach could be to reperform the optimization calculation during the operation with the latest forecasting data and real-time system status without interrupting power balancing. In further work real-time optimization will be reperformed with an hourly updated weather forecast. On the other hand, for these tests the storage is used for only 10% of its nominal capacity, which corresponds to the energy of 1.25 kWh; however, this condition is imposed to show more control of an event during a day test. A larger storage capacity could certainly be more resistant to the prediction errors.

The microgrid systems will become more and more complex and selective according to needed applications. For tertiary buildings equipped with renewable energies, which form a DC microgrid, the overall system efficiency for the local production, local consumption, and usages can be strongly improved by applying the designed supervisory control.

REFERENCES

[1] Wang BC, Sechilariu M, Locment F. Intelligent DC microgrid with smart grid communications: control strategy consideration and design. IEEE Trans Smart Grid 2012;3(4):2148–56. Special issue on Intelligent buildings and home energy management in a smart grid environment.

[2] Locment F, Sechilariu M, Houssamo I. DC load and batteries control limitations for photovoltaic systems. Experimental validation. IEEE Trans Power Electron 2012;27(9):4030–8.

[3] Sechilariu M, Wang BC, Locment F, Jouglet A. DC microgrid power flow optimization by multi-layer supervision control. Design and experimental validation. Energy Convers Manag 2014;82:1–10.

[4] Sechilariu M, Locment F, Wang BC. Photovoltaic electricity for sustainable building. Efficiency and energy cost reduction for isolated DC microgrid. Energies 2015;8(8):7945–67. Special issue on Solar photovoltaics trilemma: efficiency, stability and cost.
[5] Sechilariu M, Wang BC, Locment F. Power management and optimization for isolated DC microgrid. In: The 22nd IEEE International symposium on power electronics, electrical drives, automation and motion SPEEDAM; 2014.

General Conclusions, Future Challenges, and Perspectives

1. GENERAL CONCLUSIONS

This book presented research work on the urban direct current (DC) microgrid, on-grid and off-grid mode, by combining smart grid communication, database interaction, and energy management with power balancing. It proposes a DC microgrid integrated in buildings, which represent a large sector of energy consumption, equipped with photovoltaic (PV) sources, which are the most common renewable sources used in urban areas.

1.1 Urban Direct Current Microgrid Conclusions

In the current technical and economic context, intermittent and random production of renewable energy, such as solar and wind sources, is still a problem for their integration into the public grid. To increase their level of integration and to obtain a robust electric grid, a smart grid is needed to ensure the exchange of information on the needs of the public grid and its availability; help in balancing powers, avoiding an undesirable injection; and performing smoothing of loads during peak hours. Concerning renewable sources of low power, such as those incorporated in buildings for their public grid integration, it is possible to include them in a microgrid communicating with the smart grid. The microgrid controller must allow the connection to the public grid and ensure controls of the power, energy flow, load sharing, and load shedding and taking into account the constraints of the public grid transmitted by means of data bus communication.

Thus faced with the technical constraints of renewable sources for interfacing with the public grid, in urban areas a DC microgrid was proposed that is based on PV building-integrated sources, electrochemical storage, a diesel generator, and grid connection. The DC microgrid consists of a multisource power system, the control of which is provided by a supervisory control that is able to communicate with the smart grid and its environment (weather data, building data, and end users). Assuming that locally generated renewable electricity is consumed where, when, and in the form in which it is produced, with a public grid seen as backup source, this microgrid can be a solution for self-feeding of tertiary buildings with renewable electricity; furthermore, the excess electricity can be shared with other microgrids or sold.

The multisource power system, capable of providing power balancing through the ancillary power source (ie, storage and grid-connection or diesel generator), was developed, including modeling, simulation, implementation, and experimental tests. Considering the objective to achieve an advanced power management, a supervisory control is proposed as a local multilayer structure for power management based on forward-looking statements concerning the requests of the end users, forecasts of operation of the building, weather forecasting, energy tariff conditions, public grid availability limits, and possibilities of shedding some building appliances.

By a systemic approach, the behavior modeling of the overall system based on the interpreted Petri Net (PN) model has been proposed; the PN model is then easily translated into an adequate MATLAB-Simulink Stateflow representation for simulation. This modeling allowed for retaining the relevant microgrid operating modes of the multisource power system, the operating requirements, and the conditions of their commutations taking into account interactions with the smart grid. The main purpose of the modeling was first to better understand the problem and then to facilitate the process of identifying criteria to be taken into account, continuous and discrete variables and constraints, in the supervisory structure depending on the energy sources and structural parameters of the system. The Stateflow representations of operating modes led to numerical simulations based on a simplified control algorithm. Next, using the experimental platform, the control strategy is tested with experimental tests, and the results validated the feasibility of the microgrid and showed how the interface with the public grid affects system performance. The control strategy provided an interface parameter, $K_D(t)$, the value of which can affect the power flow. These results allowed for drawing up a specification for the design of supervisory control.

The supervisory design is based on the hybrid dynamic system approach, a system that explicitly and simultaneously involves continuous dynamic-type phenomena and events with a discrete dynamic. Thus the multisource power system is seen as a continuous system for which the supervisory control, as a discrete-event system, manages the discrete commands and other discrete events and thus forms the hybrid dynamics. This control must allow a maximum set of possible state sequences by triggering certain controllable system events: giving access permissions to the DC bus to energy sources, taking into account the availability of the public grid, constraining the load power demand, considering a production-consumption plan, and informing the public grid on short-term prospects for consumption or injection. The hybrid dynamic system representation, based on a discrete-event system built with controllers in parallel, enables the design of the supervisory control considering multiple discrete events. Thus the supervisory control becomes a hierarchical multilayer control with multiscale response time.

The DC microgrid consists mainly of a multisource power system and a supervisory control system. The multisource power system is composed of a PV array (PVA), storage, grid connection, a diesel generator, and a controllable DC load. There are four hierarchical control levels that were studied, analyzed, and implemented: the human-machine interface (HMI), prediction layer, energy management layer, and operational layer. These layers handle different functions such as power balancing, energy management, data prediction, and interaction with the smart grid and the end user.

HMI allows end users to adjust, define, and customize the operation criteria and parameters through a graphical interface; although it is not fully designed in this book, the main guidelines were given.

The prediction layer is able to calculate the PVA power prediction based on hourly solar irradiance and air temperature forecasts as weather forecasts provided by the Météo France service. For this a PV panel purely experimental model has been proposed and then extended to the PVA and validated under real operating conditions. Note that compared with the linear model and single-diode model, the experimental PV model provides higher precision. Regarding the load power demand predictions, it is considered that probability calculations could be processed from the building management system records or based on statistical data. In this study a simple load power profile was assigned to ensure enough compatibility with the system parameter values and the possibilities to illustrate the desired complex phenomena.

The energy management layer performs the optimization of the power flow based on the prediction data; it is the "intelligent" level by its role of total energy cost optimization under constraints imposed by the public grid and other metadata. Modeling the behavior of the overall system has allowed for the translation of a complex physical problem into an equivalent mathematical problem: maximize the use of renewable energy resources and the load supply at lower cost and minimize the negative impact to the public grid. Thus an optimization criterion and the constraints related to this problem have been proposed. This is the minimization of energy costs with the limits imposed by the public grid capacity, physical limits of storage and the diesel generator, PVA production possibilities, and opportunities to shed the load. Among all solutions the optimal solutions that minimize the criterion are searched. It is about finding the "most suitable solution," given a certain set of constraints, which must be optimized for several conflicting objectives. A mixed integer linear programming algorithm is used to obtain the solution to the problem. Therefore the search does not aim at finding the unique solution, but a global optimum, or an optimum Pareto set forming the "compromise surface" of the problem. Thus getting the optimum sequence of values of $K_D(t)$ (ie, the distribution coefficient reflecting the shared parts between the power grid or diesel generator and that of the storage) becomes possible. Therefore the optimization is formulated as a mixed integer linear programming problem with the objective to minimize the global energy cost on the one hand and to perform peak shaving, to avoid undesirable injection in grid-connected mode, to optimize fuel consumption in off-grid mode, and to make full use of locally produced energy with respect to element constraints on the other hand. The optimized power flow is then translated into the parameter $K_D(t)$, which is transmitted to the operational layer for real operation. With advantages provided by using $K_D(t)$, the optimized power flow can be taken into account in real-time operation while providing resistance to uncertainties.

The operational layer directly controls the multisource power system, the main control parameter of which is $K_D(t)$. The power balancing is studied for grid-connected mode and for off-grid mode, in which various constraints should be considered. The control strategy can maintain power balancing with any $K_D(t)$ value, but the optimal sequence is expected to give the best energy performance. The power flow can also be affected by power limits given by smart grid messages to offer the possibility for the smart grid to manage the grid usage. Algorithms for PVA constrained power and load shedding management are also developed to fulfill the control strategy. Numerical simulations and experimental tests were conducted allowing the multisource power system to operate with the optimal solution for production-consumption given by the top layer. However, uncertainties about the expected energy performance, no matter of related to external factors or intrinsic factors of the used technologies, require correction of parameters in real time, which means an adaptation of the optimized power scheduling taking into account the actual context.

Therefore a reactive layer should be added. One of the originalities of this supervisory design is precisely to operate the multisource power system based on the optimized sequence of values of $K_D(t)$ and with integrated algorithmic self-correction actions. This is possible because of the $K_D(t)$, which is associated with the storage power reference p_S^*. Indeed, the system is capable of operating with any $K_D(t)$ value, with $K_D \in [0,1]$, mainly because it is the public grid power, or the diesel generator power, which takes over to overcome the power balance. Secondly, there are the PVA constrained power control and load shedding action if required.

DC microgrid experimental tests are performed for grid-connected mode and off-grid mode. The results show that the proposed DC microgrid and its supervisory control are feasible for predictive energy management and can help to improve penetration of PV sources in urban areas by reducing the negative impact on the public grid. The supervisory control can be a solution for the public grid issues, such as reducing demand during peak consumption by imposing power limits and downscaling power fluctuations, and end users' requests, such as reducing the energy cost. The feasibility of implementing optimization is experimentally validated in real-time operation while respecting the rigid constraints by the proposed supervisory control, which provides an interface for optimization and parameterized power balancing strategy. The supervisory control handles various constraints at the same time, such as storage capability, grid power limitations, grid time-of-use pricing, grid peak hour, storage power limits, diesel generator working duty cycle, etc. With the same constraints a rule-based control strategy could be very complicated. Combined with a robust power balance strategy, the optimization can optimize the power flow to make full use of the produced energy and reduce load shedding and PVA constrained power. It can also help to reduce grid peak consumption, to avoid undesired grid power injection in grid-connected mode, and to minimize diesel generator fuel consumption in off-grid mode.

Regarding the grid-connected mode, from the point of view of the utility grid, no matter how the power evolves in the microgrid the grid power can be maintained within its limits. Regarding the smart grid, it can simply regulate these limits to increase or decrease the supplied power and injected power, which facilitate demand-side management for grid-operating mode.

The originality of this research lies in the following aspects:

- The supervisory control separates power balancing and energy management, but it links them through the interface parameter $K_D(t)$.

The power balancing control in the operational layer is an independent function that can work with any $K_D(t)$ value. The energy management layer is able to optimize the microgrid operation through forecast data and thus obtain the optimized power flow. The optimized power flow is then translated into the $K_D(t)$ sequence, which is a single value that represents the power flow from different sources. The communication of $K_D(t)$ does not require high-speed communication between layers. Therefore the supervisory control provides the possibility of reperforming optimization and updating the $K_D(t)$ sequence during the operation without interrupting the power balancing.

- Forecast data, smart grid messages, and rigid constraints can be considered and fully respected for optimization and real-time operation.
- Knowing that each layer provides an independent function, the multilayer-multiscale design offers a flexible structure. Several automata or microcontrollers can be used for implementation so that complex optimization and real-time power balancing control can be executed at the same time without affecting each other. The multilayer structure simplifies implementation of such a complex control strategy.
- The proposed microgrid is a DC microgrid with DC load connected directly on the DC bus. DC bus power distribution is believed to be the trend of future distribution power network development, which provides more advantages of energy efficiency for buildings equipped with native DC renewable energies and storage.
- The step-by-step experimental validation of the DC microgrid offers a comprehensible approach for readers such as researchers, students, and electrical engineers.

To summarize, the proposed DC microgrid combines smart grid interaction and energy management with power balancing. Although the microgrid only refers to building scale and involves only a few sources, the idea of parameterized power balancing and interfacing with optimization, as well as smart grid interaction, can be generalized; thus it can be used as a solution for advanced energy management for other microgrids to optimize local power flow, improve future PV penetration, and build positive energy territories.

1.2 General Conclusions on Research Approach

Electric power systems in their societal dimensions cover several research areas related to sustainable energy systems. A mid-term objective is to transform the traditional top-down managed power grid in bottom-up service, which can also imply the end users. This objective requires real-time information for the distribution grid operators aiming at better demand response and demand-side management with interaction of the end users. In this paradigm, distributed electricity generation can be installed anywhere in the public grid, including the end user, using information technology to better manage energy, because of an adequate interface coupling the distributed production, the end-user consumption, and the public grid availability. Thus the public grid of the future is the smart grid that involves microgrids and linkages based on information and communication technology. This new concept is formed on the sharing of physical data through a communication network allowing its intelligent control (optimal efficiency of production units, power flow management to avoid congestion, operation at minimal cost, etc.) taking into account the end-user needs as well as the electrical grid vulnerabilities.

Aiming at research development on the smart grid, the proposed approach has been to consider modeling tools and supervisory control considerations to propose a decentralized system of renewable electricity production that can work in stand-alone mode but also connected to the public grid and taking into account buildings equipped with renewable energy and urban areas' needs. Therefore the goal is to design, develop, and validate the urban microgrids with smart grid communication.

This research theme has benefited firstly from previous works and achievements on the analysis of hybrid dynamic system behavior, modeling, and optimal control. Moreover, through some French national projects focusing on the energy performance of electrical systems, studies were performed focusing on the low-consumption equipment of buildings, energy management, and sharing of data via a physical communications network. The global system's overall performance, seen as a dual approach, has led to a common concern of all energy reflections currently undertaken, which could be formulated as how to implement energy self-sufficiency in built environments and especially in urban areas and how to make and manage the pooling of energy resources in urban areas. From these questions research works on the intelligent control and advanced power management of a microgrid have started.

Such a complex system is seen as a hybrid dynamic system. The hybrid nature is related to the realization of the control (numerical control implying information transfer between the system and its control by a communication network). Therefore two subsystems were considered:

- the continuous part of the system as the operation of the multisource power system (ie, energy production and power flow transfer) and
- the discrete part of the system as controlling the switching of an operation mode to another and external events to the system and perturbations.

Therefore the microgrid supervisory control deals with access permissions to one or more energy sources according to their operating states, preliminary conditions of the public grid, predetermined priorities, profiles of energy consumption, etc.

The main difficulty in the design of such a system is that the hybrid nature of the issue is not generally satisfied with the purely discrete or purely continuous approach; both approaches should be combined to optimize the overall performance and to lead to the interaction between the discrete scheduling algorithms and continuous control algorithms. The hybrid dynamic system approach has led to a model that includes a set of techniques to have a graphical and mathematical representation of the system. The temporal structural and behavioral properties were analyzed using hybrid PN and Stateflow representations, and the design of supervisory control is based on finite state automata or hybrid automata. The transition from one mode to another is triggered by an event, and the sequence of events is treated statically for certain events and dynamically for others. Supervisory control takes into account the times of occurrence of the events and calculates the parameters needed to control the continuous system (ie, the multisource power system).

Subjects explored in this study, in the form of a comprehensive analysis, are seeking optimal solutions for the design of a microgrid to minimize energy costs and risks on the one hand and the possibility to structure and/or prioritize the commands on the other hand. The objective of this study was to determine and enforce the evolution constraints of the system, including time constraints, so that the forbidden states and/or timed sequences will not be achievable.

The deployed calculation tools were those included in MATLAB-Simulink software, in particular for the implementation of the commands. Regarding the simulation, Stateflow under Simulink is the most suitable tool for hybrid phenomena. In addition, PN are transposable in Stateflow graphs, which then allow the software simulation. The used optimization algorithm is based on the mixed integer linear programming that appeals to the CPLEX solver.

The experimental results on microgrids presented in the final chapter of this book, are promising, but the proposed developments should be subject to several improvements for formal validation required by in situ tests. The subject is much broader, and only a small part of the problem has been covered; therefore work in progress and works for short- and mid-term are given in the next section.

This assessment cannot finish without evoking that this work has given rise to an experimental platform—scalable, fully equipped, and combining actual renewable sources and emulators as well as electronic devices and high-speed data acquisition performance. Because of this global and reliable tool, the research work can be validated experimentally and made credible to technology transfer. Recently, the experimental platform has been selected, together with that of the L2EP Lille laboratory (France), as one of four national platforms for smart grids. This is the PowerGrid Campus project.

Additional works are committed to analyze several different cases and further explore realistic strategies to consider and validate. Short- and mid-term perspectives on this work are given in the next section.

2. FUTURE CHALLENGES

Dealing with the intermittent nature of renewable energy by grouping power generation, load consumption, and storage together, the DC microgrid proves its capability of efficiently integrating renewable energy sources in the public grid. The power balance of the system is kept by adjusting the available power for voltage stabilization, ensuring energy quality and reliability. Nevertheless, requirements such as stringent operation conditions, increased flexibility of the public grid, and lower total energy cost lead to continuous DC microgrid supervisory improvement. Some works in progress and mid-term objectives are given in the following.

2.1 Works in Progress

The supervisory control performance is limited by uncertainties; the optimization's effectiveness is affected by the prediction precision. Actually, even if prediction uncertainties do not influence power balancing, the optimal energy cost is affected. Work in progress focus on enhancing optimization performance, especially facing low prediction precision. The following solutions are investigated.

Regarding the increasing penetration of the renewable energy sources, the precision of weather forecast data is expected to be improved by means of local weather forecasting techniques based on an advanced statistical model called "nowcasting" and associated with real-time system acquisition data. Thus knowing that the supervisory control is based on the weather forecast of a large area provided by a national weather forecast national service, the precision of the prediction layer output may be improved by a local weather forecast for a single location.

In this study the optimization is performed once before the operation. Nevertheless, because the transfer of $K_D(t)$ does not require high-speed communication between layers, the optimization can be reperformed at any instant without interrupting the power balancing. Thus hourly or more frequent optimization that updates the $K_D(t)$ sequence with the latest prediction and power system status is expected to give better supervisory control performance. However, consideration of uncertainties of the prediction has to be studied and introduced as a rule-based algorithm in the operation layer that corrects $K_D(t)$ in real time with respect to power system status. A real-time predictor-corrector seems to be an efficient method. Different prediction-based methods applied to perform an economic dispatch and operate in real time are being performed. The proposed optimized methods, with the exception of the day-ahead approach, perform an online optimization, meaning that during operation those methods are reperforming the optimization process intending to improve results in cost reduction. Regarding the weather forecast data, the PV prediction is bought from Météo France, arriving at 06:00 and being actualized at 13:00, and the real power system status dataset used during optimization comes from on-site measurements.

HMI development and implementing communication between the supervisory layers are also works in progress on the DC microgrid. A fully developed HMI with graphical interface, especially for the grid-connected operation mode, more operating criteria, or a smart grid interaction mode are being designed, such as injecting or demanding constant power in the public grid during a specific period and authorizing storage injection into the grid. In addition, implementing the automatic communication between layers and realizing a prototype of complete control in a unified device could lead to technological transfer.

Furthermore, a load shedding/restoration real-time optimization for the building-integrated DC microgrid was studied and will be implemented in the supervisory control; it will be used if the microgrid power balancing cannot be kept even with backup power sources. Taking into account the building appliance powers, priorities, and critical operating time, the proposed algorithm asks for load shedding and restoration when the available DC microgrid power is different than the load demanded power. It calculates the minimum power required to be shed and/or restored according to the actual operating generation; backup power reserve; and building appliance powers, priorities, and critical operating time. Concerning the critical operating time of appliances (ie, the minimum time required by an appliance after a shedding action that is necessary before the appliance is turned on again and the maximum time that an appliance can be shed), the load shedding/restoration controller involves a dynamic knowledge base. This controller is able to update data without interrupting the DC microgrid operation. The problem formulation is based on the binary Knapsack Problem and is solved by mixed integer linear programming integrated with the CPLEX solver. The algorithm is able to select the best combination of appliances that will satisfy these requirements. A user-defined time delay for steady-state overload conditions or high energy price following the dynamic pricing may be also considered as criteria for a load shedding action. Experimental results show the steady-state and dynamic system response and prove the technical feasibility of the proposed control.

2.2 Mid-Term Objectives

Concerning the traditional sources seen as backup power, for small-scale DC microgrids the diesel generator is the most used because of the low initial cost and simplicity. The difference between load consumption and renewable generation causes fluctuations in the microgrid DC bus voltage. Thus for off-grid operating mode or limited public grid availability a power balance is performed by adjusting the diesel generator power and storage for voltage stabilization. Although the diesel generator large capacity backup power can provide long-term support for DC microgrid operation, its response speed is slow and the starting time lasts several seconds to reach stable output power. Therefore because of the slow dynamic behavior of the diesel generator the DC microgrid power balancing requires compensation for sudden load power increasing or sudden renewable power decreasing. On the other hand, storage via a supercapacitor has a fast response and high instantaneous output power but a low capacity. Thus a supercapacitor may be integrated in a DC microgrid to compensate for dynamic power fluctuations before the steady operation of the diesel generator. This work focuses on a diesel generator's slow dynamic compensation by supercapacitors for DC microgrid power balancing during the diesel generator starting time. A cooperative control strategy for the supercapacitor and diesel generator is proposed to realize the startup of backup power, which is presented in the following. When the microgrid needs to start backup power and the output of the diesel generator is not yet stable, the supercapacitor outputs power to support the microgrid. This period of the time is called supercapacitor-dominant mode. After completion of diesel generator starting the diesel generator and supercapacitor work together in a specific coordinate strategy to stabilize output. The supercapacitor does not discharge anymore, but it absorbs power from the microgrid to replenish the lost power. This period is called the supercapacitor/diesel generator cooperative control mode. When the supercapacitor approaches the upper limit of charge, to avoid excessive charge the supercapacitor is disconnected and the power balancing is supported by the diesel generator and classical storage as electrolytic batteries. This period is the diesel generator/battery-dominant mode. This control strategy is supposed to improve the microgrid dynamic characteristics under such operating conditions, and the expected results are related to the improvement of the DC bus voltage stability.

The global energy efficiency of a DC microgrid depends on power transmission, line losses, and power converter efficiency. However, the impact of the power converter efficiency on the DC microgrid is higher than the other factors. The power converter's efficiency is defined as a power transfer function during normal operation, depending on the instantaneous power, either of the input or of the output. This power transfer function is nonlinear with a threshold input power as the starting operating point, often understood as the converter's own consumption. The ohmic effect of the output circuit induces a loss that increases quadratically with the current (or power), leading to a decrease of efficiency after a maximum. The peak efficiency of a power converter indicates the efficiency of the device at one operating point considered the optimal performing point (ie, when the power converter is operating (usually) at its rated capacity). Although this peak efficiency is almost always higher than 95% for the newest devices, it is important to note that the converter may only operate at its peak efficiency range for a very small proportion of the duty cycle or not at all. The power converter efficiency curve is rarely explicitly given by the manufacturers or measured by independent institutes. The converter efficiency depends on the output power and the applied DC voltage of the PV plant. On the one hand the DC voltage varies to impose the maximum power point at any time at the PV plant, and on the other hand the power demanded by the load also varies, especially for electric building loads. Thus the optimal sizing of the converter may become very complex. For the PV converter, its efficiency relies on the average PV power received on-site, which could be calculated by the meteorological data and estimated PV panel efficiency at such conditions. For a PV grid-connected system, the European weighted inverter efficiency benchmark is expressed as an averaged operating efficiency over a yearly power distribution corresponding to the middle-European climate and is referenced on nearly every inverter datasheet. The California Energy Commission weighted inverter efficiency is almost the same logical method, but the final formula is given according to the Californian climate with its higher solar irradiance. Unlike previous works, this study aims to analyze the power converter efficiency and the impact of the power converter efficiency on the DC microgrid. The actual studies and research results do not take into account line power losses and converter efficiency. Or, it is well known that converter efficiency could be variable because renewable energy intermittency induces strong variation of the produced power and the converter input voltage. Thus the objective is to build the converter efficiency mapping that will be used in real-time power balancing aiming to reduce running the large security power margin during the DC microgrid operations.

3. PERSPECTIVES

The concept of the microgrid and its use to facilitate the integration of intermittent renewable energy systems represents an emerging and growing thematic research worldwide. Its scientific, technological, and societal interest has led to some research works that particularly concern control structures at different levels of the public grid and are applied to different topologies of the microgrid. Today the distributed generation, at small scale or high power, and the microgrids can be considered as the new model of power generation that will certainly have to become widespread in the coming years. However, faced with the traditional architecture of the public grid, microgrids represent a real break, and their implementation requires a long period of transition during which both models will coexist.

On the other hand, the fact that microgrids will be connected to the low-voltage and medium-voltage distribution network requires reviewing control of the electrical system. Indeed, the management of conventional power systems is scheduled for the high voltage, but this model is no longer valid for distributed generation. This reality leads to studying and developing management systems specially adapted for microgrids.

Therefore this context leads to the thorough study of the intelligent control and advanced power management of an urban microgrid. On the other hand, the multisource system can and must be enriched and more secured, which calls for new contributions in this direction. Furthermore, the technology transfer should be the key point to highlight and demonstrate the expected technological innovations. For this the living laboratories are the new standard for the real-scale experimental demonstration associated with the end-user implication.

3.1 Intelligent Control and Advanced Power Management of the Urban Microgrid

The development of intelligent control and management of the advanced microgrid requires a full and detailed study of the interactions that can occur in the microgrid (internally); between the microgrid and the public grid, including the islanding operation and the isolated operating mode; and among several microgrids in terms of pooling of resources. To highlight the most relevant and realistic control strategies and power management, several numerical simulations of different strategies must be undertaken and reviewed. For this, some tracks are already well specified:

- For the grid-connected operating mode the previously performed simulations have been based on the arbitrary limit values of the state of charge (*SOC*) to illustrate most of the phenomena and interactions in the proposed period. Moreover, optimization of the total energy cost has been proposed for the full use of *SOC*, reaching the lower limit at the end of the selected period. However, it would be interesting to study the case for which the optimized operation is performed with the requirement of a certain level of *SOC* after 18:00 already preset for optimization. To do this a new constraint must be defined at the formulation of the optimization problem. This case of energy management can be applied to the urban residential microgrid, for which a peak consumption occurs almost always after 18:00.
- The concept of a microgrid and its controller as the interface with the public grid were developed for applications such as a commercial building, the consumption and implicitly the impact of which to the public

network takes place during the day. The study was conducted based on the assumption that the storage can be charged only by the PV production. Subsequently, it can be interesting to study the case of the overnight storage charging from the public grid (ie, during the period of lower public grid energy cost 22:00 to 06:00). This would take advantage of the full storage capacity during the day, without the storage oversizing. By analogy with the plug-in electric vehicle, the overnight battery charging can be seen as a predictable operation for better planning of resources in the public grid.

- Furthermore, the implementation of the dynamic change of the public grid operating mode and/or end-user control during the test day has to be designed and validated. The simulation tests and the experimental tests, which are already performed, take into account only changes that are known in advance (ie, static operating modes). Thus the dynamic pricing mechanism and implicitly the dynamic change of the public grid operating mode and/or end-user control must be studied and tested.
- One of the negative aspects of the impact that the microgrid can have on the public grid is the power fluctuation in absorption and injection. In this study the optimized calculation of the distribution coefficient was proposed, taking into account a predetermined minimum duration for injection and for absorption. The first results obtained lead to storage power fluctuations and, depending on the limits of public grid power, to load power fluctuations. Therefore the study must be more complex because on the one hand the storage aging will grow due to fluctuations and on the other hand the load often becomes too constrained. Hence the idea to introduce in the multisource system energy storage dedicated to minimizing fluctuations appears, such as that which supercapacitors could offer.
- By definition a microgrid must be capable of operating either in grid-connected mode or off-grid mode. Regarding the total grid disconnection, switching from grid-connected mode to off-grid mode, each time the control does not use the public grid connection that connection is not cut. However, the islanding may require physical disconnection. Thus this aspect should be taken into account first in simulation and then in an experimental test because it is likely to exhibit evidence issues related to the dynamic mode with respect to the stability of the DC bus; the same remark can be given for the case of grid connecting and switching from off-grid mode to grid-connected mode. In addition, the connection of two or more off-grid microgrids can be an interesting study subject as pooling excessive power production to feed the

buildings with insufficient power production to supply the load. The supervisory control needs to evaluate the best power quantity that could be exchanged between the microgrids taking into consideration the real operating conditions.

3.2 Other Renewable Energy Sources for the Multisource System

Small-scale renewable energy sources are often used in urban areas to form local DC microgrids. Therefore a small-scale wind generator can be considered as a renewable source to be integrated into an urban DC microgrid. A wind generator converts the wind energy into different forms of electrical energy by means of a wind turbine associated with a power conversion system. A small-scale wind generator represents wind generation systems with the ability to produce electrical power smaller than or equal to 30 kW (in Europe) or 100 kW (in the United States). For small wind turbines, two typically used machines are the asynchronous machine with a squirrel cage and the permanent magnet synchronous machine. Because of its easy fabrication, relatively small size, and high torque, as well as the absence of the problems of excitation in the induction machines, the permanent magnet synchronous machine has become more and more widespread. However, to increase the system overall reliability and at the same time to limit the economic cost, for the small-scale wind generator the mechanical sensor should be eliminated. Therefore to integrate a small-scale wind generator in the urban DC microgrid there are several issues to study: sensorless drive control, maximum power point tracking (MPPT) algorithms, limited wind generator power control, and wind generator power modeling aiming to calculate the wind generator power prediction. Thus with an active energy conversion structure and a sensorless permanent magnet synchronous machine a speed estimator is required to control the system. In this case the extended Kalman filter model-based estimator allows sensorless drive control in a wide speed range and estimates the rotation speed with a rapid response. Regarding the MPPT methods the indirect method based on an iterative search and knowledge of the wind generator parameters seems to operate with better performances while requiring accurate knowledge of the controlled system. The direct method, based on the perturbation of a variable system and the observation of the system in real-time, seems to be less efficient under rapid variation of wind speed.

On the other hand, the multisource system can be enriched with a fuel cell of which the hydrogen production is based on the excessive microgrid power production; therefore the limited power control of renewable energy sources will be less applied. The electrolyzer uses the excessive microgrid power to produce oxygen and hydrogen by the electrolysis of water. Then it can be used flexibly according to the microgrid needs. The supervisory control of the urban DC microgrid must consider weather forecasts and then propose operational strategies for optimized energy cost.

3.3 Living Laboratory for Real-Scale Experimental Demonstration

The interactions within the microgrid and between the microgrid and the public grid were analyzed and developed through numerical simulations and experimental tests for prevalidation of physical and algorithmic architectures. Therefore the confrontation of the proposed solutions to real operating conditions (without emulators) is required. This is to verify the performance of the solutions under the requirements of the public grid and to discriminate certain solutions to retain realistic strategies for production management, storage, and consumption on the one hand and communication with the smart grid on the other hand. As an open local innovative ecosystem, an urban DC microgrid living laboratory is a technological showcase that may serve as a demonstration for innovative microgrid concepts by integrating the end users of energy into the innovation process.

Whichever the assumptions used by the prospective scenarios on the importance of fossil reserves, the technological innovation, growth in energy demand, energy efficiency, and transforming the consumption patterns are needed as essential means for building more energy efficient networks. The concepts of energy efficiency and sobriety are also a privileged crossing point for fundamental sciences, engineering sciences, and social sciences and humanities to the extent that technology and behavior are closely intertwined.

The smart grid, smart microgrids, intelligent buildings, and new technological devices that are based on distributed intelligence are supposed to respond to problems induced by intermittent decentralized production based on renewable energy sources and also are able to perform economic optimization. The challenge is to build models of collective intelligence to technological devices and microgrids in constant interaction with human users and producers. Because of these constant interactions, these models of collective intelligence must take into account technical constraints and habits, practices, and social norms in view of the emerging optimal consumption.

INDEX

'*Note:* Page numbers followed by "f" indicate figures and "t" indicate tables.'

A

Alternating current (AC) microgrid, 7–9, 7f
Artificial intelligence, 13–15

B

Backup power resources
- bidirectional sources, 93
- capacitor, 93–94
- diesel generators, 118
  - characteristics, 118–119
  - operating cost analysis, 120–121
  - operating principle, 119–120
- electrochemical battery, 93–94
- fuel cell
  - defined, 94–95
  - hydrocarbons, 94–95
  - hydrogen production technology, 95
- lead-acid storage resource
  - dynamic phenomena, 102
  - electrochemical storage, characteristics, 97–99
  - experimental evaluation, 110–117
  - modeling, 102–109
  - operating principle, 100–102
- microturbines, 96–97
- unidirectional sources, 93
- utility grid connection, 129f
  - integral proportional corrector structure, 123, 124f
  - phase-locked loop system control, 125–127, 125f–128f
  - public grid connection, 121, 122f, 124f
  - pulse width modulation (PWM), 121–122
  - used control synopsis, 123, 123f

Battery dynamic modeling, 106f
- Laplace variable, 104
- MATLAB, 107–108
- Randles schema, 103–104, 104f–105f
- Warburg impedance, 104–105

Battery static modeling, 103, 103f
Bidirectional backup resources, 93
Big Data, 12–13, 15

C

C++, 188
Charging regime (C_r), 98
Closed batteries, 101
Concerning losses, 211
Conventional batteries, 101

D

Defuzzification, 63–65
Demand response, 139
Demand-side management, 139
Depth of discharge *(DOD)*, 98
Diesel generators, 118
- characteristics, 118–119
- operating cost analysis, 120–121
- operating principle, 119–120

Direct current (DC) microgrid, 7–9, 7f, 133f
- AC/DC separated distribution, 21
- building-integrated microgrid system, 19, 20f
- DC and AC bus distribution, 20, 133–134
- diesel generator operating mode modeling, 149–151, 151f, 151t
- grid-connected mode, 158
  - experimental tests, 159–164, 160f, 160t, 162f–163f
  - simple strategy, 158, 159f
- grid-operating mode modeling, 143–145, 145f, 145t
- load operating mode modeling, 151–152, 152f–153f, 153t

Direct current (DC) microgrid (*Continued*)
microgrid global energy efficiency, 23–26, 24f
microgrid global power transmission efficiency, 21–23, 22f
off-grid mode, 164
experimental test, 165–167, 166f
simple strategy, 164–165, 165f
Petri Net (PN) modeling, 156
autonomous PN, 142, 142f
continuous simulation, 155–156
defined, 140–141
grid-connected mode, 153–154, 154f
off-grid mode, 154–155, 155f
Simulink, 155–156
Stateflow, 155–156, 157f
timed-interpreted Petri Net, 142, 142f
photovoltaic operating mode modeling, 148–149, 149f, 149t, 150f
power balancing principle
DC bus power adjustment, 137
DC load, 136
distribution coefficient, 137
maximum power point tracking (MPPT) method, 135
multisource power system, 134–135, 135f
PV array (PVA), 134–135
Simulink, 141
smart grid interaction
demand response, 139
demand-side management, 139
Petri Net, 143, 143f
smart routers, 138–139
Stateflow, 141
storage operating mode modeling, 147t, 148f
in charge mode, 146
in discharge mode, 146
Extended Kalman Filter, 145–146
turn-off mode, 147, 147f
supervisory system. *See* supervisory system
tertiary building, 19
dSPACE 1006, 209–210, 210t
dSPACE 1103, 71, 82t
dSPACE 1104, 47–48

E

Electric double-layer capacitors, 94
Energy cost optimization problem formulation, 179
grid-connected mode, 184
grid energy tariff, 183
load shedding coefficient K_L, 181
power flow, 179, 180f
PVA power shedding, 181–182
PVA production, 180
off-grid mode, 185–186, 185f
Equivalent circuit photovoltaic model
output current, 41
parameter identification method, 42
single-diode model, 40–41, 41f
temperature and diode saturation current, 41–42

F

Fixed-step MPPT algorithms, 56, 59–60
incremental conductance, 58–59, 59f
P&O algorithm, 57, 57f–58f
Fuel cell
defined, 94–95
hydrocarbons, 94–95
hydrogen production technology, 95
Fuzzy logic MPPT approach
defuzzification, 63–65
fuzzification, 62–63, 62f, 64f
fuzzy reasoning, 63, 65t
fuzzy sets, 61–62
Fuzzy sets, 61–62

G

Grid-connected mode, 184, 197f–198f
energy cost comparison, 229–230, 231t
grid energy tariff, 183
high irradiance without fluctuations, test 1
actual power sharing ratio, storage and grid, 220, 220f
CPLEX, 216

different energy cost calculation, 220–221, 221t
experimental DC bus voltage, 217–218, 219f
experimental real-time power flow, 217–218, 219f
$K_D(t)$ and state of charge (SOC), 216, 217f
photovoltaic array (PVA) cell temperature and solar irradiance, 217–218, 218f
photovoltaic array (PVA) power prediction and measure, 215, 216f
predicted and optimized power flow, 216, 217f
prediction errors, 220
high irradiance with strong fluctuations, test 2
actual power sharing ratio, storage and grid, 225, 225f
different energy cost calculations, 225–226, 226t
experimental DC bus voltage and SOC evolution, 223, 224f
experimental real-time power flow, 223, 224f
$K_D(t)$ and SOC evolution, 221, 222f
photovoltaic array (PVA) power prediction and PVA power measure, 221, 222f
predicted and optimized power flow, 221, 222f
public grid, 223
PVA cell temperature and solar irradiance, 223, 223f
K_D, power flow simulation control, 201, 202f
$K_D(t)$, power flow simulation control, 199–201, 200f
load power prediction data, 196
load shedding coefficient K_L, 181
low irradiance with strong fluctuations, test 3
actual power sharing ratio, storage and grid, 228, 230f
actual PVA power measure, 226, 226f
different energy cost calculations, 228, 230t
experimental DC bus voltage and SOC evolution, 228, 229f
experimental real-time power flow, 228, 229f
$K_D(t)$ and SOC evolution, 226, 227f
photovoltaic array (PVA) power prediction, 226, 226f
predicted and optimized power flow, 226, 227f
PVA cell temperature and solar irradiance, 228, 228f
MATLAB-Simulink, 195–196
multisource power system electrical schema, 214, 215f
optimization and experimental parameters values, 198t, 214, 214t
power flow, 179, 180f
public grid power, 231
PVA power prediction data, 196
PVA power shedding, 181–182
PVA production, 180

H

Halogen, 45
Human-machine interface (HMI), 174–175, 174f, 253, 260
Hybrid nature, 257–258

I

IBM ILOG CPLEX solver, 188
"If-Then" rules, 63
Incremental conductance (InC), 53–55, 58–59, 59f
Internal resistance, 99
Internet of Things (IoT), 15
Isolated operating mode, 6–7

L

Lead-acid storage resource
battery dynamic modeling, 106f
Laplace variable, 104
MATLAB, 107–108
Randles schema, 103–104, 104f–105f
Warburg impedance, 104–105

Lead-acid storage resource (*Continued*)
battery static modeling, 103, 103f–104f
chemical modeling, 102
dynamic phenomena, 102
electrochemical storage, characteristics
accumulators, types of, 99, 99t
elements, 97
redox reaction, 97
variables and used terms, 98–99
experimental evaluation, 113f
12-V/130-Ah Solar Sonnenschein battery, 114, 115f–116f
current profile and voltage response, 110, 110f
errors, 113–114, 113t
mean absolute error (MAE), 112
nonlinear models, 111, 111f
parameter identification, 111, 112f
state of charge *(SOC)*, 117
state of health *(SOH)*, 117
static and dynamic models, 113–114, 114t
operating principle
components, 100
discharge-charge, 100–101
manufacturing technologies, 101–102
secondary reactions, 101
purely experimental battery model, 108–109, 109f
Linear power photovoltaic model, 43–44

M

Machine to machine (M2M), 15
Mamdani fuzzy inference system, 63
Mass energy, 98
MATLAB–Simulink software, 53, 71, 258–259
Maximum power point tracking (MPPT), 51, 135, 176
experimental comparison, 65–66, 67f
closed-loop transfer function, 69
current closed loop, 70, 70f
dSPACE 1103, 71
extracted energies, experimental tests, 73–74, 77t
FL MPPT, 74
ImP&O algorithm, 72
integral proportional (IP) controller, 69, 69f
Kirchhoff laws, 67–68
MATLAB-Simulink, 71
meteorological operation conditions, electrical power, 72, 72f–77f
pulse width modulation (PWM), 68
PV power system, 66–67
synoptic block diagram, 68, 69f
total oscillation rate, four extracted powers, 74–75, 77t
fixed-step MPPT algorithms, 56, 59–60
incremental conductance, 58–59, 59f
P&O algorithm, 57, 57f–58f
incremental conductance (InC), 53–55
international journals, number of articles, 55, 55f
perturb and observe (P&O), 53–55
variable-step size MPPT algorithms, 60
fuzzy logic MPPT approach, 61–65
perturb and observe method, 60–61, 61f
Microgrids, 9
alternating current (AC) microgrid, 7–9, 7f
control
converter interface, 9
droop control, 9
hierarchical control, 10
primary control, 10–11
secondary control, 11
tertiary control, 11
direct current (DC) microgrid, 7–9, 7f
energy management, 12–14
isolated operating mode, 6–7
protection system, 11–12
vs. smart grid, 6
utility grid, 6
Microgrid supervisory control, 258
Microturbines, 96–97
MPPT. *See* Maximum power point tracking (MPPT)

Multilayer supervisory communication
communication lines, 211, 212f
load power profile, 211, 212f
maximum power point tracking (MPPT), 213
PVA power prediction, 211
Multilayer supervisory design
continuous dynamic system, 172
discrete-event system, 172
dynamic hybrid systems, 171, 172f
hierarchical control, 173f
hybrid dynamic system, 173, 174f
operational interface, 172
Petri Net (PN) and Stateflow, 171
Multisource power system, 252

N

Nominal capacity (C_{NOM}), 98
Nominal voltage, 98
Nowcasting, 260

O

Off-grid mode, 185–186, 185f, 203, 203t, 204f, 244–248
comparison and discussion, 206, 206t
control algorithm, 193, 193f
diesel generator duty cycle, 232
high solar irradiance without fluctuations, test 1
actual diesel generator on/off signal, 237, 237f
different energy cost calculations, 237, 238t
experimental DC bus voltage and SOC evolution, 235, 236f
experimental real-time power flow, 235, 236f
$K_D(t)$ and SOC evolution, 233–235, 234f
photovoltaic (PVA) power prediction and actual PVA power measure, 233, 234f
predicted and optimized power flow, 233–235, 234f
PVA cell temperature and solar irradiance, 235, 235f
high solar irradiance with strong fluctuations, test 2
actual diesel generator on/off signal, 241, 242f
different energy cost calculations, 241, 242t
experimental DC bus voltage and SOC evolution, 240, 241f
experimental real-time power flow, 240, 241f
$K_D(t)$ and SOC evolution, 239, 239f
load shedding, 240–241
photovoltaic array (PVA) power prediction and actual PVA power measure, 238, 238f
predicted and optimized power flow, 239, 239f
PVA cell temperature and solar irradiance, 240, 240f
$K_D(t)$, power flow simulation control, 205, 205f
mixed high irradiance, strong fluctuations and low irradiance without fluctuations, test 3
actual diesel generator on/off signal, 244, 246f
different energy cost calculations, 244, 246t
experimental DC bus voltage and SOC evolution, 244, 245f
experimental real-time power flow, 244, 245f
$K_D(t)$ and SOC evolution, 242, 243f
photovoltaic array (PVA) power prediction and actual PVA power measure, 242, 243f
predicted and optimized power flow, 242, 243f
PVA cell temperature and solar irradiance, 244, 244f
multisource power system electrical schema, 232, 233f
optimization and experimental parameters values, 232, 232t
Open batteries, 101
Optimization-based approach, 13

P

PCC. *See* Point of common coupling (PCC)
Perturb and observe (P&O), 53–55, 57, 57f–58f
Petri Net (PN) modeling, 156, 252
autonomous PN, 142, 142f
continuous simulation, 155–156
defined, 140–141
grid-connected mode, 153–154, 154f
off-grid mode, 154–155, 155f
Simulink, 155–156
Stateflow, 155–156, 157f
timed-interpreted Petri Net, 142, 142f
Phase-locked loop (PLL) system control, 125–130, 125f–128f
Photovolatic array (PVA), 134–135
Photovoltaic cell, 36f
electrical characteristics, 38f
power evolution, 37, 37f
PV cell efficiency, 38
solar irradiance, 38
operating principle, 36–37
PV panel, 35, 35f
PV source, 35, 35f
Photovoltaic-constrained production control, 78, 88f
element detail, 82, 82t
meteorological conditions and photovoltaic array (PVA) power evolution, 83, 84f
photovoltaic array voltage control and current control, 83, 85f
photovoltaic-constrained power control, 79–81, 80f
photovoltaic power-constrained production strategy, 78–79
power and voltage reference transient evolution, 83, 85–86, 86f–87f
PVA control voltage reference evolution, 83, 84f
test bench, 81, 82f
Photovoltaic source modeling
equivalent circuit photovoltaic model
output current, 41
parameter identification method, 42
single-diode model, 40–41, 41f
temperature and diode saturation current, 41–42
experimental comparison, 51–53, 51f–53f
linear power photovoltaic model, 43–44
optimal operating points, 53
parameter identification problem, 39
photovoltaic cell, 36f
electrical characteristics, 37–39
operating principle, 36–37
PV panel, 35, 35f
PV source, 35, 35f
photovoltaic power prediction, 38f, 39–40, 40f
photovoltaic system efficiency, 53
purely experimental photovoltaic model, 44–51
Point of common coupling (PCC), 7
Power balancing principle
DC bus power adjustment, 137
DC load, 136
distribution coefficient, 137
maximum power point tracking (MPPT) method, 135
multisource power system, 134–135, 135f
PV array (PVA), 134–135
Power control algorithms implementation, 213–214, 213f
Programmable electronic load (PEL), 47–48
Pseudocapacitors, 94
PT100 temperature sensors, 45
Pulse width modulation (PWM) control, 213–214
Pumped-storage hydroelectricity station, 2–3
Purely experimental photovoltaic model
look-up tables (LUTs), 44–45, 46f
process measurements and data acquisition, 47–51, 48f–50f
test bench description, 45, 47f, 47t
Pyranometers, 45

R

Ragone chart, 93–94
Redox, 94
Rule-based method, 13–14

S

Sealed batteries, 101
Self-discharge, 99
Simulink, 141, 155–156, 258–259
SKM100GB063D, 210
Small-scale renewable energy sources, 266
Smart grid, 5
  building-integrated renewable generators, 2–3
  conventional production units, 1
  distributed power generation, 1–2
  electricity landscape, 3, 4f
  energy prosumer sites, 3
  energy storage, 2–3
  pumped-storage hydroelectricity station, 2–3
  urban direct current microgrid, 14
Smart microgrids, in urban areas, 16
  communication network, 17
  DC load, 18
  direct current (DC) microgrid
    AC/DC separated distribution, 21
    building-integrated microgrid system, 19, 20f
    DC and AC bus distribution, 20
    microgrid global energy efficiency, 23–26, 24f
    microgrid global power transmission efficiency, 21–23, 22f
    tertiary building, 19
  microgrid supervisory control, 18
  microgrid *vs.* smart grid, 26, 27f
  smart grid topology, 16–17, 17f
Stateflow, 141, 171
State of charge *(SOC)*, 98
Supercapacitor, 93–94
Supervisory system, 194f
  energy management layer design, 178f
    distribution coefficient K_D, 178–179
    energy cost optimization problem formulation, 179–186
    operation layer interface, 189–190
    problem solving, 187–189, 188f
  energy tariffs, 195
  grid-connected mode, 197f–198f
    comparison and discussion, 201–203, 202t
    K_D, power flow simulation control, 201, 202f
    $K_D(t)$, power flow simulation control, 199–201, 200f
    load power prediction data, 196
    MATLAB-Simulink, 195–196
    optimization and simulation parameter values, 198t
    PVA power prediction data, 196
  human-machine interface (HMI), 174–175, 174f
  multilayer supervisory design
    continuous dynamic system, 172
    discrete-event system, 172
    dynamic hybrid systems, 171, 172f
    hierarchical control, 173f
    hybrid dynamic system, 173, 174f
    operational interface, 172
    Petri Net (PN) and Stateflow, 171
  off-grid mode, 203, 203t, 204f
    comparison and discussion, 206, 206t
    $K_D(t)$, power flow simulation control, 205, 205f
  operation layer, 191f
    grid-connected mode, 192–193, 192f
    off-grid mode, 193, 193f
    power balancing, 190–191
  prediction layer design, 175f
    load power prediction, 177–178
    photovoltaic power prediction, 176, 176f

T

Total life cycle, 99

U

Ultracapacitor, 93–94
Unidirectional backup resources, 93
Urban direct current microgrid
  artificial intelligence, 14–15
  Big Data, 15
  experimental platform, 29–30, 29f
  Internet of Things (IoT), 15
  machine to machine (M2M), 15
  smart building, 16
  smart city network, 15
  smart grid, 14
  smart microgrids. *See* Smart microgrids
  urban energy management, 26–28, 28f
Utility grid connection
  integral proportional corrector structure, 123, 124f
  phase-locked loop system control, 125–127, 125f–128f
  public grid connection, 121, 122f, 124f
  pulse width modulation (PWM), 121–122
  used control synopsis, 123, 123f

V

Variable-step size MPPT algorithms, 60
  fuzzy logic MPPT approach, 61–65
  perturb and observe method, 60–61, 61f

Printed in the United States
By Bookmasters